OLFACTION IN MOSQUITO–HOST INTERACTIONS

The Ciba Foundation is an international scientific and educational charity (Registered Charity No. 313574). It was established in 1947 by the Swiss chemical and pharmaceutical company of CIBA Limited—now Ciba-Geigy Limited. The Foundation operates independently in London under English trust law.

The Ciba Foundation exists to promote international cooperation in biological, medical and chemical research. It organizes about eight international multidisciplinary symposia each year on topics that seem ready for discussion by a small group of research workers. The papers and discussions are published in the Ciba Foundation symposium series. The Foundation also holds many shorter meetings (not published), organized by the Foundation itself or by outside scientific organizations. The staff always welcome suggestions for future meetings.

The Foundation's house at 41 Portland Place, London W1N 4BN, provides facilities for meetings of all kinds. Its Media Resource Service supplies information to journalists on all scientific and technological topics. The library, open five days a week to any graduate in science or medicine, also provides information on scientific meetings throughout the world and answers general enquiries on biomedical and chemical subjects. Scientists from any part of the world may stay in the house during working visits to London.

Ciba Foundation Symposium 200

OLFACTION IN MOSQUITO–HOST INTERACTIONS

1996

JOHN WILEY & SONS

Chichester · New York · Brisbane · Toronto · Singapore

© Ciba Foundation 1996

Published in 1996 by John Wiley & Sons Ltd
Baffins Lane, Chichester
West Sussex PO19 1UD, England

Telephone National   (01243) 779777
          International (+ 44) (1243) 779777

Other Wiley Editorial Offices

John Wiley & Sons, Inc., 605 Third Avenue,
New York, NY 10158-0012, USA

Jacaranda Wiley Ltd, G.P.O. Box 859, Brisbane,
Queensland 4001, Australia

John Wiley & Sons (Canada) Ltd, 22 Worcester Road,
Rexdale, Ontario M9W 1L1, Canada

John Wiley & Sons (SEA) Pte Ltd, 37 Jalan Pemimpin #05-04,
Block B, Union Industrial Building, Singapore 2057

Suggested series entry for library catalogues:
Ciba Foundation Symposia

Ciba Foundation Symposium 200
x + 331 pages, 55 figures, 15 tables

Library of Congress Cataloging-in-Publication Data

Olfaction in mosquito–host interactions / [editors, Gregory R. Bock
  (organizer) and Gail Cardew].
      p. cm.—(Ciba Foundation symposium ; 200)
    Symposium on Olfaction in Mosquito–Host Interactions, held at the
  Ciba Foundation, London, 31 Oct.–2 Nov. 1995.
    Includes bibliographical references and index.
    ISBN 0 471 96362 3 (alk. paper)
    1. Mosquitoes—Control—Congresses. 2. Smell—Congresses.
  I. Bock, Gregory R. II. Cardew, Gail. III. Symposium on Olfaction
  in Mosquito–Host Interactions (1995 : London, England) IV. Series.
  RA640.044 1996
  614.4'323—dc20   96-15908
                                                              CIP

British Library Cataloguing in Publication Data

A catalogue record for this book is available from the British Library

ISBN 0 471 96125 6

Typeset in 10/12pt Times by Dobbie Typesetting Limited, Tavistock, Devon.
Printed and bound in Great Britain by Biddles Ltd, Guildford.
This book is printed on acid-free paper responsibly manufactured from sustainable forestation, for
which at least two trees are planted for each one used for paper production.

Contents

Participants

S. Anton Department of Ecology, Ecology Building, Feromongruppen, Lund University, S223 62 Lund, Sweden

J. Boeckh Universität Regensburg, Institut für Zoologie, Lehrstuhl Boeckh, Universitätsstrasse 31, D-93040 Regensburg, Germany

M. F. Bowen SRI International, Life Sciences Division, 333 Ravenswood Avenue, Menlo Park, CA 94025–3493, USA

J. Brady Imperial College of Science, Technology & Medicine, Department of Biology, Silwood Park, Ascot, Berkshire SL5 7PY, UK

R. T. Cardé Department of Entomology, Fernald Hall, University of Massachusetts, Amherst, MA 01003, USA

J. R. Carlson Department of Biology, Yale University, Kline Biology Tower, PO Box 208103, New Haven, CT 06520–8103, USA

A. Cork Natural Resources Institute, Central Avenue, Chatham Maritime, Kent ME4 4TB, UK

C. Costantini Imperial College of Science, Technology & Medicine, Department of Biology, Silwood Park, Ascot, Berkshire SL5 7PY, UK

C. Curtis London School of Hygiene & Tropical Medicine, Keppel Street, London WC1E 7HT, UK

E. E. Davis SRI International, Life Sciences Division, 333 Ravenswood Avenue, Menlo Park, CA 94025–3493, USA

R. De Jong Institut de Zoologie, Université de Neuchâtel, Rue Emile-Argand 11, CH-2007 Neuchâtel, Switzerland

B. Dobrokhotov Manager Molecular Entomology, TDR–OMS, World Health Organization, CH-1211 Geneva 27, Switzerland

R. Galun The Hebrew University, Hadassah Medical School, PO Box 1172, 91904 Jerusalem, Israel

M. Geier   Universität Regensburg, Institut für Zoologie, Lehrstuhl Boeckh, Universitätsstrasse 31, D-93040 Regensburg, Germany

G. Gibson   Imperial College of Science, Technology & Medicine, Department of Biology, Silwood Park, Ascot, Berkshire SL5 7PY, UK

A. Grant   Worcester Foundation for Biomedical Research, 222 Maple Avenue, Shrewsbury, MA 01545, USA

P. Guerin   Institute of Zoology, University of Neuchâtel, Rue Emile-Argand 11, CH-2007 Neuchâtel, Switzerland

J. G. Hildebrand   ARL Division of Neurobiology, 611 Gould-Simpson Building, University of Arizona, Tucson, AZ 85721–0077, USA

K. E. Kaissling   Max-Planck-Institut für Verhaltensphysiologie, D-82319 Seewiesen, Germany

M. J. Klowden   Department of Entomology, University of Idaho, Moscow, ID 83844–2339, USA

B. G. J. Knols   Department of Entomology, Wageningen Agricultural University, PO Box 8031, 6700 EH Wageningen, The Netherlands

M. J. Lehane   School of Biological Sciences, University of Wales, Bangor, Gwynedd LL57 2UW, UK

H. Mustaparta   Department of Zoology, University of Trondheim-AVH, N-7055 Dragvoll, Trondheim, Norway

J. A. Pickett   Biological and Ecological Chemistry Department, IACR-Rothamsted, Harpenden, Hertfordshire AL5 2JQ, UK

R. A. Steinbrecht   Max-Planck-Institut für Verhaltensphysiologie, D-82319 Seewiesen, Germany

W. Takken   Department of Entomology, Wageningen Agricultural University, PO Box 8031, 6700 EH Wageningen, The Netherlands

G. Ziegelberger   Max-Planck-Institut für Verhaltensphysiologie, D-82319 Seewiesen, Germany

Preface

In the autumn of 1995, the Ciba Foundation organized a symposium and open meeting in London on olfaction in mosquito–host interactions. These initiatives resulted from a joint effort with the UNDP/World Bank/World Health Organization (WHO) Special programme for Research and Training in Tropical Diseases (TDR). The open meeting, moreover, was presented in co-operation with the Royal Society of Tropical Medicine and Hygiene and the Royal Entomological Society of London.

The WHO Special Programme in TDR was created in 1975 as a globally co-ordinated effort to bring resources of modern science to bear on the control of major tropical diseases—especially malaria, the schistosomiases, the filariases, the trypanosomiases, the leishmaniases and leprosy. TDR provides a mechanism for international scientific collaboration, and plays a unique role as a co-ordinator and facilitator among the growing number of national and international tropical disease research programmes.

The principal goal of this jointly sponsored symposium was to contribute to world efforts the control of mosquito-borne diseases in general, and malaria in particular by: (a) assessing the current state of knowledge about the roles and mechanisms of olfaction in host-seeking behaviour of vector mosquitoes; (b) identifying research needs and opportunities aimed at understanding and manipulating olfaction-based anthropophyly in malaria–vector species; and (c) stimulating research in this important area.

We very much hope that this record of the wide-ranging papers presented at this symposium and the lively discussion they stimulated will help to invigorate both basic and applied research, leading to improved strategies and tactics for the prevention of transmission of disease agents to human hosts by vector mosquitoes.

John G. Hildebrand
Chairman of the symposium and open meeting

Boris Dobrokhotov
WHO Special Programme for TDR

Chairman's introduction

John G. Hildebrand

ARL Division of Neurobiology, 611 Gould-Simpson Building, University of Arizona, Tucson, AZ 85721–0077, USA

I would like to say a few words about the motivation and ideas behind this symposium because it does have a few unconventional elements. First, I do not work in this field, and it is probably unusual to have a chairman of a Ciba Foundation symposium who is not an active participant in the field under discussion. Some of the other participants also don't work on mosquito olfaction, but that is part of the essence of this symposium as I conceived it from the beginning. Although we all have an interest in the olfactory system of insects and in the behaviour that is controlled by olfactory cues, we each have our own special fields. Some of us study insect olfaction as a model to learn about the principles of olfactory function in general. Others are entomologists who are interested in insect–plant interactions and in the role of olfaction in controlling the interactions of insects with plants that they parasitize. A few of us are interested in insect reproduction and how olfactory cues bring the sexes together for mating. Finally, some of the participants are world experts on the olfactory basis of mosquito–host interactions. The thinking behind the organization of this symposium was that if we brought together people with up-to-date approaches in each of these fields, then perhaps we could summarize the current understanding of mosquito olfaction and identify ideas for future research, with respect to the control of vectors that transmit debilitating human diseases.

I am not going to say anything about the importance of blood-sucking insects as vectors of disease or about mosquitoes in particular because the introductory presentations will bring us up-to-date on these matters. Rather, I would like to pose a few questions that I had in mind during the planning of the symposium, when I was assessing whether this was a worthy enterprise and thinking about which participants to invite. What is the role of olfaction in host location and in the interaction of mosquitoes with their hosts? No one doubts that olfaction is playing a part, but we ought to take stock of the supporting evidence and to what extent olfaction may be the overriding sensory modality for host recognition and host location. What are the necessary cues within an animal's effluvium that lead to host recognition and interaction of a mosquito with its host? How are those cues detected? How are

1

they processed by the mosquito's olfactory system? How do they ultimately generate behaviour? What are the behavioural consequences of those substances? We will be making a mistake if we do not focus from the very beginning on what mosquitoes actually do in the environment in response to cues that interest them and guide them to their hosts. We need to pay attention to the 'ecophysiology', or 'ecosensory biology', of mosquito behaviour by looking at the nature of the odour cues in the environment, and the behavioural responses of mosquitoes to other cues in the sometimes chaotic conditions of the natural environment. We must think about differences among mosquito groups and pay attention to their biology. We should also pay attention, even within mosquito taxa, to the differences between those groups that are more anthropophilic and those that are more zoophilic, and try to understand whether olfaction plays a significant role in forming the distinctions among mosquito groups. Finally, we must ask ourselves if there are poorly exploited (or even unexploited) targets for chemical intervention, genetic manipulation or behavioural manipulation of mosquitoes. We would miss the mark if we didn't try to direct future research into this field.

With these questions in mind, I believe that there are some clear purposes of this symposium. The first is to summarize and assess the state of this field. Second, it is valuable to put together a coherent account of the state of the field, so that all of our colleagues and students throughout the world will have the opportunity to join us in taking stock of the subject. Finally, it is a worthy goal to try to identify and draw attention to important problems yet to be explored by investigators—and especially by those who are new to the field.

To achieve these goals, I believe that it is important to mix together the experts in this field with those of us who bring expertise in allied areas, but who are not actually working directly on mosquitoes. I hope that this will result in an outcome different from that of other meetings that have considered mosquito olfaction and behaviour. I believe firmly that hybrid vigour is desirable in science, and that it is beneficial to bring together molecular geneticists, biophysicists, physiologists and behavioural biologists, because they will have provocative ideas about mosquito research.

We must endeavour to inspire and attract new, young investigators into this area. It is no accident that some people of that description have been invited to participate in this symposium. I hope that this symposium and the published volume will not only attract other young researchers, but also stimulate interest among researchers who work with *Drosophila* and other insect species, and develop greater vigour and diversity in this field.

Introduction I: an overview of mosquito biology, behaviour and importance

C. F. Curtis

London School of Hygiene & Tropical Medicine, Keppel Street, London WC1E 7HT, UK

Every few years the question 'what is the most dangerous animal in the world?' is raised on BBC Children's Television. The young questioner presumably expects the answer to be a lion, a tiger or perhaps a spitting cobra. However, we always advise the presenters to answer 'the African malaria mosquito, *Anopheles gambiae*'. The overwhelming importance of malaria as a vector-borne disease and of African children as its victims has only been fully recognized comparatively recently. Recent estimates (WHO 1993) are that 50% of the African rural population are infected with *Plasmodium falciparum* at any one time (often without symptoms because of immunity): there are 250–450 million clinical cases in Africa per year (80% of the world's total) and 1–3 million deaths are partly or wholly due to malaria. Almost all of these deaths occur in children under the age of five, before they have had a chance to build up a sufficient level of protective immunity as a result of repeated malaria attacks.

The reasons for the predominance of tropical Africans as malaria victims are partly politicoeconomic and climatic, but are mainly because of the high efficiency of *An. gambiae* as a vector. The importance of this factor was demonstrated by the accidental introductions of *An. gambiae* into Brazil and Egypt in the 1930s and 1940s, which led to malaria epidemics far worse than the levels of malaria endemic in those countries before the invasions and after their eradication (Soper & Wilson 1943, Shousa 1948).

The exceptional danger of *An. gambiae* arises from its strong tendency to feed on humans (anthropophily) and its long life. Both of these parameters have a strong influence on the vectorial capacity of an insect population (Macdonald 1957, Garrett Jones & Grab 1965) because, to be a vector of malaria, a mosquito must pick up gametocytes from one human, live 12 more days for them to mature to sporozoites and then bite another human whom the sporozoites can infect.

There are other highly anthropophilic and long lived anophelines, such as *Anopheles dirus* in the areas of multidrug-resistant *P. falciparum* in South-East

Asia. Fortunately, *An. dirus* breeds in small forest pools and does not appear in villages at high densities, but *An. gambiae* is well adapted to breeding sites created by the activities of humans and domestic animals in and around villages, e.g. footprints and hoofprints in marshy or irrigated land. Thus, 50–100 *An. gambiae* (about 5% of which are likely to be carrying sporozoites) may bite each villager each night. *An. gambiae* rest after blood feeding on the walls of the house (i.e. they are endophilic); therefore, they are vulnerable to the standard method of malaria vector control, i.e. house spraying with a residual insecticide. By contrast, *An. dirus* exit immediately after feeding indoors and are therefore difficult to kill.

Little is known of the mechanisms by which gravid females of different species seek out different types of site for oviposition, or how blood-fed females choose their places for resting and digestion.

An. gambiae are found at lower densities in houses than *Culex quinquefasciatus*, which are found in urbanized areas all over the tropics or subtropics. They breed in wet pit latrines, incompletely sealed cesspits or open drains. We recorded a pit latrine in Zanzibar yielding 13 000 *Cx. quinquefasciatus* per night (Maxwell et al 1990) and Gubler & Bhattachariya (1974) recorded about one bite/person per minute throughout the night in Howrah, near Calcutta. These mosquitoes are also anthropophilic (but will sometimes bite birds as an alternative) and are the cause of most mosquito nuisance, as well as being the vectors that initiated most of the world's 90 million chronic filariasis infections (Knudsen & Sloof 1992). Some of these infections lead to grossly deforming elephantiasis but apparently not to death, except by suicide.

The other mosquito-borne disease of major public health importance is dengue, for which there are recent estimates of 30–60 million infections and 100 000 acute clinical cases/year (Knudsen & Sloof 1992). The main vector is *Aedes aegypti*, which is localized around its domestic breeding sites in water accumulated in storage jars and discarded tyres, for example. Here it readily encounters humans for biting, but unlike *An. gambiae* and *Cx. quinquefasciatus*, it does so during the day. We may hear more at this symposium about the mechanisms by which mosquitoes locate humans for biting and therefore become dangerous vectors and/or a nuisance. Whether the same mechanisms are involved in species that bite during the night, evening and day has not, to my knowledge, been investigated.

A proper understanding and ability to manipulate olfactory (or other) searching mechanisms for human hosts or breeding sites might be exploited in the following ways.

(1) It may be possible to use traps for blood-seeking mosquitoes that are comparable to odour- and colour-baited tsetse traps (Lavessière et al 1991). One might envisage a mosquito trap that could outcompete one's own mosquito-attractive stimuli (without itself smelling intolerable). Perhaps

more immediately realizable would be improvements in the light traps that are currently used for monitoring mosquito populations (Lines et al 1991). This is especially needed for certain species that do not respond well to light traps, and where the standard method of monitoring by teams of human baits/catchers is becoming increasingly unethical because of the spread of drug-resistant malaria.

(2) Alternatively, we may be able to divert gravid females from their normal breeding sites to ovitraps in which larval development can be prevented (McCall & Cameron 1995).

(3) At present N,N-diethyl-m-toluamide (DEET) is used in most commercially available brands of insect repellent. It is effective for several hours but not overnight. Rare cases of poisoning do occur but it is considered to be less dangerous to have around the house, where children might misuse it, than household bleach (Veltri et al 1994). However, it does attack certain plastics and paintwork. Effective plant-based repellents are now becoming commercially available, but an effective dose has a strong smell. A repellent that would remain effective overnight without being rubbed off on bedclothes and is as effective as, but cheaper than, a bednet could make a real contribution to malaria and mosquito nuisance control.

(4) A system for genetically manipulating a vector population so that it changed its food preference to animals (zoophily) would be helpful (Curtis 1994). This might be more sustainable than attempts to make vectors refractory (genetically non-susceptible) to pathogens, as the latter might lead to the evolution of a pathogen strain that could evade the refractoriness gene which had been introduced into the mosquito population. Closely related mosquito species may differ sharply in anthropophily/zoophily but may sometimes be crossed successfully in the laboratory. By an appropriate backcrossing programme, with selection of the zoophilic segregrants at each generation, one might produce a strain for release that was zoophilic, but genetically and behaviourally compatible with a dangerous anthropophilic species. The sibling species *Anopheles quadriannulatus* and *An. gambiae sensu strictu* seem to be the obvious targets for this approach.

It may be useful at this symposium to speculate about fancy ways of controlling mosquito-borne disease. However, one should not get carried away and, in particular, one should keep in mind the following two opposing constraints in the real world.

First, the problem of insecticide resistance is not yet so widespread in *Anopheles* as is sometimes implied in the introductions to grant applications and other blatantly propagandist literature. Pyrethroids for house spraying or bednet impregnation provide highly effective methods for malaria vector and nuisance control. For olfactory traps to be a useful alternative, they would have to be cheaper or otherwise more attractive than established methods so as to encourage communities to use them more widely and consistently.

Second, in the most highly malarious areas of tropical Africa, transmission may be far above the level of 'saturation'. Snow et al (1994) have compared areas that have high levels of transmission with those that have 30-fold lower levels, and they have suggested that severe malaria rates might actually be made worse by vector control, short of virtual eradication, by delaying the build up of immunity in growing babies to an age at which they are apparently more prone to cerebral malaria. In such circumstances one should be cautious about advocating vector control methods that are not known to be able to produce massive and sustainable reductions in the vectorial capacity of mosquito populations (Snow & Marsh 1995). These may need to be backed up by improvements on existing malaria vaccines that could sustain immunity levels without the need for a child to suffer malaria attacks to acquire the immunity.

References

Curtis CF 1994 The case for malaria control by genetic manipulation of its vectors. Parasitol Today 10:371–374

Garrett Jones C, Grab B 1965 The assessment of insecticidal impact on the malaria mosquito's vectorial capacity from data on the proportion of parous females. Bull WHO 31:71–86

Gubler DJ, Bhattachariya NC 1974 A quantitative approach to the study of Bancroftian filariasis. Am J Trop Med Hyg 23:1027–1035

Knudsen AB, Sloof R 1992 Vector-borne disease problems in rapid urbanization: new approaches to vector control. Bull WHO 70:1–6

Lavessière C, Vail GA, Gouteux J-P 1991 Bait methods for tsetse control. In: Curtis CF (ed) Control of disease vectors in the community. Wolfe, London, p 47–74

Lines JD, Curtis CF, Wilkes TJ, Njunwa KJ 1991 Monitoring human-biting mosquitoes in Tanzania with light traps hung beside mosquito nets. Bull Entomol Res 81:77–84

Macdonald G 1957 The epidemiology and control of malaria. Oxford University Press, London

Maxwell CA, Curtis CF, Haji H, Kisumku S, Thalib AL, Yahya A 1990 Control of Bancroftian filariasis by integrating therapy with vector control using polystyrene beads. Trans R Soc Trop Med Hyg 84:709–714

McCall PJ, Cameron MM 1995 Oviposition pheromones in insect vectors. Parasitol Today 11:352–355

Shousha AT 1948 Species eradication: the eradication of *Anopheles gambiae* from Upper Egypt. Bull WHO 1:309–348

Snow RW, Marsh K 1995 Will reducing *Plasmodium falciparum* transmission alter malarial mortality among African children? Parasitol Today 11:188–190

Snow RW, Azevedo IB, Lowe BS et al 1994 Severe childhood malaria in two areas of markedly different *falciparum* transmission in East Africa. Acta Tropica 57:289–300

Soper FL, Wilson DB 1943 *Anopheles gambiae* in Brazil 1930–1940. Rockefeller Foundation, New York

Veltri JC, Osimitz TG, Bradford DC, Page BC 1994 Retrospective analysis of calls to poison control centers resulting from exposure to the insect repellent *N,N*-diethyl-*m*-toluamide (DEET) from 1985–1989. Clin Toxicol 32:1–16

WHO 1993 A global strategy for malaria control. World Health Organization, Geneva

Vector insects and their control

M. J. Lehane

School of Biological Sciences, University of Wales, Bangor, Gwynedd LL57 2UW, UK

Abstract. This paper emphasizes the huge influence that vector-transmitted disease has on humans using plague, epidemic typhus and nagana as examples. The continuing need for vector control in campaigns against insect-transmitted disease is shown by reference to current control programmes mounted against Chagas' disease, onchocerciasis, lymphatic filariasis and nagana. These successful campaigns have not been reliant on new breakthroughs but on the forging of available tools into effective strategies widely and efficiently used by the control authorities, and the long-lasting political commitment to the success of the schemes in question. A brief mention is made of current fashions in vector control research and that great care needs to be taken by policy-makers to achieve a balance between long-term research aiming at the production of fundamentally new control technologies and operational research aiming to forge the often highly effective tools we already have into sound control strategies.

1996 Olfaction in mosquito–host interactions. Wiley, Chichester (Ciba Foundation Symposium 200) p 8–21

Importance of blood-sucking insects

At present, only 750 000 species of insect have been described; the total number of insect species, however, is likely to lie between one and 10 million species. Further, it has been suggested that as many as 10^{18} individual insects are alive at any one instant, giving 200 million insects for each man, woman and child on earth. This makes the insects the pre-eminent form of life on land. Insects make a living in a variety of ways but thankfully only a few (300–400 species) have developed the habit of blood feeding. Despite being only a small fraction of the total number, these insects transmit diseases that have played an important role in shaping the course of human history (e.g. Duffy 1953, Bruce-Chwatt 1988, Harrison 1978) and they continue to have a major influence on economic development and human welfare over large areas of the earth.

Malaria, African trypanosomiasis (sleeping sickness), Chagas' disease (American trypanosomiasis), leishmaniasis, onchocerciasis (river blindness), filariasis (elephantiasis and loaiasis), bubonic plague, epidemic typhus, yellow fever, dengue and many other diseases are transmitted to man by insects. The numbers afflicted are colossal and the numbers considered at risk are staggering (Table 1). Although mosquitoes are the most important vectors, we should not forget that other insects are also of great importance. For example, Chagas' disease is transmitted by triatomine bugs, African trypanosomiasis by tsetse flies, onchocerciasis by blackflies, leishmaniasis by sandflies, epidemic typhus by lice and bubonic plague by fleas.

Bubonic plague can be used to illustrate the huge influence vector-borne disease has on humans. It is caused by the bacterium *Yersinia pestis* and is transmitted from its normal rodent host to humans by the bite of fleas. Subsequently, it can spread in the form of pneumonic plague from human to human via droplets produced by coughing and sneezing. In the pre-antibiotic era, diseases caused by flea bites had a mortality rate of 30–90%, and in some instances pneumonic plague was reported to have a mortality rate of 100%. The disease is endemic in the Eurasian steppe and there are three major epidemics recorded in history. The first was the plague of Justinian, which was at its peak from A.D. 542 until about A.D. 590 but carried on intermittently until A.D. 750. The devastation caused appears to have been as great as in the later, better documented epidemics (McNeill 1976)—10 000 people a day were dying in Constantinople at the height of the plague and a large proportion of the urban populations in the Mediterranean were lost to it. The plague is seen as a major factor in the last decline of the Roman Empire, in the successes of the Moslem armies against its remnants in A.D. 634 (Dols 1974) and in the movement of the centre of European civilization from the Mediterranean to more northerly lands. The second great epidemic originated in central Asia. Spreading eastwards to China and south to India it spread to Europe with a Mongol army laying siege to the trading city of Caffa in the Crimea in 1346. From here the disease spread by ship through the Mediterranean ports and then rapidly northwards through Europe. The best estimates suggest that

TABLE 1 **Numbers of people infected and numbers at risk (in millions) from various insect-transmitted diseases. Estimates from a number of sources (e.g. Knudson & Sloof 1992)**

Disease	Estimated number of people infected	Estimated number of people at risk
Malaria	365	2217
Onchocerciasis	20	90
Lymphatic filariasis	90	905
African trypanosomiasis	> 0.025	50
Chagas' disease	18	90
Leishmaniasis	12	350
Dengue fever	50	100s

between 1346 and 1350 a third of the population of Europe died. Historians suggest that this initial wave of the 'Black Death' and recurrent episodes of plague through the second half of the fourteenth century led to a profound change in the course of European history driven by the socioeconomic effects of depopulation. The disease continued to ravage Europe sporadically through the next three centuries. For example, it is estimated that a million Spaniards died of plague in the seventeenth century, and this has been seen as a major factor in the decline of Spain as an economic and political power (Bennassar 1969). The next great outbreak of plague was the great Indian epidemic of 1898–1918, in which it has been estimated that between 10 and 13 million lives were lost. Originating deep in the interior of China in 1855 on the back of a military rebellion (McNeill 1976), it spread to Hunan Province in 1892, to Canton in 1894 and from there to Hong Kong, Calcutta and Bombay. Having spread through India, where the worst of its effects were felt, it passed by the shipping lanes to ports in Java, Japan, Asia Minor, South Africa, Mediterranean Africa and parts of the Americas; thankfully in these regions it remained largely confined to seaports. This third pandemic is now coming to an end: 6004 cases were reported to the World Health Organization in 1967, and there were less than 1000 reported cases in 1992 (Knudsen & Sloof 1992). In 1943, with the advent of modern antibiotics, plague became an easily treatable disease and will remain so unless antibiotic resistance once more places us at the mercy of this dreadful, history-changing zoonosis.

Insects also have an indirect effect on human welfare because they transmit diseases to domesticated animals. Nagana (sleeping sickness of animals in Africa), surra, bluetongue, African horse sickness, Akabane virus, rift valley fever and many other diseases are caused by parasites transmitted to domesticated animals by insect vectors. In addition, the massively important piroplasms causing red water fever (*Babesia*) (and the now eradicated Texas fever in the USA) and East Coast fever (*Theileria*) are transmitted by ticks (acari). These vector-borne diseases are largely confined to the tropics and subtropics. Given that the carrying-capacity has been reached for most agricultural livestock systems in the developed world, if future expansions in production are required they will need to be in these regions blighted by vector-borne disease. This will mean both increasing productivity of land already in use by minimizing disease impact and making new areas available for exploitation through disease control. One of the most celebrated examples of agricultural constraints caused by vector-borne disease is nagana. Caused by trypanosome species transmitted by tsetse flies, this disease prevents the use of about 10 million km^2 of Africa for cattle rearing (Steelman 1976). An estimate, now over 20 years old, suggested that trypanosomiasis cost the cattle industry US$5 billion per year (Goodwin 1973). Of course the arguments for control and the agricultural expansions this will permit are not straightforward because control of the tsetse and the trypanosomiasis it transmits may well, in the minds of many commentators, lead to widespread losses of African game animals and heighten African overgrazing problems. The arguments have been clearly outlined by Jordan (1986).

In addition to the diseases they transmit to livestock, blood-sucking insects also cause considerable direct losses in meat and milk yields through the annoyance of their bites (Steelman 1976). The question of the direct annoyance to man caused by blood-sucking insects is more difficult to address because it is difficult to quantify. It is best addressed as tolerance thresholds and I would contend that our tolerance in western society has decreased considerably over the last century largely because an increasingly urban and affluent population encounters blood-sucking insects increasingly rarely. Over the same period increasing leisure time and income levels have made national and international tourism one of the world's most important industries. The combination of the two trends have made nuisance insects a constraint to the development of tourism in many parts of the world—notably in the Camargue region of southern France, the Scottish highlands, the Bahamas, Florida and many parts of the Caribbean (Linley & Davies 1971)—although it is rare to find the tourism industry making public statements of this kind for obvious reasons. The Marquesas islands in French Polynesia are one specific example that has been studied recently. The day-biting midge *Leptoconops albiventris,* called the nono by local people, plagues the islands' beaches, preventing development of tourism and forming a serious impairment to economic development (Aussel 1993).

Control of blood-sucking insects

Clearly, effective vector control is essential if we are to deal with problems caused directly by blood-sucking insects. But have other control agencies rendered vector control unnecessary? Most certainly not. For most of the diseases listed above, vector control plays a significant role and for some— such as South American sleeping sickness, where no vaccines or drugs for public health use are available—it is the major control strategy. What vector control strategies are available? At the beginning of the twentieth century, when the role insects play in disease transmission was first being appreciated, the tools available for vector control were poor. For the first four decades of the century the most important advances in control were due to a rapid advance in our understanding of the biology of the insects, which permitted more efficient use of the rather limited tools available. Since 1941, with the discovery of the insecticidal properties of dichlorodiphenyltrichloroethane (DDT) by Müller in the laboratories of Ciba Geigy, we have had infinitely more powerful tools at our disposal. To illustrate this we can compare the anti-malaria operations in Mian Mir, a British army encampment in India in the early 1900s, with the campaign to rid Italy of malaria immediately after the Second World War. In the pre-insecticide era Christophers and colleagues used 500 British and Indian soldiers to control mosquito breeding within a half mile of the soldiers' living quarters in Mian Mir. Despite continuing intensive operations over a five-year period by a highly disciplined workforce and the very limited objective of preventing malaria transmission in the encampment, there was no success: the enlarged spleen rate in children remained at 66% and the number of adult mosquitoes in the encampment was undiminished. In contrast to this monumental failure, the Italian government and the health division of the United Nations relief organization, armed with the new insecticide DDT, took just two years from 1945 to sweep malaria from Italy (excluding Sardinia) (Harrison 1978). The power of these new tools was driven home to the control community by their effect on the epidemic typhus outbreak in Naples in 1943. Epidemic typhus is transmitted by lice and its huge influence on humans is famously documented by Zinsser (1935). The disease is credited with a major role in the defeat of Napoleon's armies in Russia. In Europe and Russia between 1917 and 1923 it is estimated that there were three million deaths among the 30 million who suffered from the disease. The epidemic in Naples in 1943 was the last and was brought to an abrupt end by DDT and other anti-louse measures—the first time this ancient scourge of man was halted in mid-epidemic. The immense power of these insecticidal tools was obvious to all and they were seen by many (but certainly not all) at the time as a magic bullet, capable of dealing at a stroke with long-standing problems with vector-borne disease. Many of the earlier lessons of species sanitation were set aside. This view of insecticides as a panacea lasted perhaps 10 years among the optimistic but was finally dispelled by the perceived 'failure' of the global malaria eradication campaign. (Despite its supposed failure it should be remembered that largely because of this campaign malaria was reduced from a threat to 64% of the world's population in 1950 to 38% in 1972.) The dispelling of the idea that insecticides were the magic bullets, and the realization that insecticide usage strategies must be planned carefully to be successful, leads us to the modern age in which there are several impressive success stories. All are characterized by the forging of available tools into sensible and sustainable control strategies and by political will to continue the control effort to its conclusion despite the economic pressures to switch scarce resources which inevitably arise once a degree of control has been achieved. I will outline some of these recent successes.

It is estimated that two-thirds of the 16–20 million people infected with Chagas' disease live in Argentina, Chile, Uruguay, Paraguay, Bolivia, Brazil and Peru (WHO 1991). In these seven countries the major vector is *Triatoma infestans*. For all practical purposes, this bug is only found in the domestic and peridomestic environment. It has slow population growth and is susceptible to modern insecticides; therefore, it is readily accessible to insecticidal control measures. The Southern Cone initiative, a co-ordinated effort of six of the seven countries (Peru is a non-participant) with support funding from the developed world, was established in 1991. It is attempting the elimination of *T. infestans* by house spraying with synthetic pyrethroids, and this attack on Chagas' disease is supported by efforts against transmission of the disease through blood transfusion services (PAHO 1993, Schofield 1992). Expenditure from 1991–1995 is estimated at US$156 million. Uruguay and Chile are virtually free of transmission, and Argentina, Paraguay and particularly Brazil are making good progress (C. J. Schofield, personal communication 1995). This successful campaign did not rely on a remarkable new breakthrough in control technology. It just required political will fostered and stimulated by interested and persistent individuals.

The same is true of onchocerciasis control. Onchocerciasis afflicts about 18 million people, mainly in 27 African countries. It causes blindness in about 340 000 people, including up to 40% of the adult male work force in some areas in the sub-Saharan savannah belt. In the past, onchocerciasis had a disastrous socioeconomic effect in this region because it forced people to abandon fertile land near rivers. The onchocerciasis control campaign started in 1974 and relied until the late 1980s on direct vector control through larviciding of rivers over a 1.3 million km^2 area in 11 countries to kill the vector *Simulium damnosum*. This audacious and brilliantly executed vector control programme has been remarkably successful, achieving disease control over most of the area concerned (Molyneux 1995). Since 1988, vector control has been supplemented with the drug ivermectin (Mectizan) which, like vector control, interrupts transmission but in addition helps with the treatment of the disease because it rapidly kills the microfilaria which are the cause of the eye pathology (Tsalikis 1993). Although eradication of *Onchocerca volvulus* is currently an unrealistic prospect (Duke 1990), the onchocerciasis control campaign, costing approximately US$27 million per annum, has proven that disease control is achievable. The question is will the political will be there to continue control operations indefinitely, particularly with rates of blindness falling markedly and other health priorities coming to the top of the political agenda?

Well-planned and well-executed control operations, commonly incorporating vector control, are proving exceptionally effective in dealing with lymphatic filariasis (Ottesen & Ramachandran 1995). These strategies have eliminated filariasis from Japan, Taiwan and South Korea, and China is on the road to a similar success. Control is based on the interruption of disease transmission through drug and vector control programmes. The extremely safe drug diethyl carbamazine (DEC) (probably soon to be supplemented by ivermectin) is the major tool, delivered in cost-effective ways either as an additive to table salt or in once-a-year doses. The drug kills the microfilarial stage, preventing infection of mosquitoes and the transmission of the disease. Vector control plays an effective supporting role for these drug-based campaigns and can often be integrated with anti-mosquito measures aimed at other diseases. Vector control is particularly effective where the disease is transmitted by endophilic mosquitoes, such as *Culex quinquefasciatus* and *Anopheles*, and least effective for vectors such as the *Aedes scutellaris* complex where the adults are exophilic and larval breeding sites are too small and disparate for effective control to be instigated. The remarkable success of these campaigns means that lymphatic filariasis has been identified by the International Task Force for Disease Eradication as one of only six potentially eradicable diseases (Center for Disease Control 1993).

A final example is the control of nagana in African livestock. The trypanosomes causing the disease are transmitted by tsetse flies. Despite widespread drug resistance, available trypanocidal drugs can be effective in controlling disease in sedentary, well-managed herds even when kept in areas of high challenge. But this is not typical of cattle rearing across most of Africa where pastoralism, poor herd management and lack of veterinary services suggest that prospects for the widespread success of the drug-based approach to trypanosomiasis control are poor, particularly for economically sound cattle-rearing operations (Jordan 1986). However, the tsetse fly is present at low densities and has a low reproductive rate. Removal of significant parts of the population can cause the disappearance of flies from an area. Since the early 1970s, when biconical traps were introduced for sampling tsetse populations (Challier & Laveissière 1974), trap technology for tsetse control, including the use of natural baits, has taken considerable steps forward and is now widely and successfully used for control purposes (reviewed by Green 1994). For example, a campaign run in Côte d'Ivoire since 1980 has cleared $60\,000\,km^2$, which is 19% of the total area of the country, of tsetse fly (see Green 1994).

Looking in from the outside, one might be misled into thinking that fundamentally more effective insecticides and drugs had suddenly emerged in the last decade or so to make such successes possible. Neither is true. There have been advances (such as the emergence of ivermectin as a safe drug for use in onchocerciasis control, for example) but most of the tools have been available for a long time—DEC was first used clinically in 1947 and traps were first used against tsetse in the 1930s. The insecticidal control tools are mostly long standing. Thus, the major environmental management treatments, including source reduction, house or personal screening, date back at least to the beginning of the century, and even the introduction of synthetic pyrethroids in the form of allethrin came in 1969. What has happened is that the tools available have been forged into effective strategies, which have been widely and efficiently used by the control authorities; and, of particular importance, there has been a long-lasting political commitment to the success of the schemes in question.

Future trends in vector control

Finally I would like to mention the current fashion in vector control research. Many laboratories are using molecular techniques to try to devise means for the eventual introduction of refractory genes into vector populations (Crampton 1994). Others are attempting to develop vaccination strategies against vector insects (Jacobs-Lorena & Lemos 1995) or the parasites within them. These elegant technologies hold the promise that radical, new control tools will be developed, but it is generally accepted that this is unlikely to happen within a decade. When they arrive such tools will probably fill a useful role within well-designed and well-organized control programmes, but they are unlikely to be the magic bullet some might suggest. It should be clear from the examples above that much can be achieved if available resources are concentrated on operational research designed to forge the tools we have into ever more effective control strategies suited to local conditions (e.g. Curtis 1994). Combining this with funding and training strategies, which ensure that these are implemented efficiently, and effective lobbying to gain full and lasting political backing for the control operations undertaken can achieve much for human welfare in the immediate future.

References

Aussel JP 1993 Ecology of the biting midge *Leptoconops albiventris* in French Polynesia. II. Location of breeding sites and larval microdistribution. Med Vet Entomol 7:73–79

Bennassar B 1969 Recherches sur les grandes epidemies dans le nord de l'Espagne a la fin du XVI siecle. Service d'edition et de vent des publications de l'education national, Paris

Bruce-Chwatt LJ 1988 History of malaria from pre-history to eradication. In: Wernsdorfer WH, McGregor I (eds) Malaria: principles and practice of malariology. Churchill Livingstone, Edinburgh, p 1–60

Center for Disease Control 1993 Morb Mortal Wkly Rep 42:1–38

Challier A, Laveissière C 1974 Un nouveau piege pour la capture des glossines (*Glossina:* Diptera, Muscidae): description et essais sur le terrain. Cah ORSTOM 11:251–262

Crampton JM 1994 Approaches to vector control: new and trusted. 3. Prospects for genetic manipulation of insect vectors. Trans R Soc Trop Med Hyg 88:141–143

Curtis CF 1994 Approaches to vector control: new and trusted. 4. Appropriate technology for vector control: impregnated bed nets, polystyrene beads and fly traps. Trans R Soc Trop Med Hyg 88:144–146

Dols MW 1974 Plague in early islamic history. J Am Oriental Soc 94:371–383

Duffy J 1953 Epidemics in colonial America. Louisiana State University Press, Baton Rouge, LA

Duke BOL 1990 Onchocerciasis (river blindness): can it be eradicated? Parasitol Today 6:82–84

Goodwin RFW 1973 The cost effectiveness of animal disease control. Span, London Shell International Chemical Industry 16:63–64

Green CH 1994 Bait methods for tsetse fly control. Adv Parasitol 34:229–291

Harrison G 1978 Mosquitoes, malaria and man: a history of hostilities since 1880. John Murray, London

Jacobs-Lorena M, Lemos FJA 1995 Immunological strategies for control of insect disease vectors: a critical assessment. Parasitol Today 11:144–147

Jordan AM 1986 Trypanosomiasis and African rural development. Longman, London

Knudsen AB, Sloof R 1992 Vector-borne disease problems in rapid urbanization: new approaches to vector control. Bull WHO 70:1–6

Laveissière C, Couret D, Kienon JP 1981 Lutte contre les glossines riveraines a l'aide de pieges biconique impregnes d'insecticide, en zone de savane humide. 5. Note de synthese. Cah ORSTOM 23:297–303

Linley JR, Davies JB 1971 Sandflies and tourism in Florida and the Bahamas and Caribbean area. J Econ Entomol 64:264–278

McNeill WH 1976 Plagues and peoples. Blackwell, Oxford

Molyneux DH 1995 Onchocerciasis control in West Africa: current status and future of the Onchocerciasis Control Programme. Parasitol Today 11:399–402

Ottesen EA, Ramachandran CP 1995 Lymphatic filariasis infection and disease: control strategies. Parasitol Today 11:129–131

PAHO 1993 Document PNSP/92–18. Iniciativa del cono sur. Pan American Health Organization, Washington DC

Schofield CJ 1992 Eradication of *Triatoma infestans*: a new regional programme for southern Latin America. Ann Soc Belg Med Trop 72:69–70

Steelman CD 1976 Effects of external and internal arthropod parasites on domestic livestock production. Ann Rev Entomol 21:155–178

Tsalikis G 1993 The onchocerciasis control program in west Africa: a review of progress. Health Policy & Plan 8:349–359

WHO 1991 Control of Chagas' disease. World Health Organization, Geneva (Technical Report Series 811)

Zinsser H 1935 Rats, lice and history. Atlantic Monthly Press, New York

DISCUSSION

Hildebrand: Finding a solution for these devastating diseases has important political and economic ramifications that are probably more problematic than the biology that we're discussing here. Several years ago, I read an editorial in a medical journal arguing that the majority of the most devastating and debilitating diseases of humans are vector-borne diseases. Diseases such as cancer and heart disease were characterized as barely significant in comparison.

At that time the total world investment and research on those 'great, neglected diseases of mankind' was in the order of US$40 million, or more than an order of magnitude less than the budget of the National Cancer Institute of the National Institutes of Health in the USA. My own interest in this field was initiated by reading that editorial and thinking about how distorted the priorities of our governments are. I hope that we won't forget, as we talk about the biology of olfactory-mediated, mosquito–host interactions, about economic and political issues.

Brady: Louis Miller (personal communication) has pointed out that the National Institutes of Health spends at least US$1.4 billion on AIDS research compared with about US$30 million on malaria.

Steinbrecht: If AIDS did not affect the European and industrialized Northern hemisphere, there probably wouldn't be much spent on research into AIDS either. Vector-borne diseases occur primarily in the developing countries, and they seem to be left alone to deal with them.

Hildebrand: Perhaps we could go further and suggest that so little is spent on vector-borne diseases because they usually do not afflict the families of congressmen and members of Parliament.

Lehane: It even goes beyond that because the money which is available for disease research is often not aimed directly at control: it's often aimed at supporting western laboratories, and there is a difference. We sometimes consider our best interests when we're using this money, which is an extremely difficult dilemma for a scientist. One has to do one's best for the laboratory, but we shouldn't get carried away with the idea that this is always the best route to control diseases. I'm not sure how to communicate this message to politicians. It's interesting to note that two of the most successful campaigns that I have outlined here—the onchocerciasis (river blindness) campaign and the Chagas' disease campaign—were started by the enthusiasm of just a few interested entomologists.

Curtis: I share John Hildebrand's sentiments about the importance of vector-borne diseases in developing countries, but the point may have been slightly overstated. For example, diseases such as tuberculosis, acute respiratory diseases and diarrhoea may kill more people than malaria, and they are not vector borne, although they are primarily diseases of developing countries and are therefore grossly under funded.

Hildebrand: Both Chris Curtis and Mike Lehane have stressed the importance of giving due respect to the use of pesticides. Even if traditional pesticides can still be used effectively, don't you agree that the inevitable problem of biological resistance has to be faced in the long run, so that it's also important to develop a better understanding of the biology of mosquitoes and their interactions with hosts for a long-term solution?

Lehane: But pesticide resistance has not been described for triatomine bugs or tsetse flies, although it is more of a concern for mosquitoes.

Brady: Could I make a point about tsetse flies, because they represent a model that we may refer to several times over the next couple of days. The European Economic Community was going to eradicate tsetse flies by an aerial spraying programme covering 350 000 km² of the South African bush with ULV endosulfan, but decided to switch to using odour-baited, insecticide-treated targets instead, partly because of 'green' pressures. Also, in contrast to the point which Mike Lehane mentioned that the western countries spend much of their research money on supporting their own laboratories, this target technology was developed completely endogenously within a Third World country, i.e. Zimbabwe.

Dobrokhotov: Vector control based on the use of insecticides was a key element for many World Health Organization (WHO) support disease control programmes. All of them are now facing serious problems due to the rapid increase in insecticide resistance. Many developing countries cannot afford the new, more effective insecticides. The same problems occur with the use of tsetse traps. Despite their efficiency, there is not a single example of a self-supported trap programme in Africa—all of them depend on external funding. The well-known success of the onchocerciasis control programme, which practically eliminated onchocerciasis in West Africa, was achieved by means of international multi-donor support managed by the WHO. Further sustainability of the programme will depend on the combination of black fly control with the treatment of patients with drugs (ivermectin).

Guerin: May I just raise the point regarding the consequences of controlling the behaviour of individuals on the vector population. Mike Lehane provided the example of the direct treatment of houses infested by triatomines. In the case of onchocerciasis, where rivers act as the reservoir for the vector, the control programme outlined is probably the only way of handling an acute situation using an 'environmentally acceptable' approach. In the onchocerciasis control programme, was just one species targeted or was it a series of species?

Lehane: At the start of the campaign, no one anticipated the number of problems that they would face with *Simulium* because it is a species complex, and different members of the complex may have subtle variations in their lifestyle that make control of one member of the complex more difficult than others. The aim of the campaign was not to remove the population, but to interrupt the transmission for a set period of time, i.e. until the adult worms died out in the human population, which was estimated to take about 20 years. However, it's quite clear now that it's going to have to be a continuing programme, either based on interrupted transmissions using insecticides or ivermectin if it's going to be successful.

Galun: I agree with Mike Lehane that insecticides are going to remain our major means of insect control. In a recent lecture delivered by Perry Atkinsson in Israel upon receiving the Wolf Prize, he described a 30-year effort to integrate pest management, aimed at reducing the use of insecticides in cotton. So far they have achieved a 15–20% reduction.

Even the example given by Mike Lehane concerning the development of traps for the control of tsetse flies still involves insecticides. The traps used are basically blue screens impregnated with permethrins and accompanied by a source of acetone and octenol. Thus, we are dealing with visual and chemical attractants that bring the flies to the insecticide. This method is much safer to the environment than the insecticide sprays used earlier for the control of tsetse flies.

The use of bed nets impregnated with permethrin is a similar example. I wonder if impregnation with a non-repellent insecticide will cause higher mortality of the mosquitoes attracted to the people under the nets, because more of them will land on the netting than on netting treated with repellent.

Curtis: The results of trials that my colleagues and I have been carrying out with organophosphates and carbamates are not very encouraging compared with results with pyrethroids (Miller et al 1991, Curtis et al 1996, Weerasooriya et al 1996). Repellent pyrethroids do not work at a distance: mosquitoes land on the bed nets, are irritated and fly away. Hossain & Curtis (1989) showed that mosquitoes do rest on pyrethroid-treated nets for a shorter period than on untreated nets. It may be possible, just by adjusting the pyrethroid dose, to reduce the excitorepellency whilst retaining the killing. Repellency is beneficial to the bed net user, but killing is better because the mosquito is not around to bite someone else.

Gibson: This is the perfect opportunity to make a bid for more behavioural studies similar to those of Houssain & Curtis (1989) and Miller & Gibson (1994), so that we can find out exactly what the excitorepellency effect is and how different formulations alter repellency. In some cases, it is not even clear whether the carrier chemicals of the pesticide or the pesticide itself have these effects.

Curtis: The deterrency to house entry by mosquitoes, which has sometimes been reported, does seem to be due mainly to the components of the formulation (Lindsay et al 1991), but the actual excitorepellency on contact with a treated net concerns the pyrethroid itself. One of the great dangers of repellency, especially when the programme is not highly organized or well-funded, is that mosquitoes may be diverted from the relatively rich people who can afford to buy the nets to the poorer who can't.

Hildebrand: This reminds me of an effort in the 1950s in the USA to control the Japanese beetle, which is a horticultural pest in the north-eastern States. Traps were bated with odorous material which was attractive to the beetles. This was a successful approach for trapping the beetles but, at the same time, the odour-emitting traps attracted beetles into the 'protected' gardens so that they became heavily infested. Would there be a similar danger in using the trapping techniques that you described for mosquitoes, i.e. would they attract mosquitoes into people's bedrooms?

Curtis: Yes, there are many potential problems with this approach. For example, researchers are working on human sweat and trying to establish which of its components are attractive. Unfortunately, one could imagine that a super-efficient trap might smell extremely strongly of human sweat! Mosquitoes don't target themselves directly onto humans, they come into a room and for reasons that are not understood they fly around the room before they bite. Therefore, there is a certain amount of time for the extra mosquitoes that are attracted in to get trapped if there is a super-efficient trap. But that's a big if.

Bowen: Does anyone know anything about the mechanism of bed net insecticide resistance in *Culex* found in China and why it was not present in *Anopheles*?

Curtis: In the case of *Culex*, selection has occurred in a factory in which vaporizing mats were manufactured and therefore the atmosphere was presumably contaminated with volatile pyrethroids (Kang et al 1995). That stock had been kept in the factory for about five years to provide material for daily quality control checks on the product, and simply living there for that time was sufficient to cause resistance. I heard just two days ago from Kang in China that he has taken up my suggestion of keeping an *Anopheles* anthropophagus stock in the same factory, and so far (after about six months) there have been no signs of resistance. *Culex* does seem to be more able to become resistant to pyrethroids than *Anopheles*, but I don't know why. We have got three pyrethroid-resistant *Anopheles* stocks in the London School of Hygiene & Tropical Medicine, which have mainly arisen from artificial selection.

Carlson: In your introduction you mentioned mating disruption as a possible means of controlling vectors (Curtis 1996, this volume). Can you comment on which sensory cues may be the most important in driving mating behaviour?

Curtis: From an olfactory point of view, I don't think that mating disruption has a role in controlling vectors. Male mosquitoes are attracted by the sound of female wing beats. There has been a claim that female mosquitoes can be attracted to sound traps (Kanda et al 1987), but I'm not totally convinced by those reports. I'm not aware of any reports of a volatile, long-distance sex pheromone.

Grant: I was under the impression that in *Aedes aegypti* mating occurs near the host. Is it possible that males are using the same cues to both locate the host and find the females?

Curtis: Yes. Male *Ae. aegypti* do wait around the host presumably for the females to appear.

Davis: We have found receptors for L-lactic acid on male mosquitoes. We have tried to determine whether L-lactic acid is attractive to male mosquitoes using an olfactometer, but so far this has not been successful. This suggests that other odour components are necessary (Davis 1976).

References

Curtis CF 1996 Introduction I: an overview of mosquito biology, behaviour and importance. In: Olfaction in mosquito–host interactions. Wiley, Chichester (Ciba Found Symp 200) p 3–7

Curtis CF, Myamba J, Wilkes TJ 1996 Comparison of different insecticides and fabrics for anti-mosquito bednets and curtains. Med Vet Entomol 10:1–11

Davis EE 1976 Responses of the antennal receptors of male *Aedes aegypti* mosquitoes. J Insect Physiol 23:613–617

Houssain MI, Curtis CF 1989 Permethrin-impregnated bednets: behavioural and killing effects on mosquitoes. Med Vet Entomol 3:367–376

Kanda T, Cheong WH, Loong KP et al 1987 Collection of male mosquitoes from a field population by sound trapping. Trop Biomed 4:161–166

Kang BW, Gao B, Jiang H et al 1995 Tests for possible effects of selection by domestic pyrethroids for resistance in culicine and anopheline mosquitoes in Sichuan and Hubei, China. Ann Trop Med Parasitol 89:677–684

Lindsay SW, Adiamah JH, Miller JE, Armstrong JRM 1991 Pyrethroid-treated bednet effects on mosquitoes of the *Anopheles gambiae* complex in the Gambia. Med Vet Entomol 5:477–483

Miller JE, Gibson G 1994 Behavioural response of host-seeking mosquitoes (Diptera: Culicidae) to insecticide-impregnated bed netting: a new approach to insecticide bioassays. J Med Entomol 31:114–122

Miller JE, Lindsay SW, Armstrong JRM 1991 Experimental hut trials of bednets impregnated with synthetic pyrethroid or organophosphate insecticide for mosquito control in the Gambia. Med Vet Entomol 5:465–476

Weerasooriya MV, Munasinghe CS, Mudalinge MPS, Curtis CF 1996 Comparative efficacy of permethrin, lambdacyhalothrin and bendiocarb impregnated house curtains against the vector of bancroftian filariasis *Culex quinquefasciatus* in Matara, Sri Lanka. Trans R Soc Trop Med Hyg, in press

Genetics, ecology and behaviour of anophelines

Gabriella Gibson

Imperial College of Science, Technology & Medicine, Department of Biology, Silwood Park, Ascot, Berkshire SL5 7PY, UK

Abstract. The efficiency with which mosquitoes transmit malaria is related to how closely associated they are with the human host. For example, the relative vectorial capacity of two species may be determined by differences in their degree of preference for human blood or in their degree of preference for blood-feeding indoors versus outdoors. Species complexes, such as *Anopheles gambiae sensu lato*, allow us to investigate how species differences in genetics, ecology and behaviour can lead to significant differences in vectorial capacity. The potential exists for identification of behaviour-regulating genes for exploitation by novel control measures. Close correlations have been demonstrated between certain behaviours and karyotypes in the *An. gambiae s.l.* complex, but the physiological basis for these correlations has yet to be determined. Recent evidence from behavioural studies suggests that differences in host preference may reflect differences in the relative responsiveness to CO_2 and other, more specific, host odours.

1996 Olfaction in mosquito–host interactions. Wiley, Chichester (Ciba Foundation Symposium 200) p 22–45

There are more than 400 species in the genus *Anopheles*, but only about 30 are primary vectors of human malaria, and about another 30 are secondary or local vectors (White 1982). The behaviour and ecology of the vector species vary greatly, resulting in quite different degrees of vectorial importance. In general, the more closely associated they are with humans and human habitats, the greater their vectorial capacity. Vector species can be found as far north as the Arctic circle, and throughout temperate, subtropical and tropical areas, with the original distribution of malaria almost as widespread. As the map (Fig. 1) shows, malaria has disappeared from the marginal regions due to changes in climate and social practices, with virtually no intentional human intervention (MacArthur 1952), and from the temperate areas due to a combination of environmental change and human intervention (Bruce-Chwatt & De Zulueta 1980). The disease is now mainly confined to the tropics and subtropics. In some areas adequate tools are available to reduce (if not

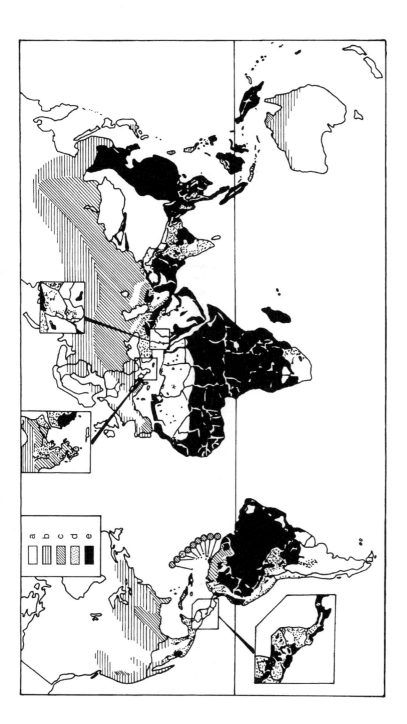

FIG. 1. Global distribution of malaria in 1968: (a) originally malaria free; (b) malaria disappeared by 1950 without specific programmes; (c) malaria eradicated by specific programmes after 1950; (d) areas in consolidation phase of malaria eradication; (e) areas in attack phase or without eradication programmes (Wernsdorfer & McGregor 1988).

eradicate) it, although economic, political and practical factors often act as constraints. In the most endemic areas, however, the current prospects for malaria control are not promising, particularly in areas where insecticide and drug resistance have developed, thus reducing the range of control measures available (Bruce-Chwatt 1988, Gramiccia & Beales 1988).

The *Anopheles gambiae sensu lato* species complex

Complexes of sibling species are common in the anophelines (White 1977), with subtle differences in the behaviour and ecology of morphologically identical species giving rise to significant differences in their vectorial capacities. Until the existence of each species complex has been identified, malaria control strategists are hard-pressed to explain how presumed vector species could be present with no transmission of malaria. This is not the forum, however, to review the extensive and excellent research into the biology of all the anopheline vector species. Rather, for several reasons, I have chosen to highlight the case of the *Anopheles gambiae sensu lato* complex in Africa to illustrate how differences in the ecology and behaviour of sibling species can have profound effects on the transmission of malaria. Of all the anopheline species complexes, the most consistent data are available on the behaviour and ecology of each species in the *An. gambiae s.l.* complex because of the ability to identify samples to the species level. In many other cases the lack of information about the distribution or even existence of sibling species has clouded research on the biology and ecology of vectors. In addition, *An. gambiae sensu stricto* is not only the most efficient and, hence, dangerous vector (Curtis 1996, this volume), but it is also one of the most difficult to control.

Since the turn of the century *An. gambiae s.l.* had been recognized as an efficient malaria vector. The range of behaviours and breeding sites found across its distribution hinted at there being more than one species present, but the sibling species are morphologically so similar that it was not until the 1960s that the six main species were identified by crossing experiments which demonstrated partial or complete sterility of the hybrids (Davidson 1964, Davidson & Hunt 1973). Pre-copulatory reproductive barriers must also exist, since hybrids are rarely found in the field. The discovery that each species has a characteristic banding pattern on their polytene chromosomes has permitted a detailed analysis of their behaviour and ecology (Coluzzi et al 1979).

Figure 2 shows the six main species that have been identified so far, although several other forms may soon be defined as distinct species. There is significant variation between the species in breeding site, distribution, host preference, resting-site preference and vectorial status. *Anopheles quadriannulatus* is thought to be the ancestral form, or most closely related to the ancestral form: it has been reported to occur only in three disconnected areas (Ethiopia, Zanzibar and in southern Africa) in fairly temperate climates, typical of

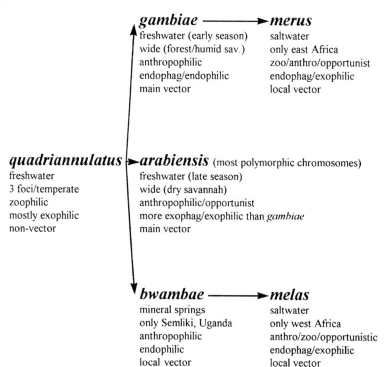

FIG. 2. Biology of the six main species of the *Anopheles gambiae* complex. Arrows indicate the most likely phylogeny based on chromosome banding patterns (Coluzzi et al 1979).

conditions which would have been more widespread in Africa during the Pleistocene, before the last ice age. It is mainly zoophagic and exophilic (although Table 1 shows that its behaviour varies to some degree) and it is not considered to be a vector. Its polytene chromosome sequence also places it centrally in the complex (Coluzzi et al 1979).

The two species of greatest medical importance are thought to have evolved independently from *An. quadriannulatus*: *An. gambiae s.s.* being originally a forest-dwelling mosquito and *Anopheles arabiensis* being more pre-adapted to arid environments. Both species are the most closely adapted to the human environment within the complex and are the most widespread. Although *An. gambiae s.s.* is generally the more efficient vector, *An. arabiensis* is also considered to be a primary vector and is the only known vector in Sudan, Ethiopia, Zimbabwe and South Africa. They are sympatric over much of their distribution across most of Africa, although *An. gambiae s.s.* generally predominates in the forest and humid savannah areas, whereas *An. arabiensis*

TABLE 1 Classification of feeding and resting behaviour patterns for four members of the **Anopheles gambiae** complex

Species	Example locality	Exophagic[a]	Endophagic[b]			
		Completely exophilic[c]	Postprandially endophilic[d]	Postprandially exophilic[e]	Partially endophilic[f]	Completely endophilic[g]
Anopheles gambiae	general	– –	– –	+ –	+ +	+ +
Anopheles arabiensis	Taveta, Kenya;	– –	– –	+ –	+ +	+ +
	Jima, Ethiopia;	– –	+ –	– –	+ +	+ +
	Pare, Tanzania;	+ +	+ +	+ –	+ +	+ –
	Manyara, Tanzania	+ +	– –	– –	– –	– –
Anopheles quadriannulatus	Ethiopia;	– –	+ –	– –	+ +	+ +
	Zimbabwe	+ +	– –	+ –	+ –	not known
Anopheles bwambae	Bwamba, Uganda	+ +	– –	+ –	+ +	+ –

[a]Outdoor feeding.
[b]Indoor feeding.
[c]Outdoor resting, two or more days outdoors.
[d]Indoor resting, unknown number of days indoors.
[e]Outdoor resting, no days indoors.
[f]Indoor resting, one day indoors.
[g]Indoor resting, two or more days indoors.
– –, less than 5%; + –, 5–25%; + + more than 5%. Taken from White (1974).

tolerates drier habitats, including part of the Arabian peninsula, and continues to breed further into the dry season.

Otherwise, it is difficult to identify significant differences in their biology, ecology or behaviour, except as differences of degree: they both breed mainly in small, temporary pools of sunlit water found in areas created by human activity, such as mud tracks, hoofprints and irrigated fields. In fact, it is difficult to find breeding sites that are not made by humans. Both species are highly anthropophagic, although if cattle are available *An. arabiensis* will also feed on them. As Table 1 shows, the behaviour of *An. gambiae s.s.* is generally less variable and more closely associated with humans. They feed most often indoors and remain there to complete egg development.

The other three species—*Anopheles merus*, *Anopheles bwambae* and *Anopheles melas*—appear to have evolved from these two species, generally adapting to breeding sites that are not so closely associated with human activity. They are allopatric, although their distributions usually overlap with *An. gambiae s.s.* and/or *An. arabiensis*. They can all be vectors, but malaria transmission is not always maintained by *An. merus* without the presence of *An. gambiae s.s.* or *An. arabiensis*.

Polytene chromosomes

Polytene chromosomes are found in the larval salivary glands and adult ovarian nurse cells of some anopheline species. Each chromosome can be characterized by its length and its pattern of light and dark bands. Segments of chromosome can become inverted, leading to a change in the band sequence. Figure 3 shows the inversions that have been found so far on the chromosomes of the *An. gambiae* complex (the banding pattern itself is not shown). An arbitrary lettering system has been established to define standard inversions. This complex system of overlapping and contiguous inversions can give rise to arrangements that are difficult to distinguish from one another, and the details of the analysis are beyond the scope of this paper. The crucial point, however, is that some of these inversions are fixed, and can therefore be used to distinguish one species from another, as shown in Fig. 4. *An. quadriannulatus* is considered to be the standard form, i.e. with no fixed inversions. *An. gambiae s.s.* and *An. arabiensis* are the most polymorphic, showing marked geographical variations in distribution and frequency of polymorphic inversions. Some differences in chromosomal types may have no adaptive significance, but simply reflect the slow spread of new arrangements. In other cases, however, chromosomal differences occur over limited areas within large panmictic populations. These are mostly clinal and relatively stable geographic variations, and appear to result, at least in part, from selective forces acting on the alternative arrangements.

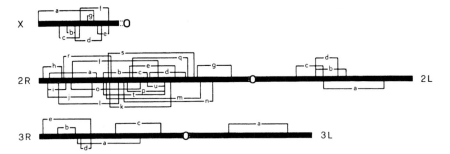

FIG. 3. Polytene chromosome complement and representation of the fixed and polymorphic inversions observed in the main six species of the *Anopheles gambiae* complex. The three chromosomes are X, 2 and 3. R and L identify the right and left arms of each chromosome. Lower case letters have been assigned to inversions found on each arm. Hence an inversion is uniquely described by its position, e.g. 2La is inversion 'a' on the 2L chromosome arm (Coluzzi et al 1979).

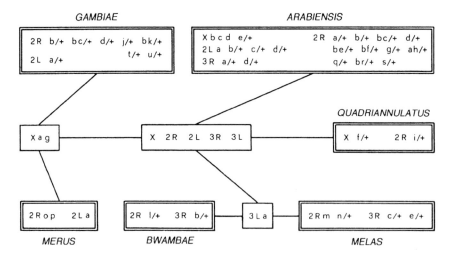

FIG. 4. Polytene chromosome relationships among the six species of the *Anopheles gambiae* complex (Coluzzi et al 1979). The inversion formula for each species is presented additively by following the line leading from the standard chromosome sequences for the five polytene arms (X, 2R, 2L, 3R, 3L). Letters appearing singly represent fixed inversions, while polymorphic inversions are indicated by the symbol for heterozygosity (a/+).

One area where such a cline in chromosomal forms has been found includes one of the steepest and most continuous ecological clines in the world, ranging from mangrove swamps and tropical rainforest in the south to sudan and sahelian savannahs in the north (Fig. 5). Figure 5 also shows, as expected, that *An. arabiensis* predominates in the drier savannah areas, and that *An. gambiae*

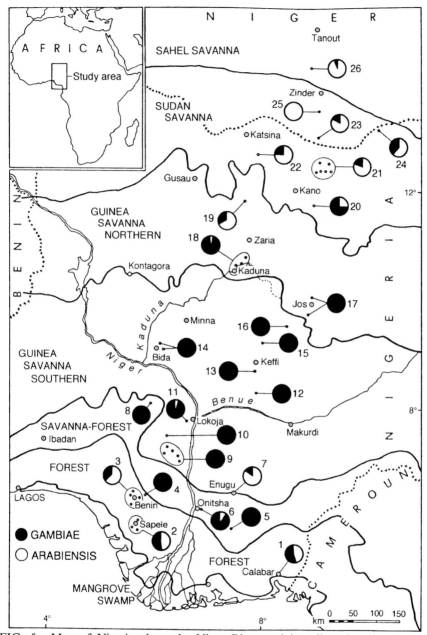

FIG. 5. Map of Nigeria along the Niger River and its tributaries, showing the frequency of *Anopheles gambiae sensu stricto* and *Anopheles arabiensis* in 26 sampling localities in different ecological zones. The data refer mostly to samples of blood-fed, indoor-resting adult females collected during the rainy season (Coluzzi et al 1979).

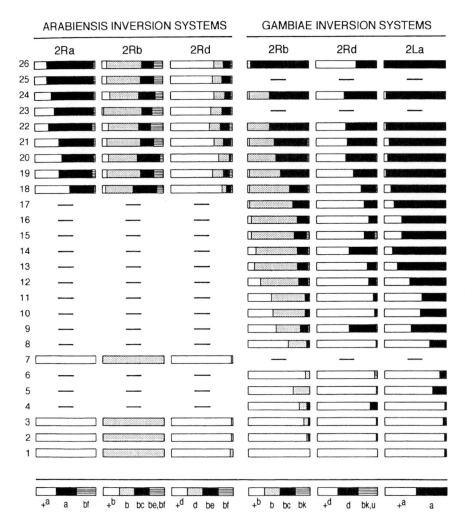

FIG. 6. The frequencies of intraspecific polymorphic inversions in *Anopheles gambiae sensu stricto* and *Anopheles arabiensis* found at the sites shown in Fig. 5 (Coluzzi et al 1979). The lettering system is the same as that described in Fig. 3. The most common forms found are the uninverted forms (shown by the symbol for wild-type '+' at the bottom of the figure), inversions 'a', 'b' and 'd' on the R side of chromosome 2 (2Ra, 2Rb and 2Rd), and inversion 'a' on the L side of chromosome 2 (2La). Occasionally, more complicated inversions are found involving overlapping or contiguous segments, such as 'bf' and 'bc' (see Fig. 3). These can be difficult to identify because of their complexity, and they have been scored with the nearest straightforward inversion, as shown at the bottom of the figure.

s.s. is widespread in both forest and guinea savannah and can be found in some areas of the sahel. There is also a remarkably high degree of inversion polymorphism here, with as many as eight common inversions. Figure 6 shows that a cline of variations on chromosome 2 appears to be related to the ecological cline. Samples of both *An. gambiae s.s.* and *An. arabiensis* from the forest zones are almost monomorphic for the standard arrangement, whereas frequencies of the inverted arrangements (2Ra, 2Rb, 2Rd and 2La) increase as one progresses towards the drier environments. In addition to the significant differences in inversion frequencies found between remote ecological zones, differences are also found in certain inversion frequencies between neighbouring villages and even between samples of mosquitoes collected indoors and outdoors in the same village. In other words, it would appear that adults of a chromosomal type better adapted to arid conditions distribute themselves locally to drier areas and are also more prone to bite and rest indoors, consistent with there being generally a higher saturation deficit indoors than outdoors (Coluzzi et al 1979).

Some of the inversions found in these two species appear to have fundamental importance in relation to the extension of the species' distributions over a wide geographical area and into a range of ecological clines. It would appear that because these inversions inhibit recombination, two alternative sets of genes can exist in a population, e.g. one set which is better adapted to humid conditions and the other to drier conditions. This allows populations to be maintained in habitats made by humans that are diverse spatially and temporally, e.g. rain-dependent and irrigated, cultivated areas (Coluzzi et al 1979, Coluzzi 1982).

Correlations between inversion frequencies and behaviour have been found elsewhere for *An. gambiae s.s.* and *An. arabiensis*, particularly in relation to endophilic and exophilic behaviour and host preference (Bryan et al 1987, Petrarca & Beier 1992) and for other anopheline vector species (*Anopheles stephensi*, Coluzzi et al 1975; *Anopheles superpictus*, Sabatini et al 1989). We may find similar dynamics in the other species complexes that have been identified amongst vector species (*Anopheles maculipennis* of Europe & Russia, *Anopheles culicifacies* of India, *Anopheles punctulatus* of Australasia and oriental *Anopheles balabacensis* [White 1982]), although these have not been so well studied.

Behavioural studies of host preference

There are at least four good reasons to investigate the genetic and behavioural basis for host preference generally and for the *An. gambiae* complex.

(1) An accurate means of assessing infective bites per person is essential, both for predicting potential outbreaks and for monitoring the progress of

control programmes. The human-biting catch is, as its name implies, the most direct source of this information, but it has come under increasing criticism mainly on ethical grounds, particularly in areas with high transmission of vector-borne diseases and/or drug-resistant parasites. Other sampling methods are available, which vary in efficacy, but there is no doubt that a standardized trapping device that specifically attracted the human-biting proportion of vector populations is highly desirable. Such a device would have to be based on the main sensory cues used by vectors to locate specifically human hosts, and it would have to elicit the appropriate differential responses of the main vector species, particularly of closely related species, such as *An. gambiae s.s.* and *An. arabiensis.*

(2) Evidence that subpopulations exist which differ in their host-seeking behaviour is of particular interest to control strategists. There are already several examples where the efficacy of control programmes based on house spraying has been reduced due to selection for subpopulations of mosquitoes which feed outdoors and/or earlier in the evening than the general population (Bown et al 1985, Coluzzi et al 1979, Gillies 1988, Hii 1985). Better methods of assessing 'host-seeking' behaviour (e.g. host preference, endophagic/exophagic behaviour and circadian timing of biting activity) of individual mosquitoes of known genotype (and even karyotype) is urgently needed.

(3) There has been much debate recently about the potential for altering the vectorial capacity of mosquitoes by genetic manipulation. The genes for extreme zoophily are expressed in one member of the *An. gambiae* complex (*An. quadriannulatus*). Depending on the nature of the genes, it may be possible to reintroduce them into *An. gambiae s.s.*, thereby deflecting this most important vector from humans and reducing their vectorial capacity (Curtis 1996, this volume).

(4) An analysis of the evolution of host preference within the *An. gambiae* complex would complement current investigations on the evolution of other aspects of their ecology and behaviour.

Mechanisms of host preference

Mosquito species are categorized as being anthropophagic/zoophagic or endophagic/exophagic, for example, but these terms describe only the outcome of a complex behaviour pattern, and they are based on evidence such as the identity of blood meals and the relative numbers caught feeding indoors and outdoors. Little is known about the mechanisms by which they arrive at a particular host. The 'ultimate causes' are related to the evolution of optimal feeding strategies, but the 'proximate causes' are determined by the stimuli available and the nature of the mosquitoes' responses to these stimuli. How do mosquitoes locate the host, and what differences in stimuli and response lead one species to be a 'generalist' feeder and another a 'specialist' feeder?

The flight path of *An. gambiae s.l.* that leads it to enter houses and feed on sleeping humans entails a sequence of stimuli and responses: activation at a particular phase of the day; orientation to human odour plumes carried by large-scale air movement outdoors ('wind', albeit less than 0.5 m/s); entry into houses; and orientation to host cues carried by convection currents and other small-scale air currents (probably less than 0.1 m/s). The aim of our research is to obtain an objective, standard, representative sample of the mosquitoes that would normally bite a human. The approach is to devise trapping systems that collect mosquitoes at specific stages in their host-location sequence without altering their behaviour significantly.

Odour-baited entry trap

The first step was to determine whether or not it is possible to attract mosquitoes into a device baited with cues associated only with the first stage of host location, i.e. orientation to host odours carried by the wind outdoors. Is it then possible, using only human odours, without warm, moist air currents, to intercept mosquitoes as they enter a house to feed? The odour-baited entry trap (OBET) aims to do this (Costantini et al 1993). Its design is based on that of a standard entry trap, placed in a window to catch mosquitoes as they enter a house. In the OBET, however, the air flowing out of the trap is driven by a fan and carries standard test odours at set release rates.

We tested the attractiveness of whole-human odours by drawing air from a tent in which a human was sleeping. The levels of CO_2 released by different people varies significantly, so the amount of CO_2 that we released at the trap entrance was kept constant at the highest concentration normally released by the human subjects used (0.15%, the background concentration being 0.04%). The OBET successfully caught similar numbers of *An. gambiae s.l.* as a CDC (Center for Disease Control, Atlanta, USA) light trap next to a human bait under a bed net, and it caught about 33% of an all-night, human-biting catch nearby.

Carbon dioxide dose response

CO_2 released on its own from the OBET at a human-equivalent rate caught 40% as many *An. gambiae s.l.* as a whole human bait. Clearly, CO_2 is a major attractive component in human odour. A dose–response experiment was therefore conducted to investigate further the response to CO_2 (Costantini et al 1996). The three main species caught vary in their host preferences: *An. gambiae s.l.* is the most anthropophagic; *Anopheles pharoensis* is less anthropophagic; and *Mansonia uniformis* is anthropophagic, but often feeds on other animals. Nevertheless, all three species showed the same linear relationship between log catch and log dose (i.e. there was no significant difference in the slopes of the dose–response curves).

However, when we compared the results of the experiment with contemporary human-biting catches nearby, we observed significant differences in the response to CO_2. For *An. gambiae s.l.*, CO_2 on its own was not attractive enough, even at the highest doses, to catch the equivalent of one human-biting catch, whereas for *M. uniformis*, the lowest dose of CO_2 was enough to catch the equivalent of one human-biting catch (at higher doses, the more CO_2, the greater the catch). The other species showed a threshold effect: at doses of CO_2 less than the human equivalent, the catch was much less than one human-biting catch, but higher doses attracted increasing numbers of mosquitoes. For *An. gambiae s.l.*, although CO_2 is an important component of whole-human odour, its efficacy is greatly enhanced by the presence of other human odours.

Relative attractancy experiment

With well over 300 chemicals emanating from the human body, identification of the key components mosquitoes respond to is rather like looking for a needle in a haystack. There is good evidence, however, that some people are significantly more attractive than others (Curtis et al 1987). If this difference is due, at least in part, to the odour 'profiles' of individuals, i.e. to the types and relative amounts of chemicals emitted, the most active components may well be those that are significantly more abundant in more attractive people. Lindsay et al (1993) have shown that significantly different numbers of *An. gambiae s.l.* were caught in five experimental huts, each of which had a different person sleeping inside under a bed net. Although these data suggest that the odours released from each hut were responsible for the differences in catch size, other close-range cues may have been involved. We therefore took the experiment further and directly compared the attractiveness of the odours alone of 12 men using the OBET system.

The 12 men first conducted human-biting catches in three teams of four. Catches were done both inside and outside four different houses. Within each team there was a highly significant ranking of mosquito catch size between the four men. The rank order of the men was not the same for all mosquito species, indicating that the ranking was not due to their catching skill (in agreement with Curtis et al 1987). Moreover, the ranking was consistent over a 12-week period, indicating that it was not just a transient phenomenon.

We reported the study with the same men, comparing the relative attractiveness of their odours alone with OBETs. Each man slept in a separate tent, with the air from each tent fed to a different OBET. The ranking of numbers of *An. gambiae s.l.* caught by each man's odour was significant (although the men did not rank in exactly the same order as for the human-biting catches). When the experiment was repeated, but topping-up each man's CO_2 output to a standard 0.15%, the differences between OBET catches disappeared. This was the case for *An. gambiae s.l.* and other mosquito species caught, but for *An. funestus* the ranking was maintained even with the standardized CO_2.

These results are consistent with the conclusion that *An. gambiae s.l.* is particularly sensitive to CO_2 in the presence of other host odours. Although the results are not promising for the identification of other active chemicals, they do confirm that some people are more attractive than others, and that this may be largely due to the levels of CO_2 they release. Environmental factors, such as the position of houses in relation to breeding sites, house design and proximity to other hosts, also affect the biting pattern within a village (Lindsay et al 1995).

It is apparent that the efficiency of the OBET could be improved. Video recordings of the mosquitoes' approach to the trap revealed mosquitoes hovering by the trap entrance rather than immediately entering. This suggests that structural or visual elements of the trap inhibit them from entering, and highlights the importance of continually evaluating the objectivity of the methods used to investigate behaviour.

Conclusion

Differences in host preference and resting site preference have evolved within the *An. gambiae* complex with profound effects on the vectorial capacity and the susceptibility to malaria control measures of each species. No doubt similar phenomena exist in other, less well-studied vector complexes. In the case *of An. gambiae s.l.*, access to polytene chromosomes has allowed us to investigate the behaviour of sibling species in some detail. Research aimed at more direct evaluation of differences in behaviour is urgently needed, however, to clarify the epidemiology of malaria in *An. gambiae s.l.* and in other species complexes. Basic research such as this can be linked to more practical projects, such as the development of odour-baited traps to replace human-biting catches, which depend on detailed knowledge of behaviour.

Acknowledgements

I thank John Brady, Chris Curtis and Carlo Costantini for their helpful comments on this manuscript, and the European Union for partial funding of this work as part of project no. TS3-CT91–0032.

References

Bown DN, Ríos JR, del Angel Cabañas G, Guerro JC, Méndez JF 1985 Evaluation of chlorphoxim used against *Anopheles albimanus* on the south coast of Mexico. 2. Use of two curtain-trap techniques in a village-scale evaluation trial. Bull Pan Am Health Organ 19:61–68

Bruce-Chwatt LJ 1988 History of malaria from pre-history to eradication. In: Wernsdorfer WH, McGregor I (eds) Malaria: principles and practices of malariology. Churchill Livingston, Edinburgh, p 1–60

Bruce-Chwatt LJ, De Zulueta J 1980 The rise and fall of malaria in Europe: an historico-epidemiological study. Oxford University Press, Oxford

Bryan JH, Petrarca V, DiDeco MA, Coluzzi M 1987 Adult behaviour of members of the *Anopheles gambiae* complex in the Gambia with special reference to *An. melas* and its chromosomal variants. Parassitologia 29:221–249

Coluzzi M 1982 Spatial distribution of chromosomal inversions and speciation in Anopheline mosquitoes. In: Barigozzi C (ed) Mechanisms of speciation. Alan R. Liss, New York, p 143–153

Coluzzi M, DiDeco M, Petrarca V 1975 Propensione al pasto di sangue in condizioni di laboratorio e polimorfismo cromosomico in *Anopheles stephensi*. Parassitologia 17:137–143

Coluzzi M, Sabatini A, Petrarca V, DiDeco MA 1979 Chromosomal differentiation and adaptation to human environments in the *Anopheles gambiae* complex. Trans R Soc Trop Med Hyg 73:483–497

Costantini C, Gibson G, Brady J, Merzagora L, Coluzzi M 1993 A new odour-baited trap to collect host-seeking mosquitoes. Parassitologia 35:5–9

Costantini C, Gibson G, Sagnon N'F, della Torre A, Brady J, Coluzzi M 1996 Mosquito responses to carbon dioxide in a West African Sudan savanna village. Med Vet Entomol 10, in press

Curtis CF 1996 Introduction I: an overview of mosquito biology, behaviour and importance. In: Olfaction in mosquito–host interactions. Wiley, Chichester (Ciba Found Symp 200) p 3–7

Curtis CF, Lines JD, Ijumba J, Callaghan A, Hill N, Karimzad MA 1987 The relative efficacy of repellents against mosquito vectors of disease. Med Vet Entomol 1:109–119

Davidson G 1964 The five mating types of the *Anopheles gambiae* complex. Rev di Malariologia 13:167–183

Davidson G, Hunt RH 1973 The crossing and chromosome characteristics of a new sixth species in the *Anopheles gambiae* complex. Parassitologia 15:121–128

Gillies MT 1988 Anopheline mosquitoes: vector behaviour and bionomics. In: Wernsdorfer WH, McGregor I (eds) Malaria: principles and practices of malariology. Churchill Livingston, Edinburgh, p 453–485

Gramiccia G, Beales PF 1988 The recent history of malaria control and eradication. In: Wernsdorfer WH, McGregor I (eds) Malaria: principles and practices of malariology. Churchill Livingston, Edinburgh, p 1335–1378

Hii JLK 1985 Evidence for the existence of genetic variability in the tendency of *Anopheles balabacensis* to rest in houses and to bite man. Southeast Asian J Trop Med Pub Health 16:173–182

Lindsay SW, Adiamah JH, Miller JE, Pleass RJ, Armstrong JRM 1993 Variation in attractiveness of human subjects to malaria mosquitoes (Diptera, Culicidae) in the Gambia. J Med Entomol 30:368–373

Lindsay SW, Schellenberg JRMA, Zeiler HA, Daly RJ, Salum FM, Wilkins HA 1995 Exposure of Gambian children to *Anopheles gambiae* malaria vectors in an irrigated rice production area. Med Vet Entomol 9:50–58

MacArthur W 1952 A brief story of English malaria. Trans R Soc Trop Med Hyg 46:359–370

Petrarca V, Beier JC 1992 Intraspecific chromosomal polymorphism in the *Anopheles gambiae* complex as a factor affecting malaria transmission in the Kisumu area of Kenya. Am J Trop Med Hyg 46:229–237

Sabatini A, Coluzzi M, Boccolini D 1989 Field studies on inversion polymorphism in *Anopheles superpictus* from southern Italy. Parassitologia 31:69–87

Wernsdorfer WH, McGregor I (eds) 1988 Malaria: principles and practices of malariology. Churchill Livingston, Edinburgh

White GB 1977 The place of morphological studies in the investigation of *Anopheles* species complexes. Mosq Syst 9:1–24

White GB 1982 Malaria vector ecology and genetics. Br Med Bull 38:207–212

DISCUSSION

Boeckh: When did the different species of *Anopheles gambiae* diverge?

Costantini: The speciation process is recent and is still continuing in *An. gambiae sensu stricto* from West Africa. Coluzzi and his collaborators have shown that, within the general frame of chromosomal inversion polymorphisms as presented by Gay Gibson, in West Africa there are several distinct chromosomal arrangements not conforming to the Hardy–Weinberg equilibrium. The forest chromosomal form of *An. gambiae* is thought to be the primitive one, based on evidence of its standard chromosomal arrangement. Coluzzi's hypothesis is that this form evolved from a *quadriannulatus*-like savannah pre-*gambiae* taxon that was able to occupy the forest niche only by linking its ecology and behaviour to humans; in fact this habitat in its original 'natural' state does not provide exploitable larval breeding sites for *An. gambiae* (e.g. small shallow sunlit pools), whose presence in the forest is dependent on human activities. This phase should have coincided with the introduction in the forest of the practice of shifting cultivation in burnt forest areas, i.e. by creating savannah-like ecological islands within the forest. So, in the case of the most anthropophilic *An. gambiae*, divergence might have started as recently as 5000 years ago. Subsequently, the anthropophilic and forest-adapted *An. gambiae* could have re-invaded the savannahs by evolving specific adaptations to more arid environments. Paracentric inversions were presumably instrumental for such adaptive processes, from which originated the chromosomal forms 'Savanna', 'Mopti', 'Bamako' and 'Bissau' in West Africa (Coluzzi et al 1985). These taxa appear to be still diverging.

Gibson: One of the main take-home messages of my presentation should be that the genes for zoophily and anthropophily are present in these closely related species, which means that we should be able to study how one behaviour may have evolved from the other.

Boeckh: A relevant question is why do closely related species that have the same sensory mechanisms have different behaviours? In my opinion, you have jumped too quickly from behaviour to genetics. There are variations in behaviour between species, but this may not necessarily be caused by genetic factors. Environmental factors may also play a role.

Takken: I have a general comment about the host preferences of *Anopheles arabiensis*, whose distribution overlaps that of *An. gambiae*. In West Africa it is largely anthropophilic, but further east it loses much of its anthropophily and becomes more opportunistic. Coluzzi has observed *An. arabiensis* populations from Madagascar that are entirely zoophilic (M. Coluzzi, personal communication 1995). Therefore, even within one species there seems to be a wide variation of host preference, which must be under genetic control.

Curtis: Another point to add is that one can cross different members of the *An. gambiae* complex that differ in their degree of anthropophily in the lab. The female hybrids are fertile, so it's perfectly possible, at least in principle, to backcross the gene or genes that are responsible for zoophily into the genetic background of *An. gambiae*. The resulting mosquitoes would presumably mate freely with *An. gambiae* if they were released into the field.

Cardé: Presumably, there are environmental isolation mechanisms in these sibling species, but do behavioural isolation mechanisms also exist?

Gibson: Laboratory studies show that crosses between *An. gambiae s.s.* and *An. arabiensis* are viable, but they are rarely found in the field. This means that there must be efficient behavioural isolation mechanisms to keep *An. gambiae s.s.* and *An. arabiensis* from interbreeding. It is assumed that the two species must mate at different times of day and/or at different sites. What is even more incredible is the evidence that some of the chromosomal inversion types within these species (which occur sympatrically) are also reproductively isolated from each other, thus allowing subpopulations of alternative inversions to co-exist (Bryan et al 1982).

Curtis: White (1974) reported that there are about 1 : 1000 hybrids in an area where *An. gambiae* and *An. arabiensis* are approximately in equal numbers, so one would expect there to be plenty of opportunity to cross-mate. They can be cross-mated easily in the lab.

Galun: The main message that I get from your presentation is that there is a tremendous diversity of sibling species of mosquitoes. For convenience, we use *Aedes aegypti* as a model for laboratory work and we tend to extend our conclusions to other species of mosquitoes. If such a diversity exists within a single species, it may be dangerous to generalize between genera.

Brady: I would like to reinforce this point using the tsetse fly as an example. The species that everyone works with in the lab is *Glossina morsitans*, because it is the easiest to breed. However, the species that everyone studies in the field in southern and eastern Africa is *Glossina pallidipes*, which responds wonderfully well to synthetic ox odours. However, the *Glossina palpalis* group of anthropophilic tsetse flies in West Africa aren't nearly as impressed with these imitation ox odours.

Knols: There are also major differences in the responses to CO_2, even for populations of single sibling species of the *An. gambiae* complex from West or East Africa. In West Africa, 40–50% of the relative attraction of this species to humans can be attributed to CO_2 (Costantini et al 1996), whereas in Tanzania this figure is only 9% (L. E. G. Mboera & B. J. G. Knols, unpublished data 1995). This shows that the geographical location of the species studied is also important.

Carlson: The genetics of these behavioural differences could turn out to be extremely complicated. They may not map to a single locus.

Gibson: I agree. I have used *An. gambiae* as an example because we know more about the genetic background of this species than that of other vectors.

Mustaparta: You mentioned that you standardized the level of CO_2 in your experiments, which would eliminate intra-individual differences. Would you have observed a different result using a different standard level of CO_2?

Gibson: It is possible. Our approach was to choose the person with the highest level of CO_2 and top the others up to that level because there are many practical difficulties with reducing the levels of CO_2.

Steinbrecht: CO_2 is produced not only by breathing but also by burning fuels and using cars. Is anything known about the attraction of mosquitoes to the latter sources of CO_2?

Gibson: There is no direct answer to your question because it's a difficult area to study. We have some new collaborators (Franz Meixner and colleagues at the Max Planck Institute for Chemistry in Mainz), who have been studying CO_2 emissions in different places and different environments. They have an extremely sensitive method for detecting variations in CO_2 levels. In the dry season in Zimbabwe, when there's not much vegetation around, the background CO_2 levels can be so constant that they can detect the breath of two cows 60 m downwind (F. Meixner, personal communication 1995).

Kaissling: This fits nicely with the calculations of Stange (1992), who calculated, from the sensitivity of CO_2 receptors in moths and the amount of CO_2 released by a tree, that a moth should be able to detect a tree in a desert at a distance of 10–100 m.

Also of interest is a study by Stange et al (1995) about the function of the CO_2 detection in the moth *Cactoblastis cactorum*. They have shown that two types of CO_2 gradients can be detected by this moth. The first is a vertical gradient, which could be used for maintaining the flight altitude; and the second, which extends from the surface of the food plant, could be used to monitor the physiological condition of the plant prior to oviposition.

Steinbrecht: CO_2 receptors of moths show little adaptation to sustained stimulation, thus they are able to monitor absolute levels of CO_2 (Bogner 1990). At the same time these receptors can follow fluctuations in CO_2 density of only 0.5 ppm around the ambient concentration (Stange 1992). Thus, given that CO_2 receptors of tsetse flies and mosquitoes perform equally well, CO_2 might well be a major cue for host orientation.

Gibson: But in a village CO_2 may be a disruptive cue because of its multiple sources.

Klowden: I have a comment about some of the generalizations that are made about lab populations in terms of the relationship between biting and the gonotrophic cycle. We tend to think that the gonotrophic cycle is affiliated with only one blood meal. For *Ae. aegypti* in the lab that is certainly the case, although many moderating factors are involved. In the case of *An. gambiae* there is no relationship whatsoever between the gonotrophic cycle and feeding, they feed at every opportunity and they're not inhibited in the same way.

Gibson: Even after the first cycle?

Klowden: Yes, even when they're gravid they continue to come back for more. However, I wondered whether there was any field evidence for multiple feeding in *An. gambiae*?

Costantini: Apart from pre-gravid females, i.e. those requiring a second blood meal to complete their first gonotrophic cycle, we rarely collect *An. gambiae sensu lato* and *An. funestus* gravid females when doing biting catches in the field.

Takken: Lyimor (1993) found that in Tanzania, up to 40% of female *An. gambiae* take two or three blood meals during the first cycle. However, we have no evidence for multiple blood meals after the first cycle, so the general statement that *An. gambiae* takes multiple blood meals may be a laboratory artefact.

Davis: Also, the number of meals that the adult takes often depends on how well fed it was during its larval stage. Sometimes it may have to take a second meal in order to be able to produce eggs.

Boeckh: Once *An. gambiae* has located a good blood meal, how does it decide what to do next? Is there some sort of learning process involved?

Knols: There is a report from Southeast Asia that *Anopheles balabacensis* employs some sort of learning process whilst host seeking (Hii et al 1991).

Bowen: Is this olfactory conditioning or host range learning?

Takken: It seems to be post-olfactory learning.

Hildebrand: This doesn't surprise me because other insects exhibit olfactory learning.

Mosquitoes are also nectar feeders, and male mosquitoes presumably also have to feed. Where do the strictly endophagous mosquitoes that spend most of the time in houses feed when they are not taking blood meals? Do they sneak outside to get their snacks?

Gibson: First, let me clarify some terminology that is often not used precisely. Endophagous means to take a blood meal indoors. After a blood meal there is a variable length of time before oviposition when some species prefer to rest indoors, outdoors or a combination thereof. This is usually referred to as endophilic/exophilic behaviour. The most 'strictly endophilic' species, such as some forms of *Ae. aegypti*, feed, rest and even oviposit indoors, thus spending their whole lives indoors. Many species that are referred to as endophilic, however, must oviposit outdoors, and therefore spend a variable amount of time outdoors when nectar feeding is possible.

Hildebrand: Do they mate indoors?

Gibson: Species that breed indoors can mate indoors as well, but most species mate outdoors, within a day or two of emerging from their breeding sites. Little is known about mating behaviour, except that it can occur in swarms and may occur elsewhere when mature males encounter virgin females. Males detect the wing-beat frequency of conspecific females.

Pickett: Are there any examples of intrinsically different oviposition sites for *An. arabiensis*?

Curtis: There are some anomalies in the behaviour of urban *An. arabiensis* in southern Nigeria (Coluzzi et al 1979). I've seen pictures of breeding sites in Nigerian cities that look very much like those of *Culex*, e.g. open drains. Consequently, I suspect that *An. arabiensis* is relatively tolerant of pollution.

Bowen: Are there any results which demonstrate that the distribution of the breeding site is a function of olfactory choice for the gravid female? We assume that the female chooses an oviposition site, but breeding site distribution may be a function of other factors, such as larval competition.

Costantini: It is currently difficult to answer this question, but in the field we sometimes find two breeding sites close together that look exactly the same to us, but for some unknown reason one is occupied by *An. gambiae s.l.* larvae and the other is not. This seems to imply that egg-laying females prefer one site over the other.

Bowen: This is also true for snowpool *Aedes* in California (B. F. Eldrige, personal communication 1995). However, to my knowledge there are no good data suggesting that the choice of site is due to site discrimination by the gravid female.

Costantini: The problem with *An. gambiae s.l.* is that, generally speaking, the study of larval ecology has been neglected. Of the studies I am aware of, apart from early ones by Christie (1958), and then by Service (1971, 1973) on larval presentation and population dynamics, and more recent ones by Le Sueur & Sharp (1988) on larval habitats, all the research seems to be concentrated on the adult.

Brady: Marston Bates did experiments on this in the 1940s. He transferred eggs from the breeding sites selected by various species of mosquitoes to other larval habitats. The larvae seemed to manage perfectly well in the 'wrong' sites, which he took to imply that different types of breeding sites were equally suitable for development, and that the choice of site was made by the female.

An interesting analogy to the question of endophagic behaviour is the behaviour of *Anopheles smithi* in West Africa. This feeds on bats in small caves and breeds in muddy water at the bottom of the cave. It seems to have no reason to leave the cave.

Klowden: There was a recent paper which demonstrated that in Thailand female *Ae. aegypti*, in contrast to the males, did not feed on nectar (Edman et al 1992). Blood was their main source of nutrition.

Costantini: There may also be seasonal changes in nectar-feeding behavioural patterns. We usually observe in sudan savannahs, where *An. gambiae s.l.* population dynamics are strictly seasonal (as determined by the onset and departure of rains), that, when we collect the indoor-resting anopheline fauna from houses at the end of the rainy season, there is an increased number of females that have taken mixed blood and sugar meals. This is rare to observe in such conditions during other periods of the breeding reason.

Knols: We cannot say that a particular mosquito species is entirely endophilic or endophagic. There are high levels of plasticity in this behaviour. For instance, *An. gambiae s.s.*, which is reportedly endophagous, will bite humans outdoors in northeast Tanzania. I would also like to add that Gillies (1957) found that more than 50% of *An. gambiae* left houses after the first day of the gonotrophic cycle, so it is possible that small nectar meals are taken during the latter part of egg development.

Takken: I would like to remind everyone that the experiments that Gay Gibson has described here, where malaria vectors are trapped in a house with a concentrated odour, are the first experiments of this nature that have been undertaken in Africa. It's extremely encouraging to learn that it is possible to attract such an important vector with this system. This system may also be used to improve the kairomones responsible for attraction, so that we may develop an effective bait.

Costantini: It is possible that odour technology will also be useful for maybe less-appealing but still important purposes, such as for sampling anophelines in the field. For example, at the moment we use human-biting catches as the most reliable measure of human–vector contact for calculating the entomological inoculation rate, i.e. the number of infective bites received on average by a person in a given time unit. This is probably not the most ideal method in terms of ethics because of the exposure of humans to the risk of contracting vector-transmitted diseases and the increasing numbers of drug-resistant parasites.

Carlson: Is it possible that heat is another variable which may be involved? Are there any convection gradients extending from houses?

Gibson: The assumption has been that, at least at a distance, convection gradients dissipate. Certainly, the closer a mosquito gets to the host, the more it will have to start taking convection gradients into account.

Pickett: Does anyone know anything about the heat-detecting mechanisms of the mosquitoes? Do they detect warm air as convection currents or as radiant heat?

Davis: Several years ago we spent some time studying the pair of temperature receptors at the tip of the antenna (Davis & Sokolove 1975). We did not find any behavioural or electrophysiological evidence to suggest that radiant heat was being detected. The wavelength region wasn't complete, but over the range that we had available there was no response. We could measure the spike activity of those receptors after exposing them to some radiant heat. As the substrate heated up due to the radiant energy, there was an increase in spike frequency, but only as the radiant energy heated the substrate. It is possible that the moisture in the air carries thermal energy rather than heat *per se*.

Lehane: My understanding of attraction is that it's a complex process in which olfaction plays a large part. However, it seems as though the olfactory component operates at a distance, and this is supplemented first by visual cues as the mosquito gets nearer and then by other cues, such as heat detection. Many years ago there was a suggestion that the increased temperature of infected hosts caused by the parasite acted as a 'come-and-get-me' signal to other parasites (e.g. Gillett & Connor 1976). However, the experiments that were done in America didn't support an increased attraction of mosquitoes to infected hosts (Day & Edman 1984).

Gibson: In the case of malaria, human hosts are most infective to mosquitoes when the parasites are in the gametocyte stage of development in the human's bloodstream. The human host is not necessarily feverish at this time, and therefore there is not a strong correlation between body temperature and infectivity of malaria parasites. There is no doubt that the list of possible cues used by mosquitoes to locate the host is long and complex, involving many potential interactions and synergisms. However, we must not lose sight of our goal. Although intellectually it would be satisfying to identify the complete set of cues, we need to identify only those which are necessary to accomplish the tasks. For example, the successful tsetse targets employ only a few of the known host attractants. A few odours and the appropriate visual cues are sufficient to attract flies to the area of the targets, which they circumnavigate with few landing. The target kills because it is bordered by fine netting, impregnated with insecticides, which the fast-flying insects do not see and collide with. One collision on the impregnated netting is enough to kill the fly. At this stage we do not know how many host cues we will need to attract a 'useful' number of mosquitoes. This will depend on the efficacy of the cues we identify and on the definition of an 'adequate' catch. For malaria-monitoring purposes we do not need to trap as many mosquitoes as possible. On the contrary, we only want to trap the equivalent number of mosquitoes that would normally be attracted to an average human host in a night, i.e.

mosquitoes of the same species and same age range, and a similar proportion that are infected with malaria parasites. A trap that catches the house-entering proportion of the population may be sufficient, in which case we do not need to identify all the cues used by a mosquito to, for example, locate the host inside a house and identify a suitable landing site on the host. If the aim is to kill as many mosquitoes as possible, then it may be necessary to develop a device that includes more of these short-range cues as well.

References

Bogner F 1990 Sensory physiological investigation of carbon dioxide receptors in Lepidoptera. J Insect Physiol 36:951–957

Bryan JH, DiDeco MA, Petrarca V, Coluzzi M 1982 Inversion polymorphism and incipient speciation in *Anopheles gambiae s.s* in the Gambia, West Africa. Genetica 59:167–176

Christie M 1958 Predation on larvae of *Anopheles gambiae* Giles. J Trop Med Hyg 61:168–176

Coluzzi M, Sabatini A, Petrarca V, Dideco MA 1979 Chromosomal differentiation and adaptation to human environments in the *Anopheles gambiae* complex. Trans R Soc Trop Med Hyg 73:483–497

Coluzzi M, Petrarca V, DiDeco MA 1985 Chromosomal inversion intergradation and incipient speciation in *Anopheles gambiae*. Boll Zool 52:45–63

Costantini C, Gibson G, Sagnon N, della Torre A, Brady J, Coluzzi M 1996 Mosquito responses to carbon dioxide in a West African Sudan savanna village. Med Vet Entomol, in press

Day JF, Edman JD 1984 The importance of disease-induced changes in mammalian body temperature to mosquito blood feeding. Comp Physiol Biochem 77:447–452

Davis EE, Sokolove PG 1975 Temperature response of the antennal receptors of the mosquito, *Aedes aegypti*. J Comp Physiol 96:223–236

Edman JD, Strickman D, Kittayapong P, Scott TW 1992 Female *Aedes aegypti* (Diptera: Culicidae) in Thailand rarely feed on sugar. J Med Entomol 29:1035–1038

Gillett JD, Connor J 1976 Host temperature and the transmission of arboviruses by mosquitoes. Mosq News 36:472–477

Gillies MT 1957 Age-groups and biting cycle of *Anopheles gambiae*. A preliminary investigation. Bull Entomol Res 48:553–559

Hii JLK, Chew M, Sang VY et al 1991 Population genetic analysis of host-seeking and resting behaviours in the malaria vector *Anopheles balabacensis* (Diptera: Culicidae). J Med Entomol 28:675–684

Le Sueur D, Sharp BL 1988 The breeding requirements of three members of the *Anopheles gambiae* Giles complex (Diptera: Culicidae) in the endemic malaria area of Natal, South Africa. Bull Entomol Res 78:549–560

Lyimo EOK 1993 Bionomics of *Anopheles gambiae* in Tanzania. PhD thesis, Wageningen Agricultural University, The Netherlands

Service MW 1971 Studies on sampling larval populations of the *Anopheles gambiae* complex. Bull WHO 45:169–180

Service MW 1973 Mortalities of the larvae of the *Anopheles gambiae* Giles complex and detection of predators by the precipitin test. Bull Entomol Res 62:359–369

Stange G 1992 High resolution measurement of atmospheric carbon dioxide concentration changes by the labial palp organ of the moth *Heliothis armigera* (Lepidoptera: Noctuidae). J Comp Physiol A 171:317–324

Stange G, Monro J, Stowe S, Osmond CB 1995 The CO_2 sense of the moth *Cactoblastis cactorum* and its probable role in the biological control of the CAM plant *Opuntia stricta*. Oecologia 102:341–352

White GB 1974 *Anopheles gambiae* complex and disease transmission in Africa. Trans R Soc Trop Med Hyg 68:278–298

General discussion I

Guerin: If one exposes an insect population to an insecticide and some of them escape with only sublethal effects, these escapees may have major effects on the vector population one is trying to control. We should take note of this when trying to manipulate certain behaviours using olfactory stimulants. Should we do so only when we can be sure that an all-or-nothing response will occur? A partial response may do more harm than good. For example, take a house where some people are covered by bed nets and others are not. Because of variations in the mosquitoes' behaviours, some mosquitoes are presented with the opportunity to blood feed whilst being exposed to sublethal effects of the insecticide. Such mosquitoes then depart to breed, passing on any effects due to the sublethal exposure to the next generation.

Lehane: This reflects my own feelings from working in Africa. I have never worked on malaria-transmitting mosquitoes but I've been bitten by many of them, and it seems they're very plastic in the way that they respond. Most of them come to you at night when you're asleep. But others do come in the early evening or early morning, i.e. at times when you would not expect them to from reading the text books. This situation does concern me because if you do knock out parts of the population, with for example bed nets, are you not merely going to replace one problem with another problem?

Pickett: This is a rather dangerous philosophy to adopt because it is the naïve answer to resistance. During a bacterial infection of a single host, it is possible to destroy all the bacteria. However, if the animal is in a semi-natural ecosystem, then it is most unlikely that all the individuals in the pest population could be destroyed by insecticide treatment. If a strong insecticide is used, the few remaining pests would have a higher incidence of insecticide-resistant genes than if a weaker one is used. We are in the habit of using strong insecticides and getting enormous resistance problems as a result. If we are going to try to exploit mosquito olfaction in control, we shouldn't step in with a 99.9% killing approach. We should follow an enlightened approach that uses the olfactorially mediated aspect only to an extent. The influence of 40% CO_2 is in itself good enough. At the same time we should exploit other cues involving essential ecological aspects of the mosquitoes' life cycles. Together, these may provide a satisfactory means of control. We have misused good insecticides to the extent that the public is frightened to use them and we have serious problems of resistance. If we can obtain an effective approach to exploiting mosquito olfaction, we will have a chance to play things differently.

Brady: I would like to emphasize that holoendemic malaria is inherently stable and therefore difficult to force into decline. For *Anopheles gambiae*, the relationship between mosquito numbers and the intensity of the disease rises rapidly and then reaches a long, stable plateau. Any intervention at the right-hand end of this curve (i.e. on vector density) therefore has negligible effect on transmission.

Curtis: It may even be worse than that. The recent data of Snow et al (1994), comparing an area in Kenya with 30-fold less biting than an area in Tanzania, showed that there were fewer deaths of babies from anaemia in the Kenyan area, but actually more deaths of one- to four-year olds from cerebral malaria. This is presumably because lower transmission was associated with a delay of build-up of immunity to an age at which children are apparently more susceptible to cerebral malaria.

Reference

Snow RW, Azevedo IB, Lowe BS et al 1994 Severe childhood malaria in two areas of markedly different falciparum transmission in East Africa. Acta Tropica 57:289–300

Introduction II: olfactory control of mosquito behaviour

Edward E. Davis

SRI International, Life Sciences Division, 333 Ravenswood Avenue, Menlo Park, CA 94025–3493, USA

Scope of the problem

It has been clearly, convincingly and repeatedly demonstrated that all behaviour patterns exhibited by mosquitoes are mediated, in a large part, by olfactory signals (Davis & Bowen 1994). However, before one can engage in any discussion of the olfactory control of these behaviours, we must clearly choose the mosquito, the behaviour and the point in its reproductive or feeding cycle that we are going to focus our attention on, and we must also determine its prior experiences.

Olfactory control of which mosquito?

The yellow fever mosquito *Aedes aegypti* is the laboratory model on which the most research has been done and, as a consequence, we often conveniently or without much thought base our work and interpret our results as if they are applicable to all mosquitoes. One of the most remarkable aspects of mosquitoes as a group is the great variety of lifestyles that different species exhibit. When we decide to investigate some aspect of mosquito behaviour or physiology we have several factors to consider. Is the mosquito of interest anautogenous or autogenous? If it is autogenous, is it facultative or obligatory autogeny? If it is obligatory, is it for the first gonotrophic cycle only or for all of them? Is it a non-diapausing or diapausing species and, if it is diapausing, does it exhibit gonotrophic dissociation? What is the host-locating strategy it uses? Is it an ambusher that loiters in or near human habitation or is it a patroller that must commute from its resting place to its hunting grounds? What are its diurnal patterns, if any? What host animals, nectar sources and oviposition sites are available to it? Does it exhibit any host preference or is it an opportunistic feeder? Does it have any preferences for other resources? Even though two mosquito species may have nearly the same lifestyle, they may not use the same set of signals to perform the same behaviour and accomplish the same goal. Often the study of one species may reveal important information about another.

Which behaviour?

The life cycle of a mosquito is composed of recurring cycles through the various behaviours in its repertoire. The most prominent and reproductively important behaviours are blood feeding and oviposition. Repetition of these behaviours underlies disease transmission. Mating and nectar feeding are also recurring behaviours. However, none of these behaviours occur at random, rather they occur in an organized and co-ordinated manner. In a non-gravid female, blood-feeding behaviour is assumed to be the most reproductively important behaviour, and has 'priority' over any other behaviour that might be expressed at this time (unless it is expressly inhibited, for example, by abdominal distension after a meal). Oviposition site-locating behaviour will not be expressed during the time when host-locating behaviour may be evoked. What are the controlling mechanisms that regulate and co-ordinate which behaviour can be expressed at its appropriate time? Potential control mechanisms for regulating host-locating behaviour include developmental (e.g. maturation of the host odour-sensitive neurons in post-emergent adults), physical (e.g. abdominal distension following a meal or in gravid females) and humoral (e.g. inhibition of host locating following a blood meal) mechanisms. The humoral regulation of host-locating behaviour may be mediated through an up- or down-regulation of the sensitivity or availability of receptors for the host odour, L-lactic acid. Other changes or differences in the physiological status of the female may also affect the particular behaviour of interest.

This assumes that the mosquito will receive only one set of behaviourally relevant signals at a time. What happens when signals for more than one possible behaviour are received coincidentally in time? What if the female is already engaged in one behaviour when it receives the signals for a reproductively more important behaviour? What if she is merely orienting to a set of signals for one behaviour when she detects the signal set for the different, perhaps more important behaviour? What are the mechanisms that determine which behaviour will be the dominant one? Some results from our laboratory (M. F. Bowen & E. E. Davis, unpublished results 1993) indicate that when the female is orienting to an odour associated with a reproductively less important behaviour, such as nectar feeding, when it detects host odour, it will reorient its flight to follow the host odour plume even if it has already landed at, but not yet contacted, the nectar odour source. However, once it has made contact with the nectar source, it will not leave that source and fly towards the host odour source.

When should the behaviour be studied?

At what point in the time course of a behaviour are we going to look? We should also consider the prior history of the mosquito. Is the adult from a well-fed or undernourished larva? When did it last take a sugar meal? Is it inseminated or gravid? Are we going to make our observations before or after an event that may change the physiological status of the insect? If after, immediately so or at some later time? The observed behaviour may be the same, but the mechanisms may be different at different times following a blood meal.

Fundamental questions

Two important underlying questions that may be addressed in this symposium are: (1) what must the (structural and functional) characteristics of a mosquito's system be in order for it to perform each of the various behaviours in its life cycle repertoire at the proper time and thereby achieve reproductive success; and its corollary (2) what must the underlying anatomical architecture and physiological mechanisms be to exhibit these characteristics?

Presentations

What strategies do mosquitoes use to detect and orient to an odour source?

Depending upon the habitat of a mosquito and the host animals found there, the strategies used will be different. For example, a diurnal mosquito that must fly from its daytime resting place to where its host animals are found will employ a different strategy for encountering the odour plume of its host than will a mosquito that rests in the interior of its host's dwelling place. Even with these differences, there is a common set of problems faced by each species. The female must use a flight pattern strategy that will optimize its ability to maintain or regain contact with the host odour plume in order to orient to and arrive at the host animal over a more or less long distance. Once in the vicinity of the host, it must locate and land on the host, probe and pierce the host's skin at an appropriate and perhaps preferred site, and then locate a blood vessel and ingest blood.

The search strategies and flight patterns used by moths that both increase the likelihood of encountering odour plumes and are used for orienting towards their sources, and the techniques and methods for their study, have been well described and provide a useful model for studying similar phenomena in mosquitoes (Cardé 1996, this volume). However, unlike a moth orienting to a pheromone emitted by a potential mate, the animal providing the odour signals that the mosquito orients to is not doing so to attract the mosquito and will often take defensive actions against the female as it approaches. Moths appear to employ highly predictable, self-steered, counter-turning patterns of searching (a wide zigzag flight pattern) and orienting (straight flight), in contrast to the orienting flight patterns for mosquito host locating that appear to be highly irregular, even erratic, but none the less non-random patterns with low predictability (M. F. Bowen & E. E. Davis, unpublished observations 1987). Tsetse flies also appear to employ irregular flight patterns during host location (Brady et al 1989, Gibson et al 1991).

What odours do mosquitoes respond to when engaged in locating a host?

It is well established that the host-locating behaviour of mosquitoes is mediated primarily by olfactory signals. However, we only know the identity of a few of the relevant host odours (Davis & Bowen 1994). L-Lactic acid and CO_2 are well-known host attractants. Together in a warm moist airstream they will attract about 50% of a population of avid females, 95% of which would respond to human skin odours. Furthermore, in contrast to a human hand, L-lactic acid and CO_2 will not evoke probing (Davis & Bowen 1994). Although other host odours have been reported from time to time—such as lysine, oestradiol (Bos & Laarman 1975) and 1-octen-3-ol (Kline et al 1990)—none have been verified as necessary odour components for host-locating behaviour by the major disease vector mosquitoes.

Systematic studies with *Glossina* have resulted in the identification of specific attractive odours from cattle and cape buffalo (Cork 1996, this volume). We anxiously await the results from similar studies of human volatile emanations currently underway.

Biting site selection and other close-range behaviours

Close-range flight orientation and approach to a host, and landing and probing of the host, are behaviours that to date have received little attention from scientists. Only recently has the selection of a biting site come under scrutiny (de Jong & Knols 1996, this volume). We know that some haematophagous arthropods have distinct preferences for where on the host they will settle to probe and bite. Others may have a tendency to be attracted to, and to land on and bite, certain regions of a host rather than a specific anatomical site. Still others may be mere opportunists and lack any clear preferences.

What are the factors that underlie selection of these biting sites? What sensory modalities are involved? What role does olfaction have in this process?

What is the role of olfaction in the avoidance of non-hosts and certain hosts, and in oviposition site location?

As with host location, olfactory signals have clearly been shown to play a necessary role in other behavioural modalities, such as the location of a preferred oviposition site. One important observation in the location of a preferred oviposition site by *Culex quinquefasciatus* is the discovery of an egg–raft pheromone and its interaction with other odour signals associated with the oviposition site (Pickett 1996, this volume).

Also of importance, but what have received little attention, are the findings that certain members of a host species (e.g. humans) may produce odour signals that decrease the probability of that person being bitten by a mosquito. At least one such weak deterrent signal has been identified (Skinner et al 1968), but the mechanism of its action is not known.

Summary

The following papers will tell us much about our current knowledge of certain aspects of the olfactory control of mosquito behaviour—specifically host-locating behaviour. More importantly, as we read the papers, we should pay attention to the insights they will provide about what is not known and to the questions that our lack of knowledge should raise. The authors will also provide us with some new approaches and methodologies that we can apply to the study of mosquito physiology and behaviour.

References

Bos HJ, Laarman JJ 1975 Guinea pig, lysine, cadaverine and estradiol as attractants for the malaria mosquito *Anopheles stephensi*. Entomol Exp Appl 18:161–172

Bowen MF 1991 The sensory physiology of host-seeking behavior of mosquitoes. Ann Rev Entomol 36:139–158

Brady J, Gibson G, Packer MJ 1989 Odour movement, wind direction and the problem of host-finding by tsetse. Physiol Entomol 14:369–380

Cardé RT 1996 Odour plumes and odour-mediated flight in insects. In: Olfaction in mosquito–host interactions. Wiley, Chichester (Ciba Found Symp 200) p 54–70

Cork A 1996 Olfactory basis of host location by mosquitoes and other haematophagous Diptera. In: Olfaction and mosquito–host interactions. Wiley, Chichester (Ciba Found Symp 200) p 71–88

Davis EE, Bowen MF 1994 Sensory physiological basis for attraction in mosquitoes. J Am Mosq Control Assoc 10:316–325

de Jong R, Knols BGJ 1996 Selection of biting sites by mosquitoes. In: Olfaction in mosquito–host interactions. Wiley, Chichester (Ciba Found Symp 200) p 89–103

Gibson GA, Packer MJ, Steullet P, Brady J 1991 Orientation of tsetse flies to wind, within and outside host odour plumes in the field. Physiol Entomol 16:47–56

Kline DL, Takken W, Wood JF, Carlson DA 1990 Field studies on the potential of butanone, carbon dioxide, honey extract, 1-octen-3-ol, L-lactic acid and phenols as attractants for mosquitoes. Med Vet Entomol 4:383–391

Klowden MJ 1990 The endogenous regulation of mosquito reproductive behavior. Experientia 46:660–670

Pickett JA, Woodcock CM 1996 The role of mosquito olfaction in oviposition site location and in the avoidance of unsuitable hosts. In: Olfaction in mosquito–host interactions. Wiley, Chichester (Ciba Found Symp 200) p 109–123

Skinner WA, Tong H, Johnson H, Maibach HI, Skidmore D 1968 Human sweat components: attractancy and repellency to mosquitoes. Experientia 24:679–680

Odour plumes and odour-mediated flight in insects

Ring T. Cardé

Department of Entomology, Fernald Hall, University of Massachusetts, Amherst, MA 01003, USA

Abstract. Flying insects often follow odour plumes to find resources. Some insects may employ an 'aim-and-shoot' strategy using mechanoreceptors before flight to determine wind direction. Once airborne, insects must use optomotor anemotaxis to set a course upwind. This mechanism uses a visual appraisal of how wind modifies the insect's path. A straight upwind course yields a front-to-rear image flow directly below the insect. Details of this process in male moths flying to female pheromone have emerged mainly from wind tunnel studies. Loss of the pheromone triggers 'casting', or wide lateral excursions without upwind progress, whereas contact with a plume usually induces a zigzag path upwind. The temporally regular counterturns in casting and zigzagging seem to be generated by a central programme. Brief contact with a filament of odour induces a heading towards upwind, and an optimal rate of encounter promotes a rapid, straight upwind course. Other insects, such as parasitoid wasps seeking a host and tsetse flies seeking a blood meal, seem not to have a temporally regular pattern of counterturns and often fly straight upwind. The availability of visual cues from the odour source itself, the aerial distribution of odour set by turbulent diffusion, and light and wind levels all influence the success of these manoeuvres.

1996 Olfaction in mosquito–host interactions. Wiley, Chichester (Ciba Foundation Symposium 200) p 54–70

A common strategy of many insects in locating a mate, food or a host is to fly upwind once they have encountered a wind-borne odour released from that resource. Three cases illustrate the widespread use of this searching pattern: (1) a male moth typically finds a female, perhaps at some distance, by navigating along the plume of her pheromone; (2) parasitoid wasps often discover a host by following a plume of kairomone, which may be composed of host odour mixed with volatiles from the plant on which the prospective host was feeding; and (3) haematophagous Diptera utilize host-emitted odours to guide their pursuit of a meal. The similarities in the navigational problems posed by these quests suggest that the actual manoeuvres employed by flying insects in locating an odour source in wind might rely on similar solutions, even in

54

comparing long-diverged insect groups such as moths, wasps and flies. This review will summarize what we know of these manoeuvres, pointing out common mechanisms, apparent differences and areas to be unravelled by further inquiry. The use of odour cues in wind is so pervasive in flying insects that it will only be possible to consider an invidiously small set of examples. Most of our current view of these processes stems from studies conducted in the wind tunnel, but some important discoveries have come from field tests.

Scanning the environment for odours

An organism's location of resources is typically governed by two processes. First, it either moves through the environment to contact cues that signify the presence of the resource or, alternatively, positions itself such that cues from the resource are likely to reach it passively. Second, the organism should adopt a course that will bring it to the resource. In cases of odour cues borne in wind, an example of a strategy that appears to enhance the probability of locating odour plumes has been uncovered in the heights at which cockroaches perch on tree trunks in the rainforest. A species-specific stratification assures that males will be at the appropriate elevation to intercept a plume of pheromone emanating from a conspecific female (Schal & Bell 1986). Similarly, *Drosophila* fruit flies adjust wind heading to increase their probability of intercepting the odour of food (Zanen et al 1994). When the wind direction holds steady, flies orient themselves crosswind, so that they adopt an optimal pattern of scanning the environment for contacting a straightened-out odour plume. When the wind direction varies, flies tend to head along the windline, again an optimal strategy to contact the plume, which in winds shifting over 60° can have a greater crosswind sweep than downwind projection (Sabelis & Schippers 1984). Determining the strategies that amplify the prospect of contacting an odour which signifies a resource is a first step in understanding how organisms locate that resource. Curiously, these strategies are a facet of chemically mediated behaviours that are little understood, mainly because they are difficult to observe directly, manipulate experimentally or even to identify—how can we be certain that a particular scanning behaviour is in 'anticipation' of a stimulus?

How are odour plumes distributed?

The dispersion of odour in wind is dominated by turbulent diffusion, and I will outline this process only briefly (reviewed by Murlis et al 1992). Odour from a small source is transported in wind and dispersed mainly by the stirring and stretching effects of turbulence, which are far more vigorous than those of molecular diffusion. The minimum eddy size of a centimetre or so is limited by its inherent instability. Eddies in the size range of hundreds of millimetres set the fine-scale structure within a plume, whereas those of metres or more in scale cause the plume itself to meander and undulate. Although in some cases the source of the odour may be less than a centimetre in breadth, its effective initial size is set by the substrate on which the source is situated and the local airflow around the source. Thus, even a small moth calling from a leaf would have a initial plume expanse that at least matches the size range of the smallest eddies. The initial size of many other odour sources can be considerably larger: plants and many mammals would comprise comparatively large plumes at the outset.

When the odour source is on the scale of centimetres or so, the resulting odour signal in the plume comprises filaments of odour generated by turbulence. A stationary sampling device set downwind would detect the filaments as odour 'bursts'. The means of odour concentration in these bursts decline with distance from the source, but many metres downwind some filaments contain relatively undiluted odour. The signal's 'intermittency'—when the signal is present at a fixed site downwind—also declines with distance from the source, even if it is measured near the putative centreline of the plume.

Another factor setting the temporal and spatial patterning of odour is temporal patterning of odour release. This is an infrequent phenomenon in pheromone communication, but some arctiid moths, for example, emit pheromone rhythmically at approximately one second intervals, although this may merely be an anatomical consequence of 'pumping' the pheromone out through two tiny pores into minute droplets (Yin et al 1990), rather than it volatilizing from a glandular surface. The parallel for blood-seeking Diptera is the pattern of CO_2 release from prospective hosts, which through the host's breathing rhythm is also periodic, typically on the order of a few seconds. In the latter example, it is doubtful that the intermittency of this signal aids the host-seeking insect—indeed, it is perhaps more likely to add to signal intermittency and therefore the difficulty of maintaining contact with the odour.

Odour plumes can be spatially complex in another way. The odour cues in a plume may be generated from different sites. For example, the volatiles influencing orientation of the mosquito *Anopheles gambiae* to humans may originate from foot odour and exhaled CO_2 (de Jong & Knols 1996, this volume). Similarly, parasitoid wasps may 'home' into a composite plume of odours from the insect and the host plant. How the internal structure of such plumes influences orientation remains to be established.

A final meteorological challenge to finding an odour's source in wind is posed by the combined effects of changing wind direction and velocity (Fig. 1). Instantaneous wind direction at a given position and the plume's long axis (its downwind projection) are aligned only infrequently (Elkinton et al 1987, Brady et al 1989). Thus, an insect proceeding upwind in a 'snaking' plume may lose the odour frequently. Recontacting the odour can occur when the wind swings around to its former position. In the case of moths, recontacting the odour is facilitated by 'casting', that is ever-widening lateral excursions to the windline (Kennedy 1983). Other insects may employ a 'looping' manoeuvre to regain the scent. Sustained progress along the plume should occur when the upwind direction and the plume's long axis are aligned. Field experiments (Elkinton et al 1987) with male gypsy moths (*Lymantria dispar*) released up to 80 m downwind of a source of pheromone demonstrate that detection of pheromone occurs much further downwind than a male's ability to follow a pheromone routinely to its source. The discordance between being able to detect an odour and find its source is thus attributable to frequent discontinuities between wind direction and the plume's long axis.

Possible mechanisms to guide an organism to the source of an odour

The solution to finding a wind-borne odour would seem straightforward: fly upwind when in contact with the odour and make crosswind course corrections if needed. However, the task has been outlined in a deceptively simple fashion. An insect sitting on a substrate could head upwind accurately by detecting upwind direction with mechanoreceptors such as those invested in the antennae. Once launched, that heading could be maintained by setting a course with visual cues, essentially an 'aim-and-shoot' strategy. For the

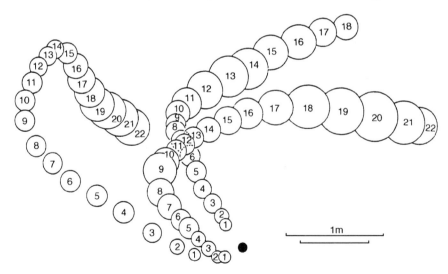

FIG. 1. The combined effects of shifts in wind direction and changes in wind speed upon the path that an odour plume takes in woodland. Odour plumes are visualized by video recording the path of smoke issuing from a small source positioned 15 cm above ground level. Three plumes are shown as consecutively numbered 'puffs' of odour. (From Brady et al 1989.)

distance over which the selected visual cues remains informative and as long as the wind direction holds steady, this strategy produces in-flight progress towards the odour. Landing should occur if the odour is 'lost'. Should the odour plume swing around to recontact the insect, another aim-and-shoot flight should bring the insect closer to the odour's source. Tsetse flies may use such a series of steps (Bursell 1984, 1987), although the local wind flow at probable landing sites on tree trunks or near the ground is not a good predictor of overall wind direction at the 30–50 cm height, where tsetse flies normally cruise (Brady 1989). It is also clear that these flies detect wind direction whilst airborne and make adjustments in their trajectory to achieve upwind progress in shifting winds (Colvin et al 1989).

The task of determining wind direction whilst airborne is accomplished by detecting progress relative to the organism's visual surround, not by wind flow detectors. For example, in the case of flight directly up the windline, the visual field directly below flows front-to-rear if the organism's airspeed exceeds the wind speed. This general principle was first verified experimentally by Kennedy (1939), who used CO_2-stimulated *Aedes aegypti* mosquitoes, and it is termed optomotor anemotaxis. Upwind displacement in an odour stream, however, entails more than charting an uncomplicated straight upwind course by achieving a front-to-rear flow of the visual field below. The actual paths of insects flying to an odour range from rapid, relatively straight-line dashes, represented by tsetse flies seeking a meal (Gibson & Brady 1988), to a zigzag course, exemplified by male moths flying to female pheromone (Kennedy 1983), and to convoluted paths with loops or vertical excursions, characteristic of some parasitoid wasps 'seeking' a host (Drost & Cardé 1992, Kaiser et al 1992, Zanen 1993). These three types of flight to an odour illustrate different strategies of navigation.

Male moth flight to female-emitted pheromone

Optomotor anemotaxis enables moths to determine the upwind heading. Their flight tracks are usually described as a zigzag executed predominantly in the horizontal plane. The counterturns of the zigzag occur several times a second and most current views of this manoeuvre assume that turns are initiated by a centrally generated programme of counterturns (Kennedy 1983). When a moth loses contact with the plume, counterturning rapidly turns into casting (Kuenen & Cardé 1994), that is, with progressively wider counterturns generally without upwind progress. Baker (1990) proposed that the moth response was governed by a tonic–phasic system. The tonic response to the general presence of pheromone is a counterturning programme, which produces upwind zigzagging, and, following the recent loss of pheromone, casting. The phasic response to the flicker of the signal, caused by the wispiness of the plume, is to head more towards the upwind direction. The rate of encountering pheromone filaments directs the degree to which the moth attains an upwind heading. The importance of a flickering signal in sustained moth flight to pheromone had been established in earlier wind tunnel trials (see Baker 1990).

The visual feedback generated by a zigzag path can be separated into longitudinal (L) and transverse (T) components, with the L image flow occurring front-to-rear of the insect for the upwind element and the T image flow occurring because of the effect of sideslip when the path is headed crosswind. It appears that the insect measures T and L separately (Willis & Cardé 1990), rather than as some approximation of T + L, although much remains to be learned about how moths process image flow and how it feeds back to motor control of counterturning and velocity of flight.

It now appears that the fine-scale structure of plumes, generated by the smallest filaments, dictate the counterturning manoeuvres (Mafra-Neto & Cardé 1994, 1995a,b, 1996). In our wind tunnel studies of almond moth (*Cadra cautella*), males flying along a thin ribbon plume in a wind tunnel follow a zigzag course (Fig. 2, upper panel). In plumes of a known filament structure (pulses at 5 Hz from a mechanical device), or broad turbulent plumes, however, males fly routes ranging from shallow zigzags to essentially straight upwind dashes; in both cases moths accelerate in plumes with such fine-scale structures (Fig. 2, middle and lower panels). Evidence on how these differences in track arise came from presentations of a single brief puff of pheromone to a casting almond moth male: males reacted by surging directly upwind (Fig. 3). We have interpreted this single puff response as a template (Mafra-Neto & Cardé 1996) that explains the variability in *Cadra* tracks, ranging in structure from casting to zigzag, to shallow zigzag and to straight upwind. Provided pheromone and gaps of pheromone-free air are encountered with the 'optimal' timing, males head upwind and accelerate; other combinations of filament encounter promote flight more towards the crosswind. Vickers & Baker (1994) found in *Heliothis virescens* essentially the same reactions to rapidly encountered filaments. Because these two moths are from long-diverged lineages, the parallel findings that the rate of filament encounter modulates the path of upwind flight are suggestive of a widespread mechanism in moths.

This interpretation of moth flight to pheromone is not universally accepted. Preiss & Kramer (1986) view zigzagging as a consequence of moths being unable to aim upwind accurately. Reiterative corrections yield the counterturning. Witzgall & Arn (1990) posit that an incomplete blend of pheromone generates counterturning and the natural pheromone evokes a straightened-out trajectory. They point out that nearly all wind tunnel recordings of moth flight to pheromone use a synthetic, possibly incomplete, copy of the female's pheromone as the stimulus. Evidence for and against these competing hypotheses is debated elsewhere (Cardé & Mafra-Neto 1996).

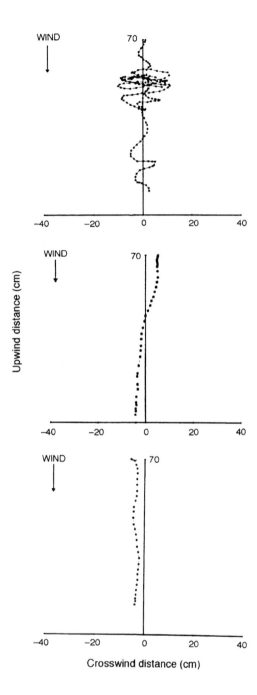

Upwind distance (cm)

Crosswind distance (cm)

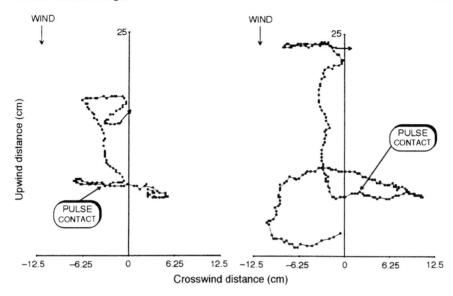

FIG. 3.   Effect on *Cadra cautella* of encountering a single brief pulse of pheromone. Top view of videorecorded tracks in the wind tunnel. Each dot in the track represents a moth's position at 0.017 s intervals. Males that were casting following the truncation of a pheromone plume were given a brief (0.25 s) encounter with pheromone. Males surged nearly straight upwind and then returned to casting. (From Mafra-Neto & Cardé 1994.)

Female parasitoid wasps seeking a host

The selective forces moulding pheromone communication in moths comprises a system wherein a sender seeks to ensure that its signal is detected, and a responder whose success is dependent on rapidly finding the sender. A prospective host of a parasitoid wasp, in contrast, seeks not to be detected. Indeed, some moth caterpillars thwart location by moving away from chemical signals they generate: they walk away from their faeces (Zanen & Cardé 1991). This 'hide-and-seek' game can be more complex from two additional perspectives. Because odour from the plant on which the host feeds is often a component of the odour plume used by wasp parasitoids in locating their hosts, the odour plume can be quite large at its origin and the imbedded odour cue from the host itself may not match the dimensions of the plant's plume.

FIG. 2.   (*opposite*) Top view of video recorded flight tracks of *Cadra cautella* males encountering three types of pheromone plumes in a wind tunnel. Each dot in the track indicates the moth's position at 0.03 s intervals. Upper panel: a thin ribbon plume. Middle panel: a plume with pulses presented at 5 Hz. Bottom panel: a wide plume with internal turbulence. (From Mafra-Neto & Cardé 1994.)

Second, wasp parasitoids routinely learn features of the chemical plume useful in host location, particularly when parasitoids shift between host species or the host shifts between different plants (reviewed by Vet et al 1995).

In a wind tunnel the tracks of *Cotesia rubeculla*, a braconid parasitoid of the cabbage white butterfly (*Pieris rapae*), towards a stimulus of the caterpillar on a cabbage leaf, range from rapid, straight-line dashes, to a zigzag path, and to a track with counterturns and loops. Experience with the host, which can alter the probability of attempting to find a particular chemical stimulus, did not alter the form of the track in any discernible way (Kaiser et al 1992). The braconid wasp *Microplitis croceipes* also displays a variety of upwind manoeuvres, including straight flights, counterturning and loops which are evidently aimed at recontacting a lost plume (Zanen 1993). Another flight pattern is seen in the chalcid wasp *Brachymeria intermedia*. Pupae of the gypsy moth, its favoured host, are typically situated on tree trunks or in leaf litter on the forest floor. The foraging track of this wasp in the presence of host kairomone is either vertical, tree-oriented scanning within several centimetres of the bark or near the forest floor (Drost & Cardé 1992). This wasp can, however, also fly directly towards a small source of kairomone from gypsy moth pupae in a wind tunnel. The counterturns seen in the wind tunnel with these three wasps, however, did not exhibit a metronomic quality and thus are unlikely to be under the control of a central programme. In most cases, parasitoid orientation, unlike moth navigation, is in daylight when visual cues may aid in scanning the environment for the host's precise location, although in some moths visual cues also aid search.

Orientation of female tsetse flies to their host for their blood meal

In obtaining blood meals the tsetse fly must often find a host at some distance. Tsetse is particularly instructive as a model of mechanisms, because the experiments undertaken are a mixture of wind tunnel studies under defined conditions and field observations, where the olfactory cues are presented to flies in simulations of the natural conditions of host finding. As noted, an aim-and-shoot flight appears responsible for some rapid direct flight towards an odour source (Bursell 1984, 1987, Griffiths et al 1995). Optomotor anemotaxis is also employed as a navigation strategy (Colvin et al 1989). Although looping flights appear important to the flies regaining contact with the host odour when shifts in wind direction cause its loss, it is a rather straight anemotactic flight that seems to account for the successful location of the odour (Gibson et al 1991). The lack in many observed flights of any evident crosswind component, and experimental manipulation of the visual field in wind tunnel trials, suggests that the flies can detect a 1–2° difference between their track and the direction upwind. They evidently do so by measuring transverse image flow. The ability to detect transverse image flow is remarkable because the fly is achieving approximately 5 m/s velocity in a background wind of less than 0.5 m/s.

Orienting without optomotor anemotaxis

All of the cases considered involved upwind movement towards an odour. What mechanisms allow insects to locate the source of an odour plume when either the wind velocity falls below the level detectable by the optomotor reaction or the light level falls below the threshold for detecting movement? (Most of the insects studied in detail—even the moths—are day fliers.) Three species of moths have been shown to continue along a plume in zero wind (Farkas & Shorey 1972, Kuenen & Baker 1983, Willis & Cardé 1990). Male moths were allowed to fly along a narrow plume in wind and then the airflow was abruptly stopped, leaving the plume suspended in the tunnel; many males continued along the plume over approximately a metre. They appeared to direct their course by sampling the plume as they progressed along its long axis (a longitudinal klinotaxis), but they also may have continued in the same direction using visual cues that directed their optomotor reaction before the wind was stopped.

The extent to which longitudinal klinotaxis is widely employed by moths in the field remains unanswered. It remains a plausible mechanism that supplements optomotor anemotaxis when wind or visual conditions do not permit its use. It might be a particularly valuable strategy for nocturnal mosquitoes seeking a host in wind-sheltered environments, and it seems to be used by tsetse flies in near zero wind (Brady et al 1996).

Outstanding questions and conclusions

The recent finding that moths utilize the fine-scale structure of a pheromone's plume for orientation suggests that other insect groups using optomotor anemotaxis might also employ this feature of the odour plume. In contrast to the pheromone plume from moths, however, the plumes emanating from plants harbouring caterpillars or the odours from large vertebrates may generate an initially large plume that lacks some of the wispy character of moth-generated plumes. The study of odour-mediated flight tends to camouflage the importance of other modalities, such as valuable visual cues presented by the odour source. This factor is explicitly recognized in tsetse orientation (see Colvin & Gibson 1992).

As emphasized by Kennedy (1978) and Harris & Foster (1995), it is the integration of cues from all stimulus modalities that sets a behavioural outcome, and it can be quite uninformative to study a behaviour by modulating input from only the chemical channel. How well wind tunnel conditions simulate the structure of natural plumes of odour and the visual environment of the field is rarely questioned. Perhaps our comprehension of how insects find odour sources is limited by inattention to these details.

The use of optomotor anemotaxis to locate wind-borne odours that signify a resource is widespread among insects. Among moths this trait allows mate finding over remarkable distances and it is under selective pressure from both sender and receiver perspectives. The navigational cues used by moths involve the fine-scale structure of the odour plume and a motor programme typically resulting in a zigzag path. Other examples, including parasitoid wasps and tsetse flies, encounter plumes generally of much larger expanse and presumably less fine-scale structure. These latter insects do not counterturn regularly, and they seem to locate lost plumes by looping manoeuvres rather than by the casting programme of moths. The odour cues used also may vary, depending on the particular host species or even its physiological condition. This suggests some plasticity in response to odours, which in the case of wasps often involves learning these chemical cues.

Acknowledgements

The National Research Initiative Competitive Grants Program of the United States Department of Agriculture supported our studies on insect orientation to pheromones and kairomones. John Brady provided a very helpful review of this paper.

References

Baker TC 1990 Upwind flight and casting flight: complementary phasic and tonic systems used for location of sex pheromone sources by male moths. In: Døving K (ed) Proceedings of the Xth international symposium on olfaction and taste. Graphic Communication Systems, Oslo, p 18–25

Brady J, Gibson G, Packer MJ 1989 Odour movement, wind direction, and the problem of host-finding by tsetse. Physiol Entomol 14:369–380

Brady J, Griffiths N, Paynter Q 1996 Wind speed effects on odor source location by tsetse flies (*Glossina*). Physiol Entomol 20:293–302

Bursell E 1984 Observation on the orientation of tsetse flies (*Glossina pallipides*) to wind-borne odours. Physiol Entomol 9:133–137

Bursell E 1987 The problem of wind-borne odour on the direction of flight in tsetse flies, *Glossina* spp. Physiol Entomol 12:149–156

Cardé RT, Mafra-Neto A 1996 Mechanisms of flight of male moths to pheromone. In: Cardé RT, Minks AK (eds) Pheromone research: new directions. Chapman & Hall, New York, in press

Colvin J Gibson G 1992 Host-seeking behavior and management of tsetse. Ann Rev Entomol 37:21–40

Colvin J, Brady J, Gibson 1989 Visually-guided, upwind turning behaviour of free-flying tsetse in odour-laden wind: a wind tunnel study. Physiol Entomol 14:31–39

de Jong R, Knols BGJ 1996 Selection of biting sites by mosquitoes. In: Olfaction in mosquito–host interactions. Wiley, Chichester (Ciba Found Symp 200) p 89–103

Drost YC, Cardé RT 1992 Use of learned visual cues during habitat location by *Brachymeria intermedia*. Entomol Exp Appl 64:217–224

Elkinton JS, Schal C, Ono T et al 1987 Pheromone puff trajectory and the upwind flight of male gypsy moths in a forest. Physiol Entomol 12:399–406

Farkas SR, Shorey HH 1972 Chemical trial following by insects. A mechanism for orientation. Science 178:67–68

Gibson G, Brady J 1988 Flight behaviour of tsetse flies in host odour plumes: the initial response to leaving or entering odour. Physiol Entomol 13:29–42

Gibson G, Packer MJ, Steullet P, Brady J 1991 Orientation of tsetse to wind, within and outside host odour plumes in the field. Physiol Entomol 16:47–56

Griffiths N, Paynter Q, Brady J 1995 Rates of progress up odour plumes by tsetse flies: a mark-release video study of the timing of odour source location by *Glossina pallidipes*. Physiol Entomol 20:100–108

Harris MO, Foster SP 1995 Behavior and integration. In: Cardé RT, Bell WL (eds) Chemical ecology of insects, vol 2. Chapman & Hall, New York, p 3–46

Kaiser L, Willis MA, Cardé RT 1992 Flight maneuvers used by a parasitic wasp to locate host-infested plant. Entomol Exp Appl 70:285–294

Kennedy JS 1939 The visual responses of flying mosquitoes. Proc Zool Soc Lond 109:221A–242A

Kennedy JS 1978 The concepts of olfactory arrestment and attraction. Physiol Entomol 3:91–98

Kennedy JS 1983 Zigzagging and casting as a response to wind-borne odour: a review. Physiol Entomol 8:109–120

Kuenen LPS, Baker TC 1983 A non-anemotactic mechanism used in pheromone source location by flying moths. Physiol Entomol 8:277–289

Kuenen LPS, Cardé RT 1994 Strategies for recontacting a lost pheromone plume: casting and upwind flight in the male gypsy moth. Physiol Entomol 19:15–29

Mafra-Neto A, Cardé RT 1994 Fine-scale structure of pheromone plumes modulates upwind orientation of flying moths. Nature 369:142–144

Mafra-Neto A, Cardé RT 1995a Influence of pheromone plume structure and pheromone concentration on upwind flight of *Cadra cautella*. Physiol Entomol 20:116–133

Mafra-Neto A, Cardé RT 1995b Effect of the fine-scale structure of pheromone plumes: pulse frequency modulates activation and upwind flight of almond moth males. Physiol Entomol 20:229–242

Mafra-Neto A, Cardé RT 1996 Dissection of the pheromone-modulated flight of moths using single-pulse response as a template. Experientia 52:373–379

Murlis J, Elkinton JS, Cardé RT 1992 Odor plumes and how insects use them. Ann Rev Entomol 37:505–532

Preiss R, Kramer E 1986 Mechanisms of pheromone orientation in flying moths. Naturwiss 73:555–557

Sabelis M, Schippers P 1984 Variable wind directions and anemotactic strategies of searching for an odour plume. Oecologia 63:225–228

Schal C, Bell WJ 1986 Interspecific and intraspecific stratification of tropical cockroaches. Ecol Entomol 11:411–423

Vet LEM, Lewis WJ, Cardé RT 1995 Parasitoid learning and foraging. In: Cardé RT, Bell WL (eds) Chemical ecology of insects, vol 2. Chapman & Hall, New York, p 65–101

Vickers NJ, Baker TC 1994 Reiterative responses to single strands of odor promote sustained upwind flight and odor source location by moths. Proc Natl Acad Sci USA 91:5756–5760

Willis MA, Cardé RT 1990 Pheromone-modulated optomotor response in male gypsy moths, *Lymantria dispar* L.: upwind flight in a pheromone plume in different wind velocities. J Comp Physiol A 167:699–706

Witzgall P, Arn H 1990 Direct measurement of the flight behavior of male moths to calling females and synthetic sex pheromones. Zeit Naturfor 45:1067C–1069C

Yin LRS, Cardé RT, Schal C 1990 Sex pheromone gland of the female tiger moth, *Holomelina lamae* (Lepidoptera: Arctiidae). Can J Zool 69:1916–1921

Zanen PO 1993 A comparative study of the role of vision and olfaction during in-flight maneuvers by four species of insects to semiochemicals. PhD thesis, University of Massachusetts, Amherst, MA, USA

Zanen PO, Cardé RT 1991 Learning and the role of host-specific volatiles during in-flight host finding in the specialist parasitoid *Microplitis croceipes*. Physiol Entomol 16:381–389

Zanen PO, Sabelis MW, Buonaccorsi JP, Cardé RT 1994 Search strategies of fruit flies in steady and shifting wind in the absence of food odors. Physiol Entomol 19:335–341

DISCUSSION

Boeckh: You remarked that it is not necessary for moths to have high levels of absolute sensitivity because a single molecule can have an impact on behaviour. Therefore, why have they developed such a high level of absolute sensitivity?

Cardé: It is possible that at low population densities, a male has to rely on a low threshold, in part, to locate a female. John Murlis and Chris Jones (see Murlis et al 1992) have suggested, however, that well downwind of an odour source there are occasional packets of moderately high odour density, and that there are also large gaps without odour. It appears that male moths are not capable of following pheromone plumes over long distances to locate a female, so the general notion that moths routinely find females over distances of many kilometres is questionable. To relate the threshold of response at the single-cell level to response in the field, we would need to know the airborne concentration or flux downwind.

Carlson: What is the greatest distance at which you've measured a behavioural response to a release of pheromone?

Cardé: A large proportion of individually caged male gypsy moths will begin wing fanning within about 10–15 s of a pheromone reaching them a distance of 120 m from a source comparable in strength to a female. If these males are then released most of them head upwind, but few of them find the source.

Carlson: What do you know about the ability of males to find the source from greater distances?

Cardé: Unfortunately, the frequently cited experiments involve first the release of male moths and their recapture several hours later at either caged females or a synthetic source of attractant. It's difficult to determine when, during these long intervals, the behaviour of males was under active guidance of pheromone.

Carlson: Is it possible that there are other mechanisms that come into play at distances greater than 100 m? These mechanisms may involve the higher levels of absolute sensitivity that Jürgen Boeckh mentioned.

Cardé: The only known way to fly upwind continually is by optomotor anemotaxis. There is no evidence for the involvement of other strategies, although if a large number of moths are released and you give them enough time, some of them are able to locate attractants over distances of a few kilometres. For example, in the 1970s we did some experiments (Kochansky et al 1975) where we released hundreds of marked polyphemus moths (a large common American saturniid moth). We found that only several of these travelled distances of 3.5 km to the females. We tried to release the males downwind of the females but we obviously couldn't control the wind direction. These kinds of experiments are cited as evidence of very long distance communication, but they suffer from the problems of methodological interpretation: how much of the movement was upwind anemotaxis induced by pheromone and how much was random movement?

Brady: Is it possible that the high level of sensitivity is not so much to do with distance detection or the detection of low levels of pheromone, but to do with distinguishing between the odours of closely related species?

Cardé: This is an area that is not well worked out. There are many model systems in which some minor components of attractants have been identified, but we do not have a clear idea of how males estimate the ratio of components, or if they make those calculations far away from or close to the odour source. There's one school of thought which says that they're always measuring the ratio from the very first moment that they respond to the odour, and there is good evidence of this in the oriental fruit moth (Linn et al 1987). However, some neurophysiological analyses suggest that the males don't have enough receptors sensitive to the minor components to be able to do this (Mustaparta 1996, this volume).

Mustaparta: The sensitivity of the receptor neurons to the major pheromone component varies by at least a factor of 1000 in some species of moth. It is therefore possible that the recruitment of receptor neurons is a mechanism by which the males determine the distance from the pheromone source.

Takken: Charlwood has done some research in Papua New Guinea, and he has found that within one night mosquitoes can cover a distance of 2 km (Charlwood et al 1988).

Cardé: But that doesn't tell us how they managed to get there.

Guerin: Is there any circumstantial evidence which suggests that mosquitoes use odour-mediated chemoanemotaxis? Do they do so only in certain conditions? For example, do they do more host searching when they are guided by wind?

Cardé: One can demonstrate that mosquitoes fly upwind in wind tunnels. It would be useful to determine their behaviour under essentially zero wind conditions and whether they are able to detect convection currents. I have watched mosquitoes in the forest when I can't detect any wind, and yet they are still able to orient themselves towards me. Perhaps a combination of odour and visual stimuli is involved, and I believe that mosquitoes are able to follow the plume through still air. However, this ought to be explored experimentally.

Brady: We have some thin, although statistically sound, evidence that tsetse flies can arrive at an odour source in unmeasurably low winds. Presumably, this orientation cannot involve anemotaxis, but instead uses some kind of kinetic strategy (Brady et al 1996).

Grant: The chemical signals that are modulated either by convection currents in relatively still air or by winds in the forest should be different in structure, and consequently different behaviours may result from their processing. Also, it is likely that signals with similar structures may elicit different behaviours if other visual and environmental cues are involved.

Kaissling: How did you create conditions of zero wind velocity? Were the zigzag patterns obtained in a wind tunnel?

Cardé: Yes. These experiments were performed in a wind tunnel. A large leather belt was employed to stop the fan very quickly. We ended up with a plume (which we could visualize with titanium tetrachloride 'smoke') that was suspended in the tunnel. After a period of about 10 s, it began to break apart, but prior to this males continued to fly up along the plume as if they were flying along a plume in wind.

Kaissling: Is it possible that they are just continuing the type of behaviour that they were doing before the wind stopped?

Cardé: This was an interpretation of Harry Shorey's original experiment of this type, i.e. that males could be using the visual cues that had allowed them to maintain their course in wind (Farkas & Shorey 1972).

Kaissling: In principle, the animal could control and monitor its movements idiothetically, i.e. by relying solely on internal signals. By scanning behaviour it could build up an internal map of an odour plume or of a quasi-static odour distribution, as under conditions without wind, and this way find the point of highest odour concentration, i.e. the odour source.

Cardé: That's possible. We've done other experiments with moths where we've changed the visual field half way through a course down a wind tunnel. At the position when either the retinal velocity of the pattern changes, or the angle that is subtended by the pattern changes, the moth stops upwind progress for several seconds. It seems as though it's recalibrating its optomotor system. Then it continues upwind. It would be interesting to go into a tropical rainforest to find out how nocturnal male moths use optomotor cues, given the low levels of light and calm winds that are typical at night in that environment.

Hildebrand: Edward Davis in his introduction mentioned the terms 'ambushers' and 'patrollers'. Are vector mosquitoes predominantly ambushers or patrollers?

Davis: The mosquitoes that live outside and enter huts through windows fly over a relatively long distance to reach the hut. Therefore, for that period of their approach they may be termed 'patrollers'. Once inside the hut, they have to change their strategies because all the odour, wind and light considerations have changed. Sometimes mosquitoes rest on a couple of surfaces. It is possible that they behave more like an 'ambusher', that is they alight on an interior wall and wait until a host comes near. Alternatively, they may employ a search strategy similar to those observed in circular arenas rather than a 'linear' upwind plume orientation strategy. A mosquito might do one or the other depending on the set of circumstances that it finds itself in. *Ae. aegypti* which lives its whole life cycle in a human dwelling, will employ a set of strategies associated with that environment. In contrast, those mosquitoes that are more zoophilic and live outside in more wooded areas, but which must go towards more open areas to find animals upon which to feed, will have adopted a totally different set of strategies.

Cardé: That the behaviour could be changing at some point is an important consideration. Therefore, in wind tunnels we may be overlooking the whole realm of behaviours that occur inside a hut.

Ziegelberger: The strategy of odour source finding depends on whether wind is present or not. In the absence of wind, the possible orientation mechanisms are limited. Besides an accidental discovery or idiothetic searching, I can only think of a mechanism involving the following of a stimulus diffusion gradient. In contrast, in the presence of wind stimulus molecules are transported in filaments and gradients play a minor role. For the upwind plume orientation a 'go-and-wait' mechanism is sufficient, at least for moths (Kaissling & Kramer 1990), where 'go' represents walk or fly upwind when the receptor cells receive stimulus molecules, and 'wait' represents stop moving upwind when odour stimulation ceases. For this strategy measurements of the odour concentration are of minor importance.

Boeckh: I would like to emphasize the same point using cockroaches as an example. Cockroaches seem to be able to switch from anemotactic to patrolling behaviour, irrespective of the wind conditions. We should bear in mind these different options and not just focus on one stereotypical situation.

References

Brady J, Griffiths N, Paynter Q 1996 Wind speed effects on odor source location by tsetse flies (*Glossina*). Physiol Entomol 20:293–302

Charlwood JD, Graves PM, Marshall TF de C 1988 Evidence for a 'memorized' home range in *Anopheles farauti* females from Papua New Guinea. Med Vet Entomol 2:101–108

Farkas SR, Shorey HH 1972 Chemical trail-following by flying insects: a mechanism for orientation to a distant odor source. Science 178:67–68

Kaissling K-E, Kramer E 1990 Sensory basis of pheromone-mediated orientation in moths. Verh Dtsch Zool Ges 83:109–131

Kochansky J, Tette J, Taschenberg EF, Cardé RT, Kaissling K-E, Roelofs WL 1975 Sex pheromone of *Antherea polyphemus*. J Insect Physiol 21:1977–1983

Linn CE Jr, Campbell MG, Roelofs WL 1987 Pheromone components and active spaces: what do moths smell and when do they smell it? Science 237:650–652

Murlis J, Elkinton JS, Cardé RT 1992 Odor plumes and how insects use them. Ann Rev Entomol 37:505–532

Mustaparta H 1996 Introduction IV: coding mechanisms in insect olfaction. In: Olfaction in mosquito–host interactions, Wiley, Chichester (Ciba Found Symp 200) p 149–157

Olfactory basis of host location by mosquitoes and other haematophagous Diptera

Alan Cork

Natural Resources Institute, Central Avenue, Chatham Maritime, Kent ME4 4TB, UK

Abstract. The behavioural role of odours released by mosquito hosts is poorly understood, indeed for many species it is still uncertain whether olfactory cues play a significant part in host location. Generalist attractants, such as CO_2, have found application in mosquito trapping systems, and yet more host-specific attractants, such as L-lactic acid, remain of questionable value. Recent work with other haematophagous Diptera, notably *Glossina*, has shown that by a coordinated multidisciplinary approach it is possible to develop odour-baited trapping systems with a high level of attractiveness and specificity. Many of the compounds shown to attract *Glossina* have been tested with mosquitoes, and one of these, 1-octen-3-ol, attracts female mosquitoes of a number of species, but only in the presence of CO_2. The behavioural significance of other compounds identified as host attractants of haematophagous Diptera, such as phenols, indoles and carboxylic acids, are currently under investigation. Efforts to produce a host odour attractant for the highly anthropophilic species *Anopheles gambiae* have been hindered by the chemical nature of the compounds associated with its human host, although a number of short-chain fatty acids identified in sweat samples have been shown to be electrophysiologically active.

1996 Olfaction in mosquito–host interactions. Wiley, Chichester (Ciba Foundation Symposium 200) p 71–88

Host-seeking mosquitoes are exposed to a wide variety of visual, olfactory, gustatory and physical stimuli, many of which could potentially act as cues for host identification and location. The response of mosquitoes to these cues is dependent on their age, sex, physiological state and host preferences. Of the stimuli studied to date, olfactory cues appear to offer the greatest potential for eliciting anemotactic host-seeking responses that can form the basis of trapping systems for epidemiological surveys or population suppression. Although odour-baited traps have been developed that utilize generalist attractants, such as CO_2, more specific olfactory signals, based on feeding or oviposition attractants, are still in their infancy. This paper will consider the current status of knowledge on the chemical nature of these olfactory stimulants and, by reference to related work with other haematophagous Diptera, will consider ways of incorporating these into odour-baited trapping devices for Culicidae.

Role of host breath and carbon dioxide on host location

The role of CO_2 in mosquito host location has been studied both in the laboratory and field for many species (Clements 1963, Takken 1991) since the first report of its potential as a mosquito attractant (Rudolfs 1922). Early laboratory studies suggested that CO_2 activated rather than attracted mosquitoes (Daykin et al 1965, Khan & Maibach 1966), yet other workers contended that CO_2 is also an attractant (Van Thiel & Weurman 1947, Brown et al 1951) as suggested by the increased trap catches obtained when CO_2 is incorporated in a trap, either with (Huffaker & Back 1943, Newhouse et al 1966) or without (Reeves 1951, Defoliart & Morris 1967) a light source.

Hocking (1963) proposed that CO_2 and host odour are long-range attractants, with elevated temperature and moisture contributing to close-range orientation. This view is supported by field work reported by Snow (1970), who showed that removal of 95.5% of the CO_2 released in expired human air significantly reduced the number of mosquitoes attracted to the host but did not reduce the proportion of mosquitoes attempting to feed once in the proximity of the host. The relative reduction in catch associated with removing the CO_2 is greater for *Culex thalassius* and *Culex tritaeniorhynchus* than *Anopheles gambiae* and *Anopheles melas*, suggesting that CO_2 is more important for host location in *Culex* than *Anopheles* spp, and that *Anopheles* spp use other olfactory cues for host location. This is supported by the work of Costantini et al (1996), who found that increasing CO_2 concentration in an odour-baited trap increased the catch of most generalist feeders but that the catch of *An. gambiae* did not increase above that expected with a whole human bait. Indeed, de Jong & Knols (1995) found that human breath has no demonstrable effect on the behaviour of *An. gambiae* in a wind tunnel bioassay. However, the possibility remains that compounds in human breath, other than CO_2, are involved in host location by *Culex* spp because the soda lime filter used to remove CO_2 from breath is not specific and would have removed all other acidic compounds such as carboxylic acids and phenols.

In related work by Gillies & Wilkes (1972) the range of attraction of two calf baits was compared with that of an equivalent amount of CO_2 (500–700 ml/min). They showed that *Cx. thalassius* and *Cx. tritaeniorrhynchus* orient to both sources at distances of up to 36 m, whereas *An. melas* responded to CO_2 at distances of up to 36 m but could orient to the calf at distances of up to 55 m. They concluded, in common with Snow (1970), that CO_2 was the principal olfactory attractant used by the *Culex* spp, but that for *An. melas* other, as yet unidentified, host odours are important for host location.

Omer (1979) found that *Anopheles arabiensis* and *Culex pipiens fatigans* responded to 0.5% CO_2 by upwind orientation in a wind tunnel, but the response was only observed when the concentration of CO_2 was changing (see also Gillies 1980). Extracellular recordings made by Kellogg (1970) support this finding by showing that CO_2 receptors only responded to changes in concentration. The ability of mosquitoes to respond to rapid changes in odour concentration would enable them to discriminate between constant background concentrations of chemicals, such as CO_2 and water, and host odours that would be perceived in an intermittent manner under natural conditions (Murlis & Jones 1981).

Role of moisture and heat in mosquito host location

Temperature and humidity gradients are thought to influence the close-range orientation and landing behaviour of mosquitoes onto hosts (Gillies 1880). This is supported by electrophysiological studies, which confirmed the presence of both thermoreceptors (Davis & Sokolove 1975) and hygroreceptors (Kellogg 1970) on the antennae of *Aedes aegypti*. However, in behavioural studies conducted in the laboratory, Price et al (1979) found that increased humidity alone did not attract *Aedes quadrimaculatus*, although increased humidity and temperature did elicit a response. More recently, Eiras & Jepson (1994) independently tested the effect of humidity and temperature on the behaviour of female *Ae. aegypti* in a bioassay (Feinsod & Spielman 1979). They found that by increasing the temperature from 27 °C to 30 °C, the proportion of *Ae. aegypti* entering a collection chamber increased significantly from 4% to 22%. However, there was no corresponding effect when they increased the humidity by up to 5%. These results taken together suggest that humidity gradients alone do not elicit a response from mosquitoes. This result was unexpected since Kellogg (1970) showed that *Ae. aegypti* is able to discern 2% changes in relative humidity at the peripheral sensilla level and W. Takken (personal communication 1995) found that female *An. gambiae* only respond in a wind tunnel bioassay (Knols et al 1994) to host-related odours when the humidity is changing.

Role of 1-octen-3-ol in host location

Vale (1974) showed that mammalian host odours are attractive to *Glossina* in the absence of visual cues. Ox odours were subsequently collected and analysed by gas chromatography linked to electroantennography (Cork et al 1990) in order to identify electrophysiologically active compounds that might be responsible for the observed attractiveness. This work resulted in the identification of a number of electrophysiologically active compounds, one of which, 1-octen-3-ol, was shown to increase odour-baited trap catches significantly (Vale & Hall 1985). The behavioural role of 1-octen-3-ol, despite increasing trap catches, has been a matter of some debate because it was not until recently that 1-octen-3-ol was shown to elicit upwind flight in a laboratory wind tunnel (Paynter & Brady 1993) and even now there still appears to be some uncertainty about its behavioural significance in the field (Torr 1990, Brady & Griffiths 1993).

1-Octen-3-ol attracts a wide range of haematophagous Diptera, including species of Tabinidae (French & Kline 1989), Oestridae (Anderson 1989), Stomoxyinae (Holloway & Phelps 1991, S. Schofield, personal communication 1995) and Ceratopogonidae (Kline 1994, Blackwell & Wadhams 1995). It also elicits electrophysiological responses from the Calliphoridae, *Cochliomyia hominivorax* (Cork 1994) and *Lucilia cuprina* (K. C. Park & A. Cork, unpublished results 1995), although no corresponding behavioural responses have yet been demonstrated for either species.

The first reported use of 1-octen-3-ol as an attractant for mosquitoes was by Takken & Kline (1989), who found that significant numbers of *Aedes taeniorhynchus*, *Anopheles crucians*, *Ae. quadrimaculatus* and *Wyeomyia mitchellii* could be trapped in modified CDC (Center for Disease Control, Atlanta, USA) light traps with 1-octen-3-ol released at between 1.6 and 2.3 mg/h. The catches were comparable to those obtained with CO_2-baited CDC light traps released at 200 ml/h, although in each case 1-octen-3-ol synergized the attractiveness of CO_2. In a subsequent publication, however, Kline et al (1990) were unable to confirm the response of *Ae. taeniorhynchus* to 1-octen-3-ol alone but did confirm the attractiveness of 1-octen-3-ol combined with CO_2. *Culex* did not respond to 1-octen-3-ol alone or in combination with CO_2 (Kline et al 1990). The lack of response from *Culex* was thought to relate to their host preferences because 1-octen-3-ol is a mammalian component of host odour and *Culex* are ornithophilic. Further work conducted by Kline (1994) in a range of ecological niches showed that 35 species of mosquito could be attracted by the odour baits tested and of these, 1-octen-3-ol in combination with CO_2 caught species of *Aedes*, *Anopheles*, *Psorophora*, *Coquillettidia* and *Mansonia*. Interestingly, he found that the responses varied geographically, seasonally and according to the physiological state of the mosquitoes, especially for *Anopheles*. 1-Octen-3-ol has been identified in human sweat (A. Cork, unpublished results 1993) and collected from human volatiles (Sastry et al 1980). It remains to be seen whether 1-octen-3-ol, either alone or in combination with other human volatiles, can be used to form the basis of an attractant bait for anthropophilic species such as *An. gambiae*.

Role of phenols and indoles in mosquito attraction

Attraction of mosquitoes to oviposition sites is thought to be mediated by a combination of environmental odour cues and, in the case of aggregating species such as *Culex quinquefasciatus*, by oviposition pheromones (McCall & Cameron 1995). Oviposition traps baited with fermenting biomass infusions have gained in popularity recently as a means of monitoring mosquitoes, particularly because they attract primarily gravid females. However, traps that rely on infusions vary in attractiveness and their replacement with synthetic equivalents is seen as desirable (Millar et al 1992). Many of the behaviourally relevant compounds present at natural oviposition sites are phenols. Ikeshoji (1975) isolated phenol and a series of monomethylphenols, dimethylphenols and trimethylphenols from extracts of wood creosote and demonstrated that several of these compounds elicited oviposition behaviour from *Aedes*, *Armigeres* and *Culex*. In subsequent studies Bentley et al (1979) isolated 4-methylphenol from decaying birch infusions and demonstrated that it was attractive to both male and female *Aedes triseriatus* Say. In an effort to identify systematically the environmental cues responsible for eliciting oviposition in *Cx. quinquefasciatus*, Millar et al (1992) identified phenol, 4-methylphenol, 4-ethylphenol, indole and 3-methylindole in fermented Bermuda grass infusions, but attributed most of the biological activity to 3-methylindole alone.

Phenol and its alkylated derivatives are frequently associated with mammalian waste products. The possibility that they might act as host odour attractants of haematophagous Diptera first received attention in work on tsetse flies. Chorley (1948) and Owaga (1985) demonstrated that mammalian urine was attractive to tsetse flies and this activity was subsequently attributed to phenolic compounds in urine (Hassanali et al 1986, Bursell et al 1988). Of the eight phenols identified in urine, Vale et al (1988) showed that only two (4-methylphenol and 3-*n*-propylphenol) were essential for maximum attraction of *Glossina pallidipes*.

Kline et al (1990) tested six of the phenols identified in urine (phenol, 3-methylphenol, 4-methylphenol, 3-ethylphenol, 4-ethylphenol and 2-propylphenol) as putative mosquito attractants. A combination of the phenolic blend and 1-octen-3-ol attracted more *Ae. taeniorhynchus* and *Culex furens* than either the phenolic blend or 1-octen-3-ol tested alone, although the differences were not statistically significant. At least some of these phenols are known to act as oviposition attractants, so it is uncertain whether the mosquitoes were responding to an oviposition or host attractant.

Indole and 3-methylindole were also found in mammalian waste products and they have been found to be attractive to a number of Muscidae and Calliphoridae that oviposit on mammals (Cragg & Ramage 1945, Mackley & Brown 1984, Mulla & Ridsdill-Smith 1986). However, despite electrophysiological studies demonstrating the sensitivity of a number of haematophagous Diptera, such as Glossinidae (A. Cork, unpublished results 1986), Calliphoridae (Cork 1994, K. C. Park & A. Cork, unpublished results 1995) and Stomoxyinae (Schofield et al 1996), to these compounds their role in host-locating behaviour remains obscure. It is conceivable that, for some species at least, they could be involved in both oviposition and host-locating behaviour, particularly because such highly anthropophilic species as *Ae. aegypti* are attracted to hay infusions (Reiter et al 1991) that contain 3-methylindole.

Human sweat: the role of carboxylic acids and steroids in host location

Samples of human sweat have been bioassayed by many workers with varying results. Howlett (1910), Rudolfs (1922) and Reuter (1936) reported that sweat was unattractive to mosquitoes. On the other hand, Parker (1948) found that female *Ae. aegypti* aggregated near sweat samples in a cage, Brown et al (1951) concluded that dilute axillary sweat was attractive in an olfactometer, Rahm (1956) found that sweat was more attractive to *Ae. aegypti* than warm moist air and Skinner et al (1965) concluded that lyophilized sweat was attractive in a dual-port olfactometer. Khan et al (1969) showed that the level of attraction was positively correlated with the amount of sweat produced, and that this was independent of the method by which sweating was induced. Schreck et al (1967) found that acetone washes of human skin were attractive to *Ae. aegypti* in the presence of 0.1% CO_2. The major chemical component of that extract was subsequently identified as L-lactic acid (Acree et al 1968). Smith et al (1970) confirmed the attractiveness of L-lactic acid to *Ae. aegypti* in the presence of CO_2 but they concluded that other chemicals may also be involved in host location. Price et al (1979), working with stored human arm emanations on *Ae. quadrimaculatus*, stressed the importance of quantifying the effects of CO_2 and water on mosquito attraction before attempting to determine the attractiveness of other odour components. As they were able to control accurately the concentration of CO_2 and water in their sample bags, they were able to demonstrate that human-produced chemicals other than CO_2 and water were also important in attracting *Ae. quadrimaculatus*. However, whether this activity was due to L-lactic acid was not tested. Eiras & Jepson (1994) found no response from female *Ae. aegypti* to different doses of L-lactic acid or moisture in a laboratory bioassay even though the doses of L-lactic acid used were thought to be comparable with those released from human hands, i.e. typically 23–133 μg/h (Smith et al 1970). They concluded that L-lactic acid, in common with CO_2, has no effect on mosquito behaviour when it is a few centimetres away. In each case a human hand was found to elicit a significantly higher level of response than any of the other treatments tested, findings that were comparable with those of other workers (Daykin et al 1965, Khan & Maibach 1966, Smith et al 1970, Price et al 1979, Gillies 1980) and which suggested that hand volatiles contain behaviourally relevant chemicals that have yet to be identified. Eiras & Jepson (1994) went on to demonstrate that human chest sweat samples elicited responses from female *Ae. aegypti* but only at elevated temperatures. Of the most abundant compounds identified in chest sweat samples analysed by mass spectrometry linked to gas chromatography, 23 were acids, and, of those tested, 10 were found to elicit electrophysiological responses from female *Ae. gambiae* (Table 1). However, Sastry et al (1980) suggested that between 300 and 400 compounds are constantly released as by-products of metabolism, some 200 of these being carboxylic acids (Stoddart 1990). The carboxylic acids consist of families of straight, methyl-branched and dimethyl-branched saturated, mono-unsaturated and di-unsaturated aliphatic acids (Nicolaides 1974). In addition many of these compounds are subsequently modified into the equivalent alcohols, α-hydroxyacids and diols. Odorous steroids have also been identified from apocrine glands associated with axillary, anogenital, sternal and areolar body regions (Gower et al 1988), although Zeng et al (1992) suggested that the principle odorous compounds in male axillary odours are C_6 to C_{11} straight, branched and unsaturated aliphatic acids, with (Z)-3-methyl-2-hexenoic acid being the most abundant. Some of these compounds have been tested for their attractiveness to mosquitoes. Thus, Rössler (1961) obtained weak attraction with an 'artificial sweat', composed of formic, acetic, propanoic, butanoic, hexanoic, octanoic, lactic, citric, uric and amino acids, but it was only effective in the presence of CO_2. Carlson et al (1973) compared the attractiveness of L-lactic acid to a range of structurally related compounds with *Ae. aegypti* in an olfactometer (Schreck et al 1967). Many of the compounds were chosen because they had been identified in human skin emanations (Anonymous 1966). Each compound was tested independently in a range of concentrations from 25 to 100 μg in the presence of CO_2. Seventeen of the compounds tested elicited behavioural responses at least comparable with that of L-lactic acid.

Role of floral odours and honey baits in the attraction of mosquitoes

Downes (1958) suggested that female mosquitoes probably feed more often on nectar than on blood, and hungry *Ae. aegypti* are known to be stimulated into locomotory, orientation and probing activity by diethyl ether-soluble honey odours (Wensler 1972). Grimstad & DeFoliart (1974) demonstrated that Anopheline mosquitoes are attracted to and feed on *Achillea millefolium* and this was confirmed in a wind tunnel bioassay by Healy & Jepson (1988), who showed that both male and female *An. arabiensis* were attracted to inflorescences and extracts of *Ac. millefolium*. The wide range of flowers visited by mosquitoes is briefly reviewed by Foster & Hancock (1994). Although the age and physiological state of female mosquitoes likely to be attracted by floral odours may be different from that attracted to animal odour baits, it is conceivable that they may be of value in trapping systems for population suppression.

TABLE 1   Electrophysiological responses of compounds identified in human sweat

Compound	Sweat from trunk region ($\mu g/ml$)[a]	Anopheles gambiae electroantennograph (mV)[b]
Formic acid	nd	-9.04 ± 1.6 (a)
Acetic acid	9.76 ± 13.37	-3.30 ± 0.8 (bcd)
Propanoic acid	0.00 ± 0.01	-4.10 ± 0.7 (bcd)
2-Methylpropanoic acid	0.36 ± 0.46	nt
Butanoic acid	0.40 ± 0.71	-3.90 ± 0.6 (bc)
3-Methylbutanoic acid	2.36 ± 2.61	nt
Pentanoic acid	0.05 ± 0.06	-4.55 ± 0.7 (b)
4-Methylpentanoic acid	0.02 ± 0.04	nt
Hexanoic acid	0.52 ± 0.43	-3.01 ± 0.4 (cd)
2-Ethylhexanoic acid	0.50 ± 0.40	nt
Heptanoic acid	10.00 ± 0.00	-1.70 ± 0.3 (e)
Octanoic acid	0.14 ± 0.24	-0.80 ± 0.1 (f)
Nonanoic acid	0.27 ± 0.36	-0.61 ± 0.1 (fg)
Decanoic acid	0.22 ± 0.42	-0.46 ± 0.1 (g)
Dodecanoic acid	0.33 ± 0.59	-0.48 ± 0.2 (g)
Tetradecanoic acid	3.93 ± 6.38	-0.38 ± 0.1 (g)
Hexadecanoic acid	16.71 ± 20.88	-0.55 ± 0.1 (fg)
9-Hexadecanoic acid	4.52 ± 7.07	nt
Octadecanoic acid	3.43 ± 3.46	-0.56 ± 0.1 (fg)
9-Octadecanoic acid	3.66 ± 4.54	nt
Control (diethyl ether)		-0.52 ± 0.1 (g)
Benzoic acid	0.25 ± 0.39	nt
Phenylacetic acid	1.49 ± 1.98	nt
Phenylpropanoic acid	0.31 ± 0.61	nt
1-Octen-3-ol	0.49 ± 0.94	-2.47 ± 0.5 (de)
Benzyl alcohol	0.41 ± 0.88	nt
Dimethyl sulfone	0.47 ± 0.55	nt
Phenylethanol	0.77 ± 0.67	nt
Phenol	2.06 ± 4.78	nt
4-Methylphenol	0.37 ± 0.64	-1.49 ± 0.2

[a]Mean of seven replicates $\pm$ SD; nd, not detected.
[b]Mean of 10–40 replicate responses $\pm$ SD to 1 mg samples in 5 μl diethyl ether; nt, not tested. Means followed by the same letter are not significantly different at $P < 0.05$ by Duncan's multiple range test.

Conclusion

Currently available odour-baited trapping systems for mosquitoes rely on non-specific attractants such as CO_2. Despite overwhelming evidence suggesting that host-related odour cues, apart from moisture and CO_2, are involved in the host-locating behaviour of many mosquito species, synthetic baits have not been developed to replace them. This is particularly surprising for anthropophilic mosquitoes, where a large number of host-related volatiles have already been chemically characterized (Anonymous 1966, Sastry et al 1980, Stoddart 1990, Zeng et al 1992). However, the very complexity of human odour and the problems associated with distinguishing between the effects of human odour, temperature, humidity and CO_2 on mosquito behaviour in the laboratory as well as the field continue to hinder progress. Most of the electrophysiological and behavioural work already undertaken on human-derived chemicals has inevitably focused on a limited range of mosquito species, most notably *Ae. aegypti*. Yet attempts to translate these findings into odour baits for trapping systems in the field have met with limited success. The only significant advances made recently in the development of synthetic odour baits have relied on the random screening of compounds identified as host odour attractants for other Diptera (Kline 1994), but since the baits used do not mimic natural attractants it is unclear how the catches might relate to natural populations. This approach is far from satisfactory given the importance of mosquitoes to human and animal welfare.

Another approach to the identification of host odour attractants for specific species would be to adapt the systematic methodology developed for the identification of host odour attractants of tsetse flies (Torr 1994). This approach would rely on the ability to quantify the behavioural effect of host odour on the selected species of mosquito compared to other host-related cues, such as temperature, humidity and CO_2 gradients. Having identified suitable sources of host odour attractants, we can, in principle, analyse them by gas chromatography linked to electrophysiology (Cork et al 1990) in order to identify electrophysiologically active compounds that could form the basis of a synthetic mimic of the natural kairomone. In the case of the tsetse fly the process has been seen to be iterative in that the synthetic blend has been constantly refined as new electrophysiologically active compounds are identified and the ratio of components in the synthetic kairomone is modified to maximize attractiveness and specificity in field and laboratory bioassays.

However, experience with host odours of *An. gambiae* has shown that there are a number of actual problems associated with this approach. The first is the assumption that all individuals of a host species are equally attractive. Humans vary in their relative attractiveness to mosquitoes (Hocking 1971) and their relative attractiveness varies from species to species. Secondly, this approach requires that compounds of behavioural significance are amenable to analysis by gas chromatography, but unfortunately this cannot be taken for granted. α-Hydroxycarboxylic acids (Carlson et al 1973), and particularly L-lactic acid, are difficult to analyse by gas chromatography without derivatization and hence loss of electrophysiological activity. Thirdly, it is assumed that all behaviourally relevant compounds will elicit measurable electrophysiological responses at physiologically relevant concentrations. Butanone and acetone, known attractants of tsetse flies, do not elicit electrophysiological responses in gas chromatography-linked analyses of host odour samples, although they have been identified in the host odour samples by chemical analysis. The fourth assumption is that electrophysiologically active compounds will elicit a behavioural response. However, numerous compounds eliciting electrophysiological responses from tsetse flies, for example, have been identified in host odours, but only a small proportion of these have been found to cause a significant behavioural response in the field.

Work is ongoing at the Natural Resources Institute to identify the compounds in human effluvia that are used by host-seeking *An. gambiae* and related anthropophilic mosquito species. In order to circumvent the technical problems associated with the identification of electrophysiologically active compounds described above the work is currently being directed towards the identification of compounds in host secretions by various means with subsequent electrophysiological studies off-line. Synthetic blends of these compounds are then provided to colleagues for behavioural studies either in the laboratory or field. However, it remains to be seen whether this approach can produce an odour bait that would replace human biting catches for use in epidemiological surveys.

Acknowledgements

Sweat samples were generously provided by colleagues at the Centre National de Lutte contre le Paludisme, Burkina Faso. I thank K. C. Park for providing the electroantennographic recordings from *An. gambiae*. The work reported on the development of an odour bait for *An. gambiae* was undertaken as part of a collaborative project between the National Resources Institute, UK, Imperial College, UK, Wageningen Agricultural University, The Netherlands and the Istituto di Parassitologia, Università 'La Sapienza', Italy. This work was supported by the European Community under projects TS3-CT92-0101 and TS3-CT91-0032.

References

Acree F Jr, Turner RB, Gouck HK, Beroza M, Smith N 1968 L-Lactic acid: a mosquito attractant isolated from humans. Science 161:1346–1347
Anderson JR 1989 Use of deer models to study larviposition by wild nasopharyngeal bot flies (Diptera: Oestridae). J Med Entomol 26:234–236
Anonymous 1966 Beckman Instruments, Inc, Advanced Research Department, Advanced Technology Operations, Fullerton, California, Contract No. DA-49-092-ARO-103

Bentley MD, McDaniel IN, Yatagai M, Lee H-P, Maynard R 1979 *p*-Cresol: an oviposition attractant of *Aedes triseriatus*. Environ Entomol 8:206–209

Blackwell A, Wadhams LJ 1995 Electrophysiological and behavioural studies of the biting midge, *Culicoides impunctatus*: interactions between some plant derived repellent compounds and a host-odour attractant. In: Animal and cell abstracts. Society of Experimental Biologists Symposium, St Andrews, 1995, p 68

Brady J, Griffiths N 1993 Upwind flight responses of tsetse flies (*Glossina* spp.) (Diptera: Glossinidae) to acetone, octenol and phenols in nature: a video study. Bull Entomol Res 83:329–333

Brown AWA, Sarkaria DS, Thompson RP 1951 Studies on the responses of the female *Aedes* mosquito. I. The search for attractant vapours. Bull Entomol Res 42:105–114

Bursell E, Gough AJE, Beevor PS, Cork A, Hall DR, Vale GA 1988 Identification of cattle urine attractive to tsetse flies, *Glossina spp.* (Diptera: Glossinidae). Bull Entomol Res 78:281–291

Carlson DA, Smith N, Gouck HK, Godwin DR 1973 Yellow fever mosquitoes: compounds related to lactic acid that attract females. J Econ Entomol 66:329–331

Chorley TW 1948 *Glossina pallidipes* Austen attracted by scent of cattle dung and urine (Diptera). Proc R Entomol Soc Lond A 23:9–11

Clements AN 1963 Physiology of mosquitoes. Macmillan, New York

Costantini C, Gibson G, Sagnon N'F, della Torre A, Brady J, Coluzzi M 1996 The response to carbon dioxide of the malaria vector *Anopheles gambiae s. l.* and other sympatric mosquito species in Burkina Faso. Med Vet Entomol 10, in press

Cork A 1994 Identification of electrophysiologically-active compounds for New World Screwworm, *Cochliomyia hominivorax*, in larval wound fluid. Med Vet Entomol 8:151–159

Cork A, Beevor PS, Gough JE, Hall DR 1990 Gas chromatography linked to electroantennography: a versatile technique for identifying insect semiochemicals. In: McCaffery AR, Wilson ID (eds) Chromatography and isolation of insect hormones and pheromones. Chromatographic Society Symposium Series, Plenum, New York, p 271–280

Cragg JB, Ramage GR 1945 Chemotropic studies on the blow-flies *Lucilia sericata* (Mg.) and *Lucilia caesar* (L.). Parasitology 36:168–175

Davis EE, Sokolove PG 1975 Temperature responses of antennal receptors of the mosquito, *Aedes aegypti*. J Comp Physiol A 96:223–236

Daykin PN, Kellogg FE, Wright RH 1965 Host finding and repulsion of *Aedes aegypti*. Can Entomol 97:239–263

de Jong R, Knols BGJ 1995 Olfactory responses of host-seeking *Anopheles gambiae s.s.* Giles (Diptera, Culicidae). Acta Tropica 59:333–335

Defoliart GR, Morris CD 1967 A dry-ice baited trap for the collection and field storage of hematophagous Diptera. J Med Entomol 4:360–362

Downes JA 1958 The feeding habits of biting flies and their significance in classification. Ann Rev Entomol 3:249–266

Eiras AE, Jepson PC 1994 Responses of female *Aedes aegypti* (Diptera: Culicidae) to host odours and convection currents using an olfactometer bioassay. Bull Entomol Res 84:207–211

Feinsod FM, Spielman A 1979 An olfactometer for measuring host-seeking behavior of female *Aedes aegypti* (Diptera: Culicidae). J Med Entomol 15:282–285

Foster WA, Hancock RG 1994 Nectar-related olfactory and visual attractants for mosquitoes. J Am Mosq Control Assoc 10:288–296

French FE, Kline DL 1989 1-Octen-3-ol, an effective attractant for Tabinidae. J Med Entomol 26:459–461

Gillies MT 1980 The role of carbon dioxide in host-finding by mosquitoes (Diptera: Culicidae): a review. Bull Entomol Res 70:525–532

Gillies MT, Wilkes TJ 1972 The range of attraction of animal baits and carbon dioxide for mosquitoes. Studies in a freshwater area of West Africa. Bull Entomol Res 61:389–404

Gower DB, Nixon A, Mallett AI 1988 The significance of odorous steroids in axillary odour. In: Van Toller S, Dodd GH (eds) Perfumery: the psychology and biology of fragrance. Chapman & Hall, New York, p 47–76

Grimstad PR, DeFoliart GR 1974 Nectar sources of Wisconsin mosquitoes. J Med Entomol 11:331–341

Hassanali A, McDowell PG, Owaga MLA, Saini RK 1986 Identification of tsetse attractants from excretory products of a wild host animal, *Syncerus caffer*. Insect Sci Appl 7:5–9

Healy TP, Jepson PC 1988 The location of floral nectar sources by mosquitoes: the long range responses of *Anopheles arabiensis* Patton (Diptera: Culicidae) to *Achillea millefolium* flowers and isolated floral odour. Bull Entomol Res 78:651–657

Hocking B 1963 The use of attractants and repellents in vector control. Bull WHO (suppl) 29:121–126

Hocking B 1971 Blood-sucking behaviour of terrestrial arthropods. Ann Rev Entomol 16:1–26

Holloway MTP, Phelps RJ 1991 The responses of *Stomoxys* spp. (Diptera: Muscidae) to traps and artificial host odours in the field. Bull Entomol Res 81:51–55

Howlett FM 1910 The influence of temperature on biting of mosquitoes. Parasitology 3:479–484

Huffaker CB, Back RC 1943 A study of methods of sampling mosquito populations. J Econ Entomol 36:561–569

Ikeshoji T 1975 Chemical analysis of wood creosote for species-specific attraction of mosquito oviposition. Appl Entomol Zool 10:302–308

Kellogg FE 1970 Water vapour and carbon dioxide receptors in *Aedes aegypti*. J Insect Physiol 16:99–108

Khan AA, Maibach HI 1966 Quantitation of effect of several stimuli on landing and probing by *Aedes aegypti*. J Econ Entomol 59:902–905

Khan AA, Maibach HI, Strauss WG, Fisher JL 1969 Increased attractiveness of man to mosquitoes with induced eccrine sweating. Nature 223:859–860

Kline DL 1994 Olfactory attractants for mosquito surveillance and control: 1-octen-3-ol. J Am Mosq Control Assoc 10:280–287

Kline DL, Takken W, Wood JF, Carlson DA 1990 Field studies on the potential of butanone, carbon dioxide, honey extract, 1-octen-3-ol, L-lactic acid and phenols as attractants for mosquitoes. Med Vet Entomol 4:383–391

Knols BGJ, de Jong R, Takken W 1994 Trapping system for testing olfactory responses of the malaria mosquito *Anopheles gambiae* in a wind-tunnel. Med Vet Entomol 8:386–388

Mackley JW, Brown HE 1984 Swormlure-4. A new formulation of the swormlure-2 mixture as an attractant for adult screwworms, *Cochliomyia hominivorax* (Diptera: Calliphoriadae). J Econ Entomol 77:1264–1268

McCall PJ, Cameron MM 1995 Oviposition pheromones in insect vectors. Parasitol Today 11:352–355

Millar JG, Chaney JD, Mulla MS 1992 Identification of oviposition attractants for *Culex quinquefasciatus* from fermented Bermuda grass infusions. J Am Mosq Control Assoc 8:11–17

Mulla MS, Ridsdill-Smith JT 1986 Chemical attractants tested against the Australian bush fly *Musca vetustissima* (Diptera, Muscidae). J Chem Ecol 12:261–270

Murlis J, Jones CD 1981 Fine-scale structure of odour plumes in relation to insect orientation to distant pheromone and other attractant sources. Physiol Entomol 6:71–86

Newhouse VF, Chamberlain RW, Johnson JG, Sudia WD 1966 Use of dry-ice to increase mosquito catches of the CDC miniature light trap. Mosq News 26:30–35

Nicolaides N 1974 Skin lipids: their biochemical uniqueness. Science 186:19–26

Omer SM 1979 Responses of females of *Anopheles arabiensis* and *Culex pipiens* to air currents, carbon dioxide and human hands in a flight-tunnel. Entomol Exp Appl 26:142–151

Owaga MLA 1985 Observations on the efficacy of buffalo urine as a potent olfactory attractant for *Glossina pallidipes* Austen. Insect Sci Appl 6:561–566

Park KC, Cork A 1995 Electrophysiological studies on the malaria vector, *Anopheles gambiae* Giles (Diptera: Culicidae) with human host odours. In: Animal and cell abstracts. Society of Experimental Biologists Symposium, St Andrews, 1995, p 68

Parker AH 1948 Stimuli involved in the attraction of *Aedes aegypti* L., to man. Bull Entomol Res 39:387–397

Paynter Q, Brady J 1993 Flight responses of tsetse flies (*Glossina*) to octenol and acetone vapour in a wind-tunnel. Physiol Entomol 18:102–108

Price GD, Smith N, Carlson DA 1979 The attraction of female mosquitoes (*Anopheles quadrimaculatus* Say) to stored human emanations in conjunction with adjusted levels of relative humidity, temperature and carbon dioxide. J Chem Ecol 5:383–395

Rahm U 1956 Zum problem der Attraktion von Stechmücken durch den Menschen. Acta Trop 13:319–344

Reeves WC 1951 Field studies on carbon dioxide as a possible host stimulant to mosquitoes. Proc Soc Exp Biol Med 77:64–66

Reiter P, Amador MA, Colon N 1991 Enhancement of the CDC ovitrap with hay infusions for daily monitoring of *Aedes aegypti* populations. J Am Mosq Control Assoc 7:52–55

Reuter J 1936 Oriënteerend onderzoek naar de oorzaak van het gedrag van *Anopheles maculipennis* Meigen bij de voedselkeuze. Dissertation, University of Leiden, Leiden, Germany. ([Summary] Acta Leid 10/11:260–267)

Rössler HP 1961 Versuche zur geruchlichen Anlockung weiblicher Stechmücken (*Aedes aegypti* L., Culicidae). Z vergl Physiol 44:184–231

Rudolfs W 1922 Chemotropism of mosquitoes. Bull N J Agr Exp Stn 367:4–23

Sastry SD, Buck KT, Janák J, Dressler M, Preti G 1980 Volatiles emitted by humans. In: Waller GR, Dermer OC (eds) Biochemical applications of mass spectrometry (suppl). Wiley, New York, p 1085–1129

Schofield S, Cork A, Brady J 1996 Electroantennogram responses of the stable fly, *Stomoxys calcitrans*, to potential components of host odour. Physiol Entomol 20:273–280

Schreck CE, Gouck HK, Smith N 1967 An improved olfactometer for use in studying mosquito attractants and repellents. J Econ Entomol 60:1188–1190

Skinner WA, Tong H, Pearson T, Strauss W, Maibach H 1965 Human sweat components attractive to mosquitoes. Nature 207:661–662

Smith CN, Smith N, Gouck HK et al 1970 L-Lactic acid as a factor in the attraction of *Aedes aegypti* (Diptera: Culicidae) to human hosts. Ann Entomol Soc Am 65:760–770

Snow WF 1970 The effect of a reduction in expired carbon dioxide on the attractiveness of human subjects to mosquitoes. Bull Entomol Res 60:43–48

Stoddart DM 1990 The scented Ape. Cambridge University Press, Cambridge

Takken W 1991 The role of olfaction in host-seeking of mosquitoes: a review. Insect Sci Appl 12:287–295

Takken W, Kline DL 1989 Carbon dioxide and 1-octen-3-ol as mosquito attractants. J Am Mosq Control Assoc 5:311–316

Torr SJ 1990 Dose responses of tsetse flies (*Glossina*) in host odour plumes in the field. Physiol Entomol 15:93–103

Torr SJ 1994 The tsetse (Diptera: Glossinidae) story: implications for mosquitoes. J Am Mosq Control Assoc 10:258–265

Vale GA 1974 The responses of tsetse flies (Diptera: Glossinidae) to mobile and stationary baits. Bull Entomol Res 64:545–588

Vale GA, Hall DR 1985 The role of 1-octen-3-ol, acetone and carbon dioxide in the attraction of tsetse flies, *Glossina* spp. (Diptera: Glossinidae), to ox odour. Bull Entomol Res 75:209–217

Vale GA, Hall DR, Gough AJE 1988 The olfactory responses of tsetse flies, *Glossina* spp. (Diptera: Glossinidae), to phenols and urine in the field. Bull Entomol Res 78:293–300

Van Thiel PH, Weurman C 1947 L'attraction excercée sur *Anopheles maculipennis atroparvus* par l'acide carbonique dans l'appareil de choix II. Acta Tropica 4:1–9

Wensler RJD 1972 The effect of odours on the behavior of adult *Aedes aegypti* and some factors limiting responsiveness. Can J Zool 50:415–420

Zeng X-N, Leyden JJ, Brand JG, Spielman AI, McGinley KJ, Preti G 1992 An investigation of human apocrine gland secretion for axillary odour precursors. J Chem Ecol 18:1039–1055

DISCUSSION

Pickett: What was the plasticizer used as the absorbant?

Cork: Dioctyl phthalate, but we have also used Cereclor.

Pickett: This would be better because low molecular weight phthalates, such as dimethyl phthalate, present as contaminants can act as repellents in their own right.

Cardé: I have a methodological query. When you say mosquitoes are attracted to these various chemical combinations, what kind of behavioural test was used?

Cork: We are only looking at what we capture at the end of the day, either with electric nets or on sticky surfaces. In the dual-port bioassay we measured the relative proportion of released mosquitoes that entered each collection chamber. The mechanisms by which they arrived at or alighted on an odour source were not considered.

Guerin: Alan Cork's presentation highlights one point that we should all be aware of. Haematophagous arthropods are responding to blends of compounds that are quite different in terms of chemical functionality and structure from the aliphatic compounds that form the pheromone blends of moths. Therefore, we have to be careful when thinking about how haematophagous species respond. Simple compounds, such as CO_2, H_2S and ammonia, may interact with ambient air in ways quite different to what happens to the long-chain compounds that form the pheromone blends of moths.

Pickett: You are quite correct, but I'm not sure what significance that would have once the compounds have left the source. The molecular properties of a compound do not determine what happens once the olfactory cue leaves the source. Obviously, such properties determine the vapour pressure, and a hydrophilic compound will leave the source at a slower rate. However, once the cue has left the source then it is the turbulent airflow that is responsible for its movement out into a plume, for example.

Guerin: I agree, airflow is involved to a certain degree, but the majority of receptors in the male moth antennae only respond to a few compounds of a similar nature. Alan Cork has been working with many different compounds, ranging from simple, polar, short-chain compounds to steroids. I find it difficult to believe that the plume, which is made up of a few long-chain compounds of a similar nature, behaves similarly to this complex mixture of products.

Pickett: I disagree. Once the mixture has left the source, the olfactory cue, however complex a mixture, just moves about as wisps of constant relative composition, so the nature of its components does not matter.

Guerin: Are you saying that no single constituent of a complex mixture would in any way affect another constituent?

Pickett: Phenols and indoles could interact because the former are acidic and the latter are basic, but Alan Cork's list of phenolic compounds are unlikely to interact chemically in the air because of the high dilution involved.

Guerin: We should not assume that they would not interact under any circumstances.

Mustaparta: Possible interactions between compounds could be tested electrophysiologically; for example, by comparing responses of receptor neurons to single compounds and mixtures.

Guerin: That approach is purely for the purposes of finding out the range of receptor types, but the receptors are functioning as a unit when the mosquito is flying upwind at a time when it is perceiving a complex mixture of compounds.

Boeckh: In my opinion, bearing in mind the numbers of molecules in the gaseous phase, it is not conceivable that these compounds would interact without an aerosol of some kind.

Cork: On the surface of an animal indoles could react with phenols if the phenols were not conjugated. However, in urine they are conjugated and are therefore protected from chemical degradation. These compounds are not released until they pass from the body, when the sterile material becomes mixed with bacteria that break down the conjugates. Once they enter the gaseous phase I see no reason why they should interact chemically.

Gibson: Does the molecular weight of the volatile chemicals affect how far they travel from the source?

Cork: The volatility of the compound does depend on its molecular weight, but it also depends on the structure itself and on the matrix in which it's held. Most of these compounds do become airborne, and even steroids are volatile.

Steinbrecht: Once the chemicals are in the plume, they should be regarded as being similar to pheromones. The antennal receptor cells will monitor the number and type of molecules present. We don't yet know whether all these compounds are necessary for a behavioural response. If, for example, a dozen compounds are necessary, then it will make a difference if the antenna catches all of these components or only some of them.

Pickett: There is one other situation where it could make a difference and that is where there is a differential affinity for sinks; for example, the cuticle on plant surfaces where lipophilic interactions are favoured. Rain or dew may also give differences because of their ability to strip hydrophilic compounds, such as phenols, more readily from the air.

Cork: Unlike lepidopteran pheromones, where the actual ratio of components is often critical to elicit a behavioural response, dipteran kairomones are very plastic, and the actual ratio of components is usually not critical for attraction. Most of the compounds that we've been looking at are probably redundant, in the sense that only a few of them are required to elicit the behaviour in which we are interested.

Lehane: I was under the impression that blood-sucking insects use differences in ratios to select their hosts at a distance. It's a waste of energy to track down a host that you don't eventually want to feed on. If they are not using ratios for this purpose, then are they using specific smells or individual chemical components from individual animals?

Cork: The presence or absence of a compound in the effluvia released by a mammal may be important to a host-seeking mosquito for identifying whether that animal could be a potential host. *An. gambiae* is highly anthropophilic, so if the compounds we have identified in human effluvia are not specific to humans then they are unlikely to provide the basis for a human-specific kairomone. Thus, we need to identify human-specific compounds or blends of compounds that would enable *An. gambiae* to discriminate between mammalian species.

Brady: You have been intimately involved in the development of the tsetse bait, and you are now at an early stage in the development of a putative mosquito bait. There are several hundred compounds that are potential human bait candidates. How does this compare with the ox compounds used for the tsetse bait? How different are they from a chemical point of view?

Cork: Whole ox odour contained a large number of compounds, probably about 300–400. Some of these—such as 1-octen-3-ol, indole and a range of phenols—were found to be highly electrophysiologically active by gas chromatography linked to electroantennography. Tsetse flies have olfactory receptors for a wide range of compounds, so that when we increased the amount of material injected into the gas chromatography column, more compounds were detected by the electroantennograph. Indeed, we have hardly found a compound that does not elicit an electrophysiological response at one dose or another. The question is not what is electrophysiologically active but at what maximum dose do we draw a line and say that compounds not demonstrating electrophysiological activity at or below that threshold are not considered in bioassays.

Mustaparta: Have you obtained electroantennograph responses to compounds that are not detected by gas chromatography?

Cork: Yes, we have obtained electroantennograph responses from compounds at levels below that detected by the gas chromatograph, such as 1-octen-3-ol and indole. However, on the other hand we are also aware of behaviourally active compounds that did not elicit an electrophysiological response from our gas chromatography-linked electroantennography system, such as acetone and butanone. This unfortunately casts doubt on the value of our technology for identifying components of kairomones.

Mustaparta: Have you looked for these compounds in odours released by other animals?

Cork: Yes, we have looked at the volatiles from a wide range of tsetse hosts, such as cows, pigs, crocodiles and warthogs.

Pickett: Critics of electroantennograph and single-cell recording techniques often make the point that anything will produce a response if it is present at a high enough concentration. The main issue is that if one is trying to understand a natural chemical/ecological situation, then one needs to look at compounds at realistic physiological levels. This means that the situation is simplified, in a sense, because one is only concerned with detecting activities of compounds and the natural levels present. This allows an understanding of the natural system, but one can perturb the situation when designing a bait. It is invaluable to link gas chromatography with electrophysiology because this allows us to look at what an insect is responding to at the kind of physiological levels with which the insect is presented. The other aspect is knowing that a compound is behaviourally active but not being able to obtain an electoantennograph response. This provides an important reason for making single-cell recordings because the individual cells are often not abundant enough to create an electroantennograph response. We had this problem with a Curculionid (weevil) pheromone. Initially, we could not obtain an electroantennograph response, although we knew there were some cells tuned to the pheromone in the antennae. When we obtained a single-cell recording we were able to use it to pick out the pheromone from the gas chromatography trace.

Cork: In the case of acetone the quantities of material collected from cattle effluvia were insufficient to elicit an electroantennographic response from the gas chromatography-linked electroantennography system, although off-line we have obtained electroantennograms and Case Den Otter has identified specific olfactory sensilla that respond exclusively to acetone. However, as the amounts of acetone released by oxen are much smaller than the quantities used in baits, it is conceivable that it can act as a mimic for other compounds.

Brady: 1-Octen-3-ol is an important component of the tsetse fly bait. However, cattle produce only small quantities of 1-octen-3-ol, so isn't it strange that tsetse flies should respond to it?

Cork: It is difficult to detect 1-octen-3-ol in ox odour. Indeed, we were lucky to pick it up in the first instance. We now know how much of each behaviourally active compound the average cow produces, and we have been able to produce a formulation that releases the compounds in physiological amounts. Field experiments in Zimbabwe showed that the synthetic blend was half as attractive as a natural ox, suggesting that we have not identified all the behaviourally relevant compounds in ox odour. On the other hand, we have identified a number of compounds that are electrophysiologically active for which we never actually obtained a behavioural response.

Brady: If you have 400 candidate compounds, as you suggest, it is tempting to study just those that produce a nice electroantennograph response. This is presumably what happened in the tsetse fly story—you said yourself that you may have been lucky with 1-octen-3-ol—but there may be a 1-octen-3-ol equivalent of mosquito baits that you may miss.

Cork: This is not a stepwise process. It's iterative, in the sense that one has to keep going back and refining the chemical blend until it is equivalent in attractiveness to the natural kairomone. The same process will presumably have to be followed with any mosquito attractants based on host odours.

Davis: 1-Octen-3-ol is attractive to host-seeking *Culex quinquefasciatus* and to gravid *Aedes triseriatus* (M. D. Bentley, personal communication 1978). It's interesting that the same compound elicits two totally different behaviours.

Selection of biting sites by mosquitoes

Ruurd de Jong[1] and Bart G. J. Knols

Department of Entomology, Wageningen Agricultural University, PO Box 8031, 6700 EH Wageningen, The Netherlands

Abstract. Most blood-feeding arthropods feed at specific sites on their hosts. There can be many reasons for such non-random distributions and not all of them are related to active selection of feeding sites by the parasites. Observations of mosquito biting on humans revealed that most species have preferred biting sites and that not all species share the same preferences. The selection of these sites may be related to several factors, depending on the mosquito species, which include the visual and chemical properties of the host. Identification of these factors can provide us with information on cues that are important in the finding and selection of a host. The choice of biting sites can be influenced by host-related cues of varying specificity. Therefore, it can in some cases reflect the host range of the mosquito species.

1996 Olfaction in mosquito–host interactions. Wiley, Chichester (Ciba Foundation Symposium 200) p 89–103

Blood-feeding arthropods are generally not evenly distributed over their host's bodies. Occasionally, these non-random distributions seem to be species specific, suggesting a selection behaviour by the parasites. Depending on the parasite–host complex, different factors, which may include olfactory orientation, can influence this behaviour. However, it is important to realize that some of these factors are unrelated to the behaviour of the parasites. Analysis of the distributions of parasites on their host's bodies can provide valuable information about the factors involved. It may also give an indication about their importance in host finding and selection by the parasite.

Landing, biting and feeding

Most blood-feeding arthropods approach their host by air, and usually feeding seems to occur on, or close to, the place where they alight. However, there are reports of landing and feeding taking place at clearly distinct sites. Sandflies,

---

[1]Present address: Institut de Zoologie, Université de Neuchâtel, Rue Emile-Argand 11, CH-2007 Neuchâtel, Switzerland

for example, do not show any preference for landing on a particular region of anaesthetized mice. They move over the host's body to find suitable locations—such as ears, eyelids, feet, the nose and tail—where low hair densities allow feeding (Coleman & Edman 1988). Townley et al (1984) investigated the landing and engorging sites of biting midges on a horse. They found that the preferential landing sites were along the mane (26%), the hind quarters (29%) and lower leg regions (17%), whereas the percentages of midges engorging on the same regions were 49%, 13% and 23%, respectively. In contrast, a similar experiment involving other species of biting midges suggested that 95% of the midges landed on the belly of the bait horse, and that this was also the feeding site (Braverman 1988).

Not all sites where feeding is attempted are suitable for obtaining a blood meal. Magnarelli & Anderson (1980) showed that the prevalence of disrupted feedings of tabanids on the legs and backs of cattle exceeded that on the head areas. The initial biting sites represent the selected sites to attempt feeding; therefore, they are less likely to be influenced by additional factors than more successful feeding sites. These factors may include a host reaction to biting, which may be different for different body regions. Although it is preferable to identify the initial biting sites rather than the feeding sites, it is not always possible to do so. For parasites such as tabanids, which start probing almost immediately after landing (Mullens & Gerhardt 1979), alighting sites might give more information about preferences than feeding sites.

Factors unrelated to parasite behaviour

The above example shows that the reasons for feeding at specific sites are not necessarily related to the behaviour of the parasites. Mechanical factors, for example predation by hosts or by other predators, can make it impossible for parasites to feed everywhere on the host's body. On cattle, which generally inhabit well-grazed pastures, large percentages of ticks are attached to the ventral and anterior body regions, including the legs. Very few ticks are found on the same regions on deer, possibly because deer travel through heavy bush, which may dislodge ticks on their anterior area and legs (Bloemer et al 1988). When blood-feeding arthropods spend a substantial time on their host, distributions are likely to be influenced by factors unrelated to an active selection of feeding sites, for example by the release of an aggregation pheromone by ticks (Sonenshine 1985). Although most insects spend only a short time on their host, these factors may still play an important role. The behaviour of the host, for example rippling of the skin and movement of the ears, head, tail and leg, influence parasite feeding. Torr (1994) found that the proportion of tsetse flies landing on the head of a warthog increased from 25% at low population densities to 60% at high population densities. This change was correlated with an increased grooming behaviour of the warthog. Tsetse flies on the head region did not appear to be disturbed by the grooming; therefore, the increased landing rate of tsetse on the warthog's head was probably, at least in part, due to the warthog's behaviour.

Factors related to parasite behaviour

A parasite can be attracted to a host by the host's visual, physical and/or chemical properties. Heat, moisture and visual cues, for example, are very important to many parasites, and their influence on the selection of biting sites is well known. An influence of chemical cues, especially CO_2, on this selection behaviour has often been assumed but little evidence actually exists.

Work on blackflies (Simuliidae) suggests that olfactory cues influence the parasite's selection of biting sites. Blackflies feed on warm-blooded vertebrates, and many species bite at the landing site (Crosskey 1990). Some species are known to feed on specific hosts. For example, *Simulium euryadminiculum* is attracted to the odour of the common loon's uropygial gland, and lands preferentially on the head–neck region of its host (Lowther & Wood 1964). Tests with loon-like decoys proved that visual stimuli are important in this process, and also that the position of an extract of the uropygial scent bait on the decoys affects the landing distribution (Bennet et al 1972).

Most blackfly species feeding on humans have preferred biting sites. Some of them bite mostly on the head region, whereas others bite mainly on the back and shoulders and relatively little on the legs below knee level (Crosskey 1990). The opposite preference occurs for *Simulium damnosum sensu lato,* where biting predominates on the feet and ankles (Duke & Beesley 1958, Renz & Wenk 1983). The results of Duke & Beesley's (1958) experiments suggest that the biting of *S. damnosum is* related to the low height above ground at which the flies approach the human host. However, Crosskey (1990) suggests that the response to a human host is more complicated because *S. damnosum* and other flies bite a certain part of the body regardless of whether the bait stands, sits or lies down. Thompson (1976, 1977) proved that these flies are attracted to worn rather than clean clothes. The attractive component is mainly present in eccrine sweat from the lower part of the body. The preference of *S. damnosum* to bite legs seems to be associated with its olfactory response to sweat from this region.

Biting by mosquitoes

Most work on mosquito biting has been done with human baits. However, Walker & Edman (1985) have studied the feeding behaviour of *Aedes triseriatus* on rodents. The selection of feeding sites by this mosquito species did not occur upon initial landing: probing and feeding took place after random foraging, and *Ae. triseriatus* then fed where the host hair was short and sparse. The effect of host hair on mosquito behaviour is presumably limited for human hosts because less than 5% of the human skin surface is densely covered with hair. Generally, biting on humans takes place on the landing site, and there are several examples of biting site preferences. Members of the small and exclusively African mosquito genus *Eretmapodites* bite close to the ground, almost entirely below the knee of a standing human (Haddow 1956). In contrast, *Aedes simpsoni* shows a marked preference for the head, followed by the shoulders and upper trunk (Haddow 1946). De Jong & Knols (1995a) and Knols et al (1994) found striking differences between some *Anopheles* species. *Anopheles atroparvus* and *Anopheles albimanus* bite on the head region of a sitting human, whereas *Anopheles gambiae* has a preference for feet and ankles. Biting site preferences of *Culex quinquefasciatus* have been reported to correlate with the body sites most affected by clinical filariasis, i.e. thighs, legs and feet (Self et al 1969, Chandra & Hati 1993). However, in these studies the behaviour of *Cx. quinquefasciatus* may well have been influenced by host movement (Self et al 1969) and by the presence and activities of mosquito collectors (Self et al 1969, Chandra & Hati 1993). Here we present results on the selection of biting sites by *Cx. quinquefasciatus* and *Aedes aegypti* on an individual human host and under essentially similar conditions as described previously for *An. gambiae*, *An. atroparvus* and *An. albimanus* (de Jong & Knols 1995a, Knols et al 1994). This experimental set-up minimizes the possible influences of host movement and the presence of a mosquito collector.

Materials and methods

Mosquitoes

The mosquito strains originated from Muheza, Tanzania (*Cx. quinque-fasciatus*) and Bilthoven, The Netherlands (*Ae. aegypti*). We kept the adults in 30 cm cubic gauze cages at 27 °C and about 80% relative humidity, and fed them on a 6% glucose solution. We used five to 10 day-old female mosquitoes that had not received a blood meal and starved them the night before experiments by confining them individually into glass vials sealed with water-moistened plugs of cotton wool.

Experimental design and analysis

We used an experimental procedure identical to the one described in detail by de Jong & Knols (1995a). The test person (male Caucasian, 27 years), who was wearing only close-fitting underwear, sat motionless and in an upright position inside a bed net. We released individual mosquitoes through a small hole in the bed net, and discarded those that did not bite within three minutes. Biting sites were confirmed by a second person who entered the net after probing was felt or three minutes had passed. We analysed the distributions of biting sites by comparing relative skin surface areas of various body parts with received and expected number of bites using a G-test (Sokal & Rohlf 1981).

Biting by *Culex quinquefasciatus* and *Aedes aegypti*

The distribution of biting sites over different body parts for *Cx. quinquefasciatus* did not differ from that expected on the basis of relative skin surface areas (Table 1). In contrast, *Ae. aegypti* bit more on the head and less on the legs than expected. Skin temperature and humidity are main factors in the short-range orientation behaviour of many mosquito species (Clements 1963). To determine whether the biting sites correlated with these factors, we divided the body into three temperature (Clarke & Edholm 1985) and three humidity regions. We used eccrine sweat gland densities (Marples 1969) to assess humidity levels above the skin because these glands affect the skin's main discharge of water. Projection of the bites on these body regions showed that biting correlated with particular combinations of eccrine sweat gland densities and skin temperature (Table 2A).

Further analysis of these results (Tables 2B and 2C) shows that for *Cx. quinquefasciatus* this correlation is caused by preferential biting on relatively dry skin areas. Biting of *Ae. aegypti* correlated with skin areas that had a high temperature and a low eccrine sweat gland density. The mere correlation of these skin properties with a high biting incidence, however, does not prove that they are important in the selection of biting sites. Previous work (de Jong & Knols 1995a, Knols et al 1994) showed that similar correlations can disappear after

TABLE 1   Distribution of biting sites of *Culex quinquefasciatus* and *Aedes aegypti* on various body regions of a human bait

Body region	Cx. quinquefasciatus		Ae. aegypti	
	Number observed	Number expected[a]	Number observed	Number expected[a]
Head[b]	13	9	24	9
Trunk	32	32	39	34
Arms	26	26	16	20
Legs	30	30	26	42
Total no. of bites	101	101	105	105
G-test[c]	n.s.		***	

[a]Relative skin surface (after Clark & Edholm 1985) expressed as expected number of bites.
[b]Head includes neck region.
[c]G-test of goodness of fit with expected frequencies (Sokal & Rohlf 1981); ***, $P < 0.001$; n.s., not significant.

TABLE 2 Distribution of *Culex quinquefasciatus* and *Aedes aegypti* bites for (A) combinations of skin temperature and eccrine sweat gland density; and (B, C) both factors separately

A

Skin temperature[a]	Eccrine sweat gland density[b]	Cx. quinquefasciatus		Ae. aegypti	
		Number observed	Number expected[c]	Number observed	Number expected[c]
Low	Low	4	5	3	5
	Medium	22	28	15	30
	High	1	4	6	4
Medium	Low	16	9	12	10
	Medium	26	34	12	35
	High	8	3	5	3
High	Low	17	6	24	6
	Medium	6	7	23	7
	High	1	5	5	5
G-test[d]		***		***	

B

Skin temperature[a]	Cx. quinquefasciatus		Ae. aegypti	
	Number observed	Number expected[c]	Number observed	Number expected[c]
Low	27	37	23	37
Medium	50	46	29	45
High	24	18	53	18
G-test[d]	n.s.		***	

C

Eccrine sweat gland density[b]	Cx. quinquefasciatus		Ae. aegypti	
	Number observed	Number expected[c]	Number observed	Number expected[c]
Low	36	16	39	16
Medium	55	73	50	72
High	10	12	16	12
G-test[d]	***		***	

[a]Categorized after Clark & Edholm (1985). Low, <30 °C; medium 30–32 °C; high, >30 °C.
[b]Categorized after Marples (1969). Low, <100 glands/cm; medium, 100–300 glands/cm; high >300 glands/cm.
[c]Relative skin surface (after Clarke & Edholm 1985) expressed as expected number of bites.
[d]G-test of goodness of fit, with expected frequencies of bites (Sokal & Rohlf 1981). n.s., not significant; ***$P < 0.001$.

modifying other host properties, for example after removal of expired breath. Nevertheless, it is interesting that a preference of *Ae. aegypti* for high skin temperatures has been shown previously by Smart & Brown (1956). They also found a decrease in the attractiveness of a human hand to *Ae. aegypti* with increased eccrine sweating, but this may contradict the work of Khan et al (1969).

In our experiments, *Cx. quinquefasciatus* did not bite preferably on the lower extremities (Table 1), which contrasts with other reports (Self et al 1969, Chandra & Hati 1993). Initially, we observed that slight movements of the head by the bait person caused *Ae. aegypti* to bite lower on the body. It is possible that catching of mosquitoes by human baits from their own bodies (Self et al 1969) and by other mosquito collectors (Self et al 1969, Chandra & Hati 1993) may have influenced *Cx. quinquefasciatus* in a similar way. In addition, we cannot rule out that differences between *Cx. quinquefasciatus* strains and/or bait were responsible for the differences in results.

Manipulating a bait

Figure 1 shows the variation in preferences of different mosquito species. For some species, the results were obtained with the same human bait and under similar conditions (see also Table 3). Manipulation of the bait can give information on the factors involved in the selection process. Removal of exhaled breath significantly reduced biting of *An. atroparvus* and *An. albimanus* on the head (de Jong & Knols 1995a, Knols et al 1994). *An. atroparvus* is strongly attracted to breath and, like *An. albimanus* (Wilton 1975), to CO_2 (Laarman 1955), which is an important constituent of breath. This attraction might explain the strong preference of both species for landing and probing on the head. *An. albimanus* required more time to bite after the host's breath was removed (Knols et al 1994), suggesting a role of breath in its host location.

Biting by *Eretmapodites chrysogaster* occurs nearly exclusively below the knee of a standing person (Haddow 1956). In a lying position, humans are bitten all over the body, but they attract fewer mosquitoes than when standing. The mosquitoes tend to circle above a lying bait without biting, but they start to bite if the bait is raised slightly. Moreover, a moving bait is more attractive than a motionless one. Haddow's experiments (1956) show that the attraction of *E. chrysogaster* to its host is mainly visual.

Manipulating the bait by washing his feet changed the *An. gambiae* preference for biting this body region (de Jong & Knols 1995a). This suggests that odours from this region are involved in the selection of biting sites by this species. Wind tunnel studies showed that *An. gambiae* is not attracted to human breath, but is attracted strongly to the odour of Limburger cheese (de Jong & Knols 1995b). Limburger cheese odour is, to the human nose, reminiscent of foot odour, and is, like foot odour, the product of bacterial processes. The attraction of *An. gambiae* to Limburger cheese odour suggests

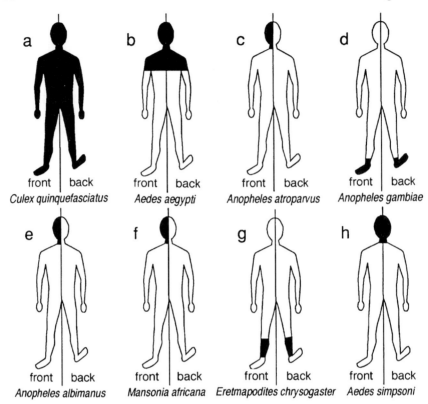

FIG. 1.   Preferred biting sites on humans for different mosquito species. a–f, bait in sitting position (c–d, de Jong & Knols [1995a]; e, Knols et al [1994]; f, B. G. J. Knols & R. de Jong, unpublished results [1994]) (see also Table 3). g, bait in a standing position (Haddow [1956]). h, bait in standing, sitting or lying position (Haddow [1946]).

that this highly anthropophilic mosquito species locates its human host by olfactory cues that have a bacterial origin (Knols & de Jong 1996).

   The observation that it is possible to divert *An. atroparvus* and *An. albimanus* from biting the head to other body regions does not imply that other species with a preference for biting the head, such as *Ae. simpsoni*, *Ae. aegypti* and *Mansonia africana*, will respond in a similar way. The two anopheline species appeared to respond to breath at a short distance from the host by a sudden reduction in flight-speed followed by orientation towards the face. We did not observe this typical behaviour in *Ae. aegypti* and *M. africana*, and it has not been described by Haddow (1946) for *Ae. simpsoni*. It is possible that cues other than those related to breath are more important for these species. A high skin temperature can be such a factor, as well as *Ae. aegypti*'s known

TABLE 3 **Biting distribution of six different mosquito species on various body regions of the same human bait**

Body region	Anopheles atroparvus[a]	Anopheles gambiae[a]	Anopheles albimanus[b]	Culex quinquefasciatus	Aedes aegypti	Mansonia africana[c]
Head[d]	50	1	49	13	24	22
Trunk	15	16	14	32	39	1
Arms	17	7	28	26	16	1
Legs	18	76	9	30	26	1
Total no. of bites	100	100	100	101	105	25
G-test[e]	***	***	***	n.s.	***	***

[a]Data from de Jong & Knols (1995).
[b]Data from Knols et al (1994).
[c]B. G. J. Knols & R. de Jong, unpublished results (1995).
[d]Head includes neck region.
[e]G-test of goodness of fit (Sokal & Rohlf 1981) with expected number of bites (based on relative skin surface areas, after Clark & Edholm [1985]). n.s., not significant; ***, $P < 0.001$.

attraction to skin-related olfactory cues (Acree et al 1968, Carlson et al 1973). Visual cues could also play a role, as Browne & Bennett (1981) found that *Aedes cantator*, *Aedes punctor* and *Mansonia perturbans* prefer the ends of rectangular shapes. However, since the preference for biting the head region by *Ae. simpsoni* remains unaffected regardless of the position of the bait (Haddow 1946), visual stimuli are probably not the most important cues for this species.

Both *Ae. simpsoni* and *An. gambiae* seem to prefer human bait (Haddow 1946). The host location by these species is probably linked with responses to human-specific cues. Once these cues are known, it will be interesting to assess their influence on the selection of biting sites by these species, for example by applying them to different sites on the host's body. We know that biting by the opportunistic feeders *An. atroparvus* (Swellengrebel & De Buck 1938) and *An. albimanus* (Breeland 1972) is strongly influenced by breath. CO_2 is a known attractant for both species (Van Thiel & Weurman 1947, Wilton 1975). It is also a well-known kairomone for many other opportunistic blood-feeding arthropods. The biting site selection of *E. chrysogaster* (Haddow 1956) reveals a visually oriented strategy for host location by this species. Haddow's experiments suggest that these visual cues are non-specific, implying a broad host range for *E. chrysogaster*. The selection of sites can be influenced by host-related cues of varying specificity; therefore, they can indirectly provide information about the host range of a mosquito species.

Conclusions

Most mosquito species, except *Cx. quinquefasciatus*, prefer biting particular sites of the host's body. *Ae. aegypti* prefers to bite the head and upper part of the trunk. The selection of biting sites can depend on many factors, and their identification could give information on the cues used by mosquitoes to locate their hosts.

Acknowledgements

We are grateful to Letizia Mattiacci for critical comments on this manuscript. The work was partially funded by the European Union as part of the project 'Behavioural studies on malaria vectors', project numbers TS3-CT92-0101 and TS3-CT91-0032.

References

Acree F Jr, Turner RB, Gouck HK, Beroza M, Smith N 1968 L-lactic acid: a mosquito attractant isolated from humans. Science 161:1346–1347

Bennet GF, Fallis AM, Campbell AG 1972 The response of *Simulium* (Eusimulium) *euryadminiculum* Davies (Diptera: Simuliidae) to some olfactory and visual stimuli. Can J Zool 50:793–800

Bloemer SR, Zimmerman RH, Fairbanks K 1988 Abundance, attachment sites, and density estimators of lone star ticks (Acari: Ixodidae) infesting white-tailed deer. J Med Entomol 25:295–300

Braverman Y 1988 Preferred landing sites of *Culicoides* species (Diptera: Ceratopogonidae) on a horse in Israel and its relevance to summer seasonal recurrent dermatitis (sweet itch). Equine Vet J 20:426–429

Breeland SG 1972 Studies on the ecology of *Anopheles albimanus*. Am J Trop Med Hyg 21:751–754

Browne SM, Bennett GF 1981 Response of mosquitoes (Diptera: Culicidae) to visual stimuli. J Med Entomol 18:505–521

Carlson DA, Smith N, Gouck HK, Godwin DR 1973 Yellow fever mosquitoes: compounds related to lactic acid that attract females. J Econ Entomol 66:329–331

Chandra G, Hati AK 1993 Correlation between the preferred biting site of *Culex quinquefasciatus* and the region of the body affected by clinical filariasis. Ann Trop Med Parasitol 87:393–397

Clarke RP, Edholm OG 1985 Man and his thermal environment. Edward Arnold, London

Clements AN 1963 Physiology of mosquitoes. Macmillan, New York

Coleman R, Edman JD 1988 Feeding-site selection of *Lutzomyia longipalpis* (Diptera: Psychodidae) on mice infected with *Leishmania mexicana amazonensis*. J Med Entomol 25:229–233

Crosskey RW 1990 Host animals, biting and bloodsucking. In: The natural history of blackflies. Wiley, Chichester, p 411–446

de Jong R, Knols BGJ 1995a Selection of biting sites on man by two malaria mosquito species. Experientia 51:80–84

de Jong R, Knols BGJ 1995b Olfactory responses of host-seeking *Anopheles gambiae ss* Giles (Diptera: Culicidae). Acta Trop 59:333–335

Duke BOL, Beesley WN 1958 The vertical distribution of *Simulium damnosum* bites on the human body. Ann Trop Med Parasitol 52:274–281

Haddow AJ 1946 Mosquitoes of Bwamba County, Uganda. Bull Entomol Res 36:33–73

Haddow AJ 1956 Observations on the biting-habits of African mosquitoes in the genus *Eretmapodites Theobald*. Bull Entomol Res 46:761–772

Khan AA, Maibach HI, Strauss WG, Fisher JL 1969 Increased attractiveness of man to mosquitoes with induced eccrine sweating. Nature 223:859–860

Knols BGJ, de Jong R 1996 Limburger cheese as an attractant for the malaria mosquito *Anopheles gambiae* s.s. Parasitol Today 12:159–161

Knols BGJ, Takken W, de Jong R 1994 Influence of human breath on selection of biting sites by *Anopheles albimanus*. J Am Mosq Control Assoc 10:423–426

Laarman JJ 1955 The host-seeking behaviour of the malaria mosquito *Anopheles atroparvus*. Acta Leiden 25:1–144

Lowther JK, Wood DM 1964 Specificity of a black fly, *Simulium euryadminiculum* Davies, towards its host, the common loon. Can Entomol 96:911–913

Magnarelli LA, Anderson JF 1980 Feeding behavior of Tabanidae (Diptera) on cattle and serological analyses of partial blood meals. Environ Entomol 9:664–667

Marples MJ 1969 Life on the human skin. Sci Am 220:108–115

Mullens BA, Gerhardt RR 1979 Feeding behavior of some Tennessee Tabanidae. Environ Entomol 8:1047–1051

Renz A, Wenk P 1983 The distribution of the microfilariae of *Onchocerca volvulus* in the different body regions in relation to the attacking behavior of *Simulium damnosum s.l.* in the Sudan savannah of northern Cameroon. Trans R Soc Trop Med Hyg 77:748–752

Self LS, Abdulcader MHM, Tun MM 1969 Preferred biting sites of *Culex pipiens fatigans* on adult Burmese males. Bull WHO 40:324–327

Smart MR, Brown AWA 1956 Studies on the responses of the female Aedes mosquito. VII. The effect of skin temperature, hue and moisture on the attractiveness of the human hand. Bull Entomol Res 47:80–100

Sokal RR, Rohlf FJ 1981 Biometry. Freeman, New York

Sonenshine DE 1985 Pheromones and other semiochemicals of the Acari. Annu Rev Entomol 30:1–28

Swellengrebel NH, De Buck A 1938 Malaria in the Netherlands. Scheltema & Holkema, Amsterdam

Thompson BH 1976 Studies on the attraction of *Simulum damnosum s.l.* (Diptera: Simuliidae) to its hosts. 1. The relative importance of sight, exhaled breath, and smell. Trop Med Parasitol 27:445–473

Thompson BH 1977 Studies on the attraction of *Simulum damnosum s.l.* (Diptera: Simuliidae) to its hosts. 2. The nature of substances on the human skin responsible for attractant olfactory stimuli. Trop Med Parasitol 28:83–90

Torr SJ 1994 Responses of tsetse flies (Diptera: Glossinidae) to warthog (*Phacochoerus aethiopicus Pallas*). Bull Entomol Res 84:411–419

Townley P, Baker KP, Quinn PJ 1984 Preferential landing and engorging sites of Culicoides species landing on a horse in Ireland. Equine Vet J 16:117–120

Van Thiel PH, Weurman C 1947 L'attraction excercée sur *Anopheles maculipennis atroparvus* par l'acide carbonique dans l'appareil de choix II. Acta Trop 4:1–9

Walker ED, Edman JD 1985 Feeding-site selection and blood-feeding behavior of *Aedes triseriatus* (Diptera: Culicidae) on rodent (Sciuridae) hosts. J Med Entomol 22:287–294

Wilton DP 1975 Mosquito collections in El Salvador with ultraviolet and CDC miniature light traps with and without dry ice. Mosq News 35:522–525

DISCUSSION

Curtis: Have you done corresponding tests with *Anopheles arabiensis* and *Anopheles quadriannulatus*?

de Jong: No, but they should be done because these species are more opportunistic. In these cases CO_2 may be a more important cue.

Takken: This is not entirely true. We've recently done some comparative studies of *An. arabiensis* and *Anopheles gambiae*. We found that *An. arabiensis* always goes for the feet and doesn't change its preference (W. Takken & E. Bouma, unpublished data 1995).

Galun: Harold Trapido (unpublished paper, XIth Int Congr Ent Vienna 1960) also studied the stratification of mosquito biting of the host. In order to differentiate between a preference of leg-biting mosquitoes to fly close to the ground or to focus on the legs, he hung people upside down so that their heads touched the ground.

Knols: A significant observation for us was that when we studied the selection of biting sites by *An. gambiae* on a host lying down, we found that he was bitten all over his body. However, we found that as soon as the subject sat upright, then mosquitoes headed for his feet. This is how we hit upon the idea that convection currents and foot odours might play a role in the selection of biting sites.

de Jong: Yes, because the convection currents also change when the position of the person changes.

Hildebrand: For *An. gambiae*, you showed that washing changed the distribution of biting sites. But did this also reduce the frequency of biting sites?

de Jong: The distribution of biting sites can be related to several factors. Identification of these factors may tell us something about the cues important for the finding and selection of a host. In our experiments we only looked at the distributions, and for that reason we just waited until we had 100 bites.

Carlson: Do you know if the soap removed attractive odours, as opposed to depositing a repellent odour?

de Jong: That's a good point. Mosquitoes are attracted to a wet bar of soap, so it is unlikely that it contains a major repellent. However, humidity is also attractive, so we tried a dry bar of soap, which was neither attractive nor repellent.

Boeckh: How often did you have to wash the subject with soap?

de Jong: We washed the subject every hour.

Boeckh: What did you use as a control?

de Jong: The control was the same person without washing. Furthermore, we tried washing just one foot with the test person standing with his feet planted wide. *An. gambiae* seemed to prefer the unwashed foot. However, we could not complete this series of experiments because the test person had difficulties feeling the bites in that position.

Carlson: Does the distribution of body hair also play a role in biting site selection?

de Jong: It is unlikely that it plays a major role because only 5% of a person's body is covered with body hair. It is possible that it depends on the species, but we have observed *Anopheles atroparvus* probing on the back of the test person's head. The hair there didn't seem to have much effect.

Knols: If a human host sits in an open field situation then *An. atroparvus* is likely to bite the ankles and feet. *Mansonia africana* will persistently bite the back of sitting individuals in the field. Under controlled laboratory conditions both species prefer to bite the head of a sitting test person. Environmental factors, especially wind, will influence the approach direction and flight height of mosquitoes orienting towards a host. These factors might therefore have influenced the selection of biting sites and were different (though constant) from the natural situation.

Ziegelberger: If I was a mosquito, I would avoid the head because of the increased risk of being heard.

Gibson: The pitch of the sound made by a mosquito is determined by its wing-beat frequency, which is relatively species specific, and is used by males to identify conspecific females for mating. It would be interesting to identify which mosquito species 'buzz' at frequencies within and outside of the auditory range of a host animal, and then to determine whether or not those species which can be heard avoid the head area so that the host does not hear them or take defensive actions.

Klowden: Did you vary the height at which you released the mosquitoes?

de Jong: We did not look at this. The mosquitoes were always released from the same level, i.e. at face height.

Cardé: What sort of tracks did the mosquitoes make as they travelled down the body?

de Jong: We did the experiments with *An. gambiae* under poor light conditions because it is a mosquito that bites at night. We noticed that they approached the host in the head region and then went downwards, touching the skin every now and then. They did not travel back up the body. *An. atroparvus* is much bigger and bites during the day, so we were able to see its tracks more clearly. *An. atroparvus* appeared to respond to exhaled breath by a sudden reduction of flight-speed at short distance from the face and then orienting to the head.

Guerin: How fast was the response?

Knols: We did not measure the time between release and biting; the subject just waited until he was bitten or until the 3 min experimental period had passed. However, we got the strong impression that *Anopheles albimanus* took longer to bite when the subject's breath was removed from the bed net than when it was present.

Curtis: In my mind these experiments are important because they should allow us to distinguish between the zoophilic *An. quadriannulatus* and the anthropophilic *An. gambiae*. The approach of back-crossing zoophilic genes into an *An. gambiae* background should now be possible, but it does depend on determining that *An. quadriannulatus* does indeed go for the head. I hope that this experiment will be done in Zimbabwe.

Gibson: I am concerned that the experiments were performed on a European who normally wears socks and shoes. The bacteria that produced his 'foot odour' would not necessarily be the same as the ones found on the feet of humans in Africa, to which *An. gambiae* has become adapted.

Knols: On the other hand if you if you consider the human toe cleft or small cracks in the sole of the foot, for instance, the environmental conditions (high temperature and humidity) are virtually similar in these areas, even if the subject is not wearing shoes or socks.

Geier: Did you always test the same person or did you test different people?

de Jong: We always tested the same person, i.e. Bart Knols.

Geier: Therefore, it is possible that there are differences in biting sites between people.

Takken: In the past year we've repeated this experiment with *An. gambiae* on five different students, and we found that all the mosquitoes went for the feet.

de Jong: We found that *Culex quinquefasciatus* bites all over the body, but there are reports which suggest that biting site selection of *Cx. quinquefasciatus* is correlated with body regions most affected with clinical filariasis, i.e. the legs (Self et al 1969, Chandra & Hati 1993).

Takken: But people who were sitting on the chair and recording these experiments were also trying to chase away mosquitoes that came to bite them. This may automatically make the mosquitoes go to the feet.

de Jong: That's a good point. We noticed this for *Ae. aegypti*. Bart Knols was sitting still, but even slight movements of the head made the mosquitoes go to the feet.

Klowden: Have you studied vascularity? Is biting correlated with the ease of getting to the blood?

de Jong: I don't know if it is for humans, but Walker & Edman (1985) looked at this for *Aedes triseriatus* in rodents, and he found that there was a correlation.

Pickett: Although we think about feet as having a distinctive smell, the ears and the face also have their own range of specific compounds. These compounds are not very odorous to us, but they could be used by the mosquitoes.

Knols: The concentrations of odours that we have been using in our wind tunnel are so low that we can't smell them.

Geier: Why did you use Limburger cheese?

de Jong: We went into a cheese shop, sniffed all of the cheeses and picked the most revolting one.

Pickett: Why didn't you pick Stilton? Stilton has a much more revolting smell!

References

Chandra G, Hati AK 1993 Correlation between the preferred biting site of *Culex quinquefasciatus* and the region of the body affected by clinical filariasis. Ann Trop Med Parasitol 87:393–397

Self LS, Abdulcader MHM, Tun MM 1969 Preferred biting sites of *Culex pipiens fatigans* on adult Burmese males. Bull WHO 40:324–327

Walker ED, Edman JD 1985 Feeding-site selection and blood-feeding behavior of *Aedes triseriatus* (Diptera: Culicidae) on rodent (Sciuridae) hosts. J Med Entomol 22: 287–294

General discussion II

Knols: I would like to mention that most of the work on moth olfaction has been done with wind tunnels in which the behavioural responses of individual insects have been recorded on video and analysed subsequently. In contrast, most of the studies on mosquito olfaction have used dual-port olfactometers, where responses of groups of mosquitoes towards test odours or controls have been compared. These studies have resulted in the identification of behaviourally active chemicals for *Aedes aegypti*, such as L-lactic acid and short-chain fatty acids (Acree et al 1968, Carlson et al 1973). In our laboratory we use a similar system with two traps in an olfactometer, so we can introduce test odour into one of them and use the other one as a control (see Knols et al 1994). This is the way in which we demonstrated the attractiveness of Limburger cheese (see de Jong & Knols 1995). These systems can thus be used to demonstrate rapidly whether particular odours induce behavioural responses in mosquitoes, but the significance of these findings can only be established under field conditions. However, to date, few sampling tools are available for testing responses of mosquitoes to odours in the field. Carlo Costantini developed an odour-baited entry trap for use in West Africa, but this system proved less effective for sampling *Anopheles gambiae* in Tanzania, which again suggests that there are large behavioural differences between West and East African populations (Costantini et al 1993). Therefore, we have tried to use so-called electric nets for mosquitoes similar to those developed for sampling tsetse flies (*Glossina* spp.). These have the advantage that they are not dependent on trap entry responses, which often decrease the effectiveness of traps substantially. Various designs have been tested, and some are as effective as the nets used for tsetse flies (B. G. J. Knols & L. E. G. Mboera, unpublished results 1995). We have also used cyclindrical nets, and have demonstrated that mosquitoes engage in positive anemotaxis upon perceiving a plume of CO_2.

Brady: This is evidence that they fly upwind.

Davis: No, only that that particular species flies upwind. Less than 5% of *Ae. aegypti* will fly upwind in a 1 m wind tunnel. This behaviour is species specific.

Cardé: In studies of moth attraction to pheromone, we release moths and they typically react to the odour within seconds, flying upwind to its source in less than a minute. I worry about situations in which the mosquitoes are left in an assay system for 'long' periods of time. How can you be sure that you're dealing with a powerful upwind attractant if you need to leave the animals in a chamber for many minutes or even hours in order to get a 60–70% catch?

Hildebrand: It seems to me that you're raising an important issue about the differences between long-range attraction and events that occur close to the source.

Cardé: It is possible that in some choice tests, we are observing the propensity of the animal to land or do something other than being attracted to a particular odour. We may be assaying their willingness to land and not their upwind anemotactic behaviour. I also worry that if the final phase of locating a human host occurs in a still air environment, wind tunnels may not be entirely suitable for unravelling all aspects of the mosquitoes' behaviours. During the final phase of host location, a mosquito may fly in a fairly random pattern before landing. The decision to land may be based on chemical cues, heat or other cues. We may be missing some of these reactions in wind tunnel studies that always use wind and may lack these additional, non-chemical sensory inputs.

Brady: Strictly speaking, Bart Knols is not using a wind tunnel. There are two streams coming out of the two ports; the rest of the air, at least at the upwind end, is still, so it is slightly different from a conventional wind tunnel. However, in his system he tests only clean air versus the odour, and I am concerned that any odour might be more attractive than no odour. Is this possible?

Knols: Human breath contains at least 102 identified compounds, and we have been testing the breath of three humans against a clean air control (Krotozynski et al 1977, de Jong & Knols 1995). I believe that if mosquitoes showed responses to any odour when tested against clean air, we would have found responses to breath. However, this was not the case.

de Jong: Bart Knols and myself (unpublished work 1995) also used different concentrations of several fatty acids, which didn't elicit a response.

Brady: So that is one answer: some odours do not give a response. But an interesting operational question remains for all of us who are interested in olfactometers, i.e. which protocol gives us the most useful information?

Cardé: I would like to see assays devised that rely on more than just odour and that can be done in the presence and absence of airflow. I'm concerned that we may be missing behavioural reactions by conducting these experiments in an airflow.

Pickett: The assay that we employed investigated only one overall aspect of oviposition behaviour (Pickett & Woodcock 1996). It is a static bioassay in which the pheromone evaporates from a floating polyethylene source on water. When Brian Laurence performed the first dose–response experiments using this assay, he found that the response levelled out, but not at 100% effect. This is probably due to the pheromone having a low volatility so that at high doses the air becomes saturated with pheromone.

Cardé: Is this assay performed with moving air?

Pickett: No, it's performed with still air. Much of our field work has also been done with still air. We performed our major field experiments in Kenya in deep concrete pits with water at the bottom.

Takken: We initially worked with a standard wind tunnel, which had a single source placed in a moving airstream. However, one problem we had with this design is that, at least for *An. gambiae* and *Anopheles stephensi*, the mosquitoes always flew upwind as soon as there was an airstream. Therefore, it was difficult to make comparisons between the results from different olfactory cues because we had to look at the flight pattern of the mosquitoes. We did do this, but it was very time consuming. We have now developed a dual-port bioassay to test putative olfactory attractants more quickly.

Cardé: I'm still not sure that this addresses the question of what behaviours would occur in still air.

Davis: With respect to the choice of olfactometer, you first have to decide which behaviour it is that you want to measure and then select the most appropriate olfactometer. If you select one with airflow, then the questions that you are addressing will be related to odour movements in an airstream over a longer distance. However, if you want find out what happens at the landing site, then you may or may not have an airstream, depending on whether you wish to look at landing sites outside or inside a hut or dwelling. In my opinion one has to choose the most suitable olfactometer.

Guerin: I would like to say a few words about orientation to host odour in the triatomines *Triatoma* and *Rhodnius* (Taneja & Guerin 1995). We looked at whether odour alone could be responsible for attraction of triatomines to hosts before proceeding with studies on host odour attractants so that we could rule out a predominating role for other modalities such as sight. The bioassay in this case was simple, consisting of a locomotion compensator and tube that served to deliver air tangentially to the north pole of the locomotion compensator (Kramer sphere) where the bugs ran. In air alone, the bugs tend to run downwind. In air plus odour, the triatomines stop and turn upwind. The bugs then run upwind to the odour source. If they veer slightly off course, they stop to correct their path. When the odour is removed but the airstream continues, the bugs do one of two things. They show an increased tendency to cut across the wind, which is an adaptation to re-contact the odour and is also observed in moths flying in air. Alternatively, the triatomines that run due upwind often continue to do so for a long period after the odour is removed, i.e. maintain course. The point is that these bugs, which are adapted to life in human habitations or near bird nests, share some basic searching strategies with moths. They are also very sensitive to the removal of odour.

Boeckh: What happens if you give the odour without wind?

Guerin: We haven't looked at that.

Carlson: I would like to return to the question of biting preferences, but from the opposite point of view. Boris Dobrokhotov once told me an anecdote about someone in a Russian village who could walk out into the tundra completely naked and not get bitten. Are there any published examples of human mutations that confer resistance to mosquito biting?

Curtis: There was some American work in this direction. One of the military laboratories tested a large number of soldiers (Maibach et al 1966). They found a large variation, but I don't know if they found a completely negative individual.

Knols: There were 838 individuals in that study. One of them received significantly fewer bites in choice tests, in which mosquitoes could choose between the arm/hand of two individuals. However, that person did not remain entirely free of mosquito attack (Maibach et al 1966).

Mustaparta: It is often claimed that there are individual differences in humans concerning mosquito bites. It would be interesting to test whether this is due to different odours released or to different immune reactions.

Kaissling: I would like to describe some work by Ernst Kramer on the effect of pheromone intensity on the anemotactic response. He repetitively offered brief (150 ms) pulses of bombykol to males of *Bombyx mori* walking on a Kramer sphere, which allowed the measurement of various behavioural parameters. The moths transiently increased their walking velocity and the velocity of turning upwind after each stimulus pulse. The anemotactic response—for example, the straightness of the path—was maximal at a repetition rate of three pulses per second, independent of stimulus intensity in the range from a walking threshold concentration to a 104-fold higher stimulus concentration (Kramer 1986). The anemotactic response was weak with permanent pheromone stimulation. No behaviour was observed when a constant impulse firing of the bombykol receptor cells was elicited by Z,E-4,6-hexadecadiene, simulating a constant 10 min bombykol stimulus. However, if this firing was modulated by repetitive stimuli (three per second) of the inhibitory terpene alcohol linalool, the moths readily walked upwind (Kramer 1992).

References

Acree F Jr, Turner RB, Gouck HK, Beroza M, Smith N 1968 L-Lactic acid: a mosquito attractant isolated from humans. Science 161:1346–1347

Carlson DA, Smith N, Gouck HK, Godwin DR 1973 Yellow fever mosquitoes: compounds related to lactic acid that attract females. J Econ Entomol 66:329–331

Costantini C, Gibson G, Brady J, Merzagora L, Coluzii M 1993 A new odour-baited trap to collect host-seeking mosquitoes. Parassitologia 35:5–9

de Jong R, Knols BGJ 1995 Olfactory responses of host-seeking *Anopheles gambiae s.s.* (Diptera: Culicidae). Acta Tropica 59:333–335

Knols BGJ, de Jong R, Takken W 1994 Trapping system for testing olfactory responses of the malaria mosquito *Anopheles gambiae* in a windtunnel. Med Vet Entomol 8:386–388

Kramer E 1986 Turbulent diffusion and pheromone-triggered anemotaxis. In: Payne TL, Birch MC, Kennedy CEG (eds) Mechanisms in insect olfaction. Oxford University Press, Oxford p 59–67

Kramer E 1992 Attractivity of pheromone surpassed by time-patterned application of two non-pheromone compounds. J Insect Behav 5:83–97

Krotozynski B, Gabriel G, O'Neill H 1977 Characterization of human expired air: a promising investigative and diagnostic technique. J Chromatogr Sci 15:239–244

Maibach HI, Skinner WA, Strauss WG, Khan AA 1966 Factors that attract and repel mosquitoes in human skin. J Am Med Assoc 196:173–176

Pickett JA, Woodcock CM The role of mosquito olfaction in oviposition site location and in the avoidance of unsuitable hosts. In: Olfaction in mosquito–host interactions. Wiley, Chichester (Ciba Found Symp 200) p 109–123

Taneja J, Guerin PM 1995 Oriented responses of the triatomine bugs *Rhodnius prolixus* and *Triatoma infestans* to vertebrate odours on a servosphere. J Comp Pysiol A 176:455–464

The role of mosquito olfaction in oviposition site location and in the avoidance of unsuitable hosts

John A. Pickett and Christine M. Woodcock

Biological and Ecological Chemistry Department, IACR-Rothamsted, Harpenden, Hertfordshire AL5 2JQ, UK

Abstract. Developments in the exploitation of mosquito olfaction are traced, in collaborative studies with various groups, from the first identification of a mosquito pheromone through to a discussion of non-host avoidance. The characterization of the oviposition pheromone for mosquitoes in the genus *Culex*, e.g. *Culex quinquefasciatus*, as a novel chiral lactone ester provided the impetus for a number of sophisticated asymmetric syntheses and economical large-scale routes to racemic products. The latter have provided material for successful field trials in three continents. During the course of this field work, we obtained evidence that semiochemicals originating directly from the oviposition site are essential for activity of the oviposition pheromone. Recent studies are elucidating the nature of these agents and their geographical variability. Initially, we used synthetic oviposition pheromone to attract mosquitoes to sites treated with a biorational larvicide. However, recyclable biological control agents offer better prospects for resource-poor regions. A biotechnological approach to pheromone production has been devised involving the generation of inexpensive starting materials by the cultivation of a higher plant. New studies on dipterous pests feeding on farm animals indicate a semiochemically based mechanism by which unsuitable individuals within the host species are avoided. There appears to be an analogous process in which mosquitoes avoid certain potential human hosts, thereby raising prospects for the development of novel, rationally identified repellents once the semiochemical/olfactory interactions have been fully elucidated.

1996 Olfaction in mosquito–host interactions. Wiley, Chichester (Ciba Foundation Symposium 200) p 109–123

Initial work on olfactory processes involved in the location of oviposition sites has given rise to new techniques and the development of existing approaches that are now available for studying other semiochemically mediated aspects of mosquito behaviour. Such approaches include the direct coupling of electrophysiological recordings with high resolution capillary gas chromatography for initial isolation and the identification of neurophysiologically active semiochemicals prior to full characterization by spectroscopy and synthesis. The objective of the work described here is to understand the role of mosquito olfaction in oviposition site location and in the avoidance of unsuitable hosts.

Oviposition pheromone

Mosquitoes in the genus *Culex*, and in particular *Culex quinquefasciatus* (*pipiens fatigens*) Say, lay rafts of eggs which, on maturation, develop droplets at their apices. These droplets release a pheromone attractive to gravid females (Bruno & Laurence 1979). This is adaptively advantageous in that it signals safety of the site, as the pheromone is only produced by eggs that have survived for at least 24 h, and may have other benefits for development of the offspring (McCall & Cameron 1995). Samples of the apical droplet placed on polyethylene disks floating in water will attract gravid females to lay eggs in the vicinity (Bruno & Laurence 1979). Gas chromatography coupled mass spectrometry (Pickett 1990) showed the presence of one major pheromone component eluting at a long retention time. The spectrum contained some unusual features that were explained by the structure I depicted in Fig. 1, with the base peak at mass to charge ratio (m/z) 99 arising from the δ-lactone ring and the rearrangement ion at m/z 142 arising from the reaction shown (Fig. 2). Synthesis of the unsaturated acid at C5 by the Wittig methodology allowed production of the *erythro*-5,6-dihydroxy acid and cyclization and acetylation to compound I (Fig. 1), comprising two enantiomers. The *erythro* product was highly active in the disk on water bioassay (Laurence & Pickett 1982). Synthetic *erythro* material was resolved into the two enantiomers using a chiral gas chromatography column in which the stationary phase comprised an isomer of chrysanthemoylmandelic acid naphthylethylamide. The two enantiomers, synthesized unequivocally by means of the Sharpless kinetically controlled asymmetric epoxidation, were employed in peak enhancement studies to show that the eggs produced only the (5R,6S)-6-acetoxy-5-hexadecanolide (compound II in Fig. 1) (Laurence et al 1985). The other enantiomer was behaviourally inactive in this study, as were the two *threo* enantiomers synthesized and tested by others (Hwang et al 1987).

To obtain synthetic and natural pheromones for chiral gas chromatography studies, we replaced the acetoxy group in compound II by a trifluoroacetoxy group, thereby increasing the vapour pressure of the product and reducing the temperature necessary for gas chromatography. This compound was, surprisingly, active in the behavioural bioassay, and indeed, there was an indication of higher activity than the natural product (Briggs et al 1986), probably because of the higher vapour pressure. It had previously been demonstrated that shortening the alkyl chain of compound II eliminated activity (Laurence & Pickett 1985). Even more surprising was that, on replacing the hydrogens on

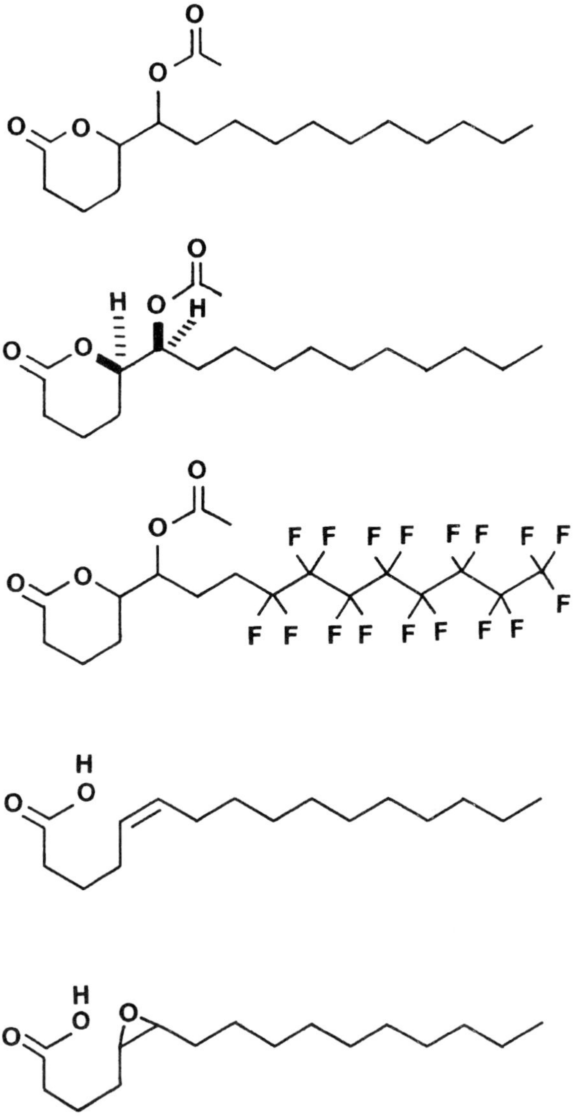

I

II

III

IV

V

FIG. 1. Structures of the *Culex quinquefasciatus* oviposition pheromone and related compounds. I, 6-Acetoxy-5-hexadecanolide; II, (5*R*,6*S*)-6-acetoxy-5-hexadecanolide, the oviposition pheromone; III, heptadecafluoro analogue of the pheromone; IV, 5-hexadecenoic acid, the pheromone synthesis precursor; and V, epoxide derivative employed in the analysis of plants for IV.

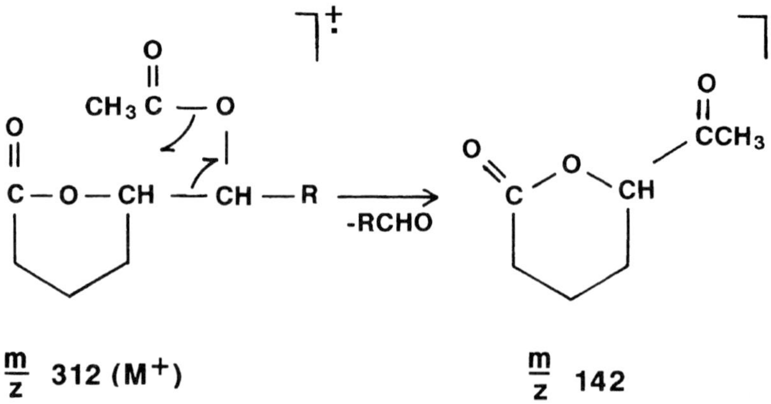

FIG. 2. Proposed mechanism for rearrangement of the *Culex quinquefasciatus* oviposition pheromone. Radicle ion from parent molecule of pheromone II at mass to charge ratio (m/z) 312 (M$^+$) rearranges in the mass spectrometer source to eliminate aldehyde, RCHO, giving a product with m/z 142.

the terminal eight carbons with fluorine, i.e. the heptadecafluoro analogue III (Fig. 1), activity remained at a high level, which was comparable with the natural product (Dawson et al 1990). The neural receptors involved in the olfactory responses of mosquitoes would be generally similar to olfactory receptors (Lancet & Pace 1987). For this pheromone, the receptor would be expected to have a site, or two closely located sites, involving hydrogen bonding and other polar interactions with the δ-lactone and acetoxy groups and an adjacent site comprising a lipophilic pocket for the decyl hydrocarbon sidechain. However, maintenance of activity after perfluorination of the terminal eight carbons indicates that the interaction cannot be of a lipophilic nature but must involve a spatial effect, although it should be borne in mind that the hydrocarbon chain would be flexible and could thereby take on a range of conformations. In contrast, fluorocarbon chains are more rigid through the higher Van der Waals radius of the fluorine atoms and would protrude in a rod-like fashion, with a 360° twist per 26 difluoromethylene groups (Banks & Tatlow 1986). Thus, the behavioural activity of analogue III gives further information on the likely structure of the protein involved in the olfactory receptor system. Such studies involving fluorine replacement of hydrogens in pheromonal molecules have subsequently been used to investigate olfactory receptors in other systems (Prestwich 1993).

The first successful field trial of the oviposition pheromone was undertaken at the International Centre of Insect Physiology and Ecology experimental site at Mbita Point, northeast of Lake Victoria in Kenya, using a number of concrete water-containing pits. For this trial, a new synthetic route was developed for large-scale synthesis of the pheromone, involving Baeyer Villiger methodology and a readily available starting material, cyclopentanone (Dawson et al 1990). The pheromone was formulated in effervescent tablets which released the active material on contact with water. An 80% increase in egg rafts was obtained by use of the pheromone, which was highly significant compared to the numbers laid in the control pits (Otieno et al 1988, Dawson et al 1989). The next step was to include, in the effervescent tablet, a means of destroying the larvae ensuing from the increased oviposition. This was achieved using the selective insect growth regulant pyriproxyfen (originally S31183), which has a mode of action related to the juvenile hormones. Again, an increase of over 80% in egg raft numbers was obtained in the pheromone-treated pits, with the larvae being destroyed by the pyriproxyfen. Alternative formulations involved slow release of pheromone from degradable metal carboxylate glasses (Blair et al 1994).

It was considered necessary to find a cheaper source of the oviposition pheromone and to this end, we examined a number of plants reported to contain the synthesis precursor, 5-hexadecenoic acid (compound IV in Fig. 1). The carboxylic acids from the triglycerides were released by saponification and were methylated. The unsaturated double bond was then oxidized to a 5,6-epoxy group (compound V in Fig. 1). The 5,6-epoxy group gave distinctive ions by α-fragmentation in the mass spectrum (i.e. at m/z 143 and m/z 183), which allowed discrimination from other fatty acids, e.g. 9-hexadecenoic, produced by most plants. The summer cypress, *Kochia scoparia* (Chenopodiaceae), was particularly promising. Therefore, we treated the fatty acid components from this plant, including the 5-hexadecenoic acid, with a catalytic (and recyclable) amount of osmium tetroxide to produce the *erythro*-5,6-dihydroxyhexadeca-noic acid. Simultaneous cyclization to the δ-lactone and acetylation by acetic anhydride with base gave the pheromone II and its enantiomer (T. O. Olagbemiro, personal communication 1991). Production of the pheromone via a renewable plant resource represents an important development in sustainable and cheap semiochemical production.

It was also desirable to find an alternative to the juvenile hormone-type insecticide for destruction of the larvae, to reduce the cost for resource-poor countries. The fungus *Lagenidium giganteum* Couch (Oomycetes) was considered to be suitable for the control of *Culex* mosquitoes because the motile zoospores, which inhabit the upper layers of the water, avidly seek out the larvae as they come to the surface for air (Brey & Remaudiere 1985, Brey et al 1988). Although this organism is highly safe to mammals (Siegel & Shadduck 1987), one problem needs to be overcome in that many *Culex* species, and certainly *Cx. quinquefasciatus*, oviposit in polluted water to ensure sufficient supplies of nutrients for their offspring. *L. giganteum*, however, does not infect mosquitoes in polluted water (Jaronski & Axtell 1982). It is therefore necessary to identify the olfactory cues responsible for the mosquitoes detecting polluted water, so that the pure compounds can be used to encourage oviposition in relatively clean sites.

Oviposition site cues

Culex mosquitoes are known to employ volatile semiochemicals, together with visual cues, to locate oviposition sites (Ikeshoji & Mulla 1970, Beehler et al 1993). A number of active components, including 3-methylindole, have been identified from Bermuda grass infusions (Millar et al 1992) and field tested (Beehler et al 1994), and the influence of ageing of the infusions has been investigated (Isoe et al 1995). For *Culex tarsalis*, non-volatile infusion components were implicated as oviposition stimuli (Isoe & Millar 1995). Other work with natural pheromones (Millar et al 1994) and field trials around the world with synthetic products have demonstrated the essential nature of oviposition site-produced semiochemicals in optimizing activity of the pheromone. Indeed, the trials conducted in various geographic locations indicated a range of different pollution sources, with an associated range of semiochemicals. Thus, in Sri Lanka it was shown that animal faeces were responsible for the generation of the pollutant semiochemicals. In East Africa, household washing waste contributed to the activity and in laboratory cultures in Britain, London tap water was responsible (B. R. Laurence, personal communication 1987). Trials conducted in France and in New York State, although successful, provided no further information on water pollution effects. In a laboratory assay, water polluted by a variety of materials, including faeces of the rabbit, *Oryctolagus cuniculus*, gave an additive effect with the oviposition pheromone, and the semiochemical activity of this water could be extracted by using ether as a solvent (Blackwell et al 1993). The ether extract could be vacuum distilled at ambient temperature to give a completely volatile product, compatible with gas chromatography, that was also extremely active in behavioural bioassays. Earlier electrophysiological studies on mosquitoes showed responses to L-lactic acid (Davis & Sokolove 1976), CO_2 and other stimuli (Bowen 1991). In the course of our work, electroantenno-grams from gravid female *Cx. quinquefasciatus* demonstrated high activity in the water polluted with rabbit faeces and, more importantly, in the vacuum distilled ether extract (Blackwell et al 1993, Mordue et al 1992, 1993). Single-cell recording studies detected olfactory cells on the *Cx. quinquefasciatus* antenna that also responded to the vacuum distillate. The single-cell recording preparations, coupled to high resolution gas chromatography (Wadhams 1990), showed two major areas of activity in the sample (Fig. 3), subsequently identified by gas chromatography coupled mass spectrometry as phenol and indole. Cells responding specifically to 3-methylindole were also found (e.g. Fig. 4), and the pure compound significantly increased oviposition in laboratory and field assays (Mordue et al 1992, Beehler et al 1994). However,

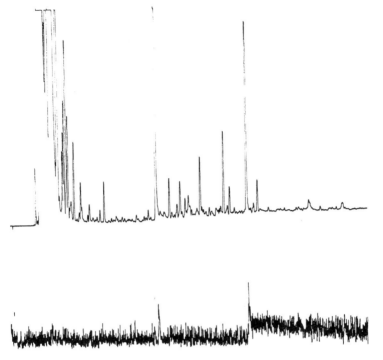

FIG. 3. Gas chromatography coupled single-cell recording on a *Culex quinque-fasciatus* female. Upper trace: gas chromatogram of volatiles from fermented aqueous faeces (rabbit, *Oryctolagus cuniculus*). Lower trace: single-cell recording showing responses to peaks from phenol and indole.

gas chromatography coupled mass spectrometry and gas chromatography coupled single-cell recordings demonstrated that this compound was not present in the extract of aqueous rabbit faeces. Electroantennogram preparations were used to construct dose–response data for a range of compounds, including *ortho-*, *meta-* and *para-*cresols, and showed that males, where tested, were less responsive than females to oviposition site stimulants (Blackwell et al 1993).

Non-host avoidance

The impetus for developments in this area is being provided by work on other Diptera. Thus, in new studies on cattle pests such as the head fly, *Hydrotaea irritans*, and the horn fly, *Haematobia irritans*, it can readily be seen that, within a herd comprising only one breed, some individuals show greater attractiveness than others (Steelman et al 1993). Preliminary evidence indicates that the less attractive cattle produce materials which mask the normal

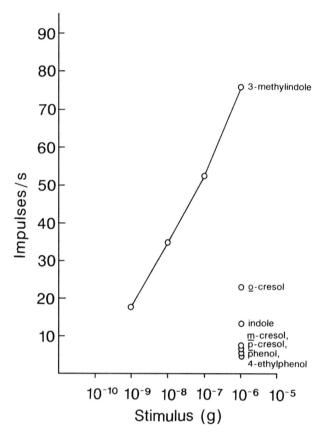

FIG. 4.   Dose–response of an olfactory cell on the antenna of a *Culex quinquefasciatus* female to 3-methylindole at 10^{-9}–10^{-6}g, and to indole, phenol, 4-ethylphenol and *ortho*-, *meta*- and *para*-cresols at 10^{-6}g.

attractiveness of such animals (G. Thomas, personal communication 1992). This is giving rise to the construction of a push–pull system in which a few highly attractive cattle protect the main herd and receive, using sophisticated electrostatic spraying equipment, selective pesticides to reduce the overall fly population and reduce stress and damage to the attractant individuals. Again, gas chromatography coupled electrophysiology is proving invaluable in identifying compounds from the unattractive individuals that may, in the long term, be used as repellents against these pests. As there is also evidence that certain humans are less attractive than others to disease-vectoring mosquitoes, the possibility exists for identifying the semiochemicals involved. As with the cattle flies, such compounds may be employed by the insects to detect potential hosts expressing inappropriate immunological responses. These semiochemicals, if identified for mosquitoes that attack humans, would represent rational and potentially highly potent repellents with mechanisms deeply embedded in mosquito ecology, thereby providing useful alternatives to currently available products such as *N,N*-diethyl-*m*-toluamide (Davis 1985, Coleman et al 1994).

Summary

Approaches and techniques arising from studies of olfactory mechanisms in mosquito oviposition behaviour are set to throw new light on host attraction and avoidance of unsuitable hosts by haematophagous insects.

Acknowledgement

IACR-Rothamsted receives grant-aided support from the Biotechnology and Biological Sciences Research Council, UK.

References

Banks RE, Tatlow JC 1986 A guide to modern organofluorine chemistry. J Fluorine Chem 33:227–284

Beehler JW, Millar JG, Mulla MS 1993 Synergism between chemical attractants and visual cues influencing oviposition of the mosquito, *Culex quinquefasciatus* (Diptera: Culicidae). J Chem Ecol 19:635–644

Beehler JW, Millar JG, Mulla MS 1994 Field evaluation of synthetic compounds mediating oviposition in *Culex* mosquitoes (Diptera: Culicidae). J Chem Ecol 20:281–291

Blackwell A, Mordue AJ, Hansson BS, Wadhams LJ, Pickett JA 1993 A behavioural and electrophysiological study of oviposition cues for *Culex quinquefasciatus*. Physiol Entomol 18:343–348

Blair JA, Mordue AJ, Duffy JA, Wardell JL 1994 Use of metal carboxylate glasses in the controlled release of bioactive molecules: *Culex quinquefasciatus* oviposition pheromone. J Controlled Release 31:145–149

Bowen MF 1991 The sensory physiology of host-seeking behavior of mosquitoes. Ann Rev Entomol 36:139–158

Brey PT, Remaudiere G 1985 Recognition and isolation of *Lagenidium giganteum* Couch. Bull Soc Vector Ecol 10:90–97

Brey PT, Lebrun RA, Papierok B, Ohayon H, Vennavalli S, Hafez J 1988 Defense reactions by larvae of *Aedes aegypti* during infection by the aquatic fungus *Lagenidium giganteum* (Oomycete). Cell Tissue Res 253:245–250

Briggs GG, Cayley GR, Dawson GW et al 1986 Some fluorine-containing pheromone analogues. Pestic Sci 17:441–448

Bruno DW, Laurence BR 1979 The influence of the apical droplet of *Culex* egg rafts on oviposition of *Culex pipiens fatigans* (Diptera: Culicidae). J Med Entomol 16:300–305

Coleman RE, Richards AL, Magnon GJ et al 1994 Laboratory and field trials of 4 repellents with *Culex pipiens* (Diptera: Culicidae). J Med Entomol 31:17–22

Davis EE 1985 Insect repellents: concepts of their mode of action relative to potential sensory mechanisms in mosquitoes (Diptera: Culicidae). J Med Entomol 22:237–243

Davis EE, Sokolove PG 1976 Lactic acid-sensitive receptors on the antennae of the mosquito, *Aedes aegypti*. J Comp Physiol A 105:43–54

Dawson GW, Laurence BR, Pickett JA, Pile MM, Wadhams LJ 1989 A note on the mosquito oviposition pheromone. Pestic Sci 27:277–280

Dawson GW, Mudd A, Pickett JA, Pile MM, Wadhams LJ 1990 Convenient synthesis of mosquito oviposition pheromone and a highly fluorinated analog retaining biological activity. J Chem Ecol 16:1779–1789

Hwang Y-S, Mulla MS, Chaney JD, Lin G-G, Xu H-J 1987 Attractancy and species specificity of 6-acetoxy-5-hexadecanolide, a mosquito oviposition attractant pheromone. J Chem Ecol 13:245–252

Ikeshoji T, Mulla MS 1970 Oviposition attractants for four species of mosquitoes in natural breeding waters. Ann Entomol Soc Am 63:1322–1327

Isoe J, Millar JG 1995 Characterization of factors mediating oviposition site choice by *Culex tarsalis*. J Am Mosq Control Assoc 11:21–28

Isoe J, Beehler JW, Millar JG, Mulla MS 1995 Oviposition responses of *Culex tarsalis* and *Culex quinquefasciatus* to aged Bermuda grass infusions. J Am Mosq Control Assoc 11:39–44

Jaronski ST, Axtell RC 1982 Effects of organic water pollution on the infectivity of the fungus *Lagenidium giganteum* (*Oomycetes:* Lagenidiales) for larvae of *Culex quinquefasciatus* (Diptera: Culicidae): field and laboratory evaluation. J Med Entomol 19:255–262

Lancet D, Pace U 1987 The molecular basis of odor recognition. Trends Biochem Sci 12:63–66

Laurence BR, Pickett JA 1982 *erythro*-6-Acetoxy-5-hexadecanolide, the major component of a mosquito oviposition attractant pheromone. J Chem Soc Chem Commun, p 59–60

Laurence BR, Pickett JA 1985 An oviposition attractant pheromone in *Culex quinquefasciatus* Say (Diptera: Culicidae). Bull Entomol Res 75:283–290

Laurence BR, Mori K, Otsuka T, Pickett JA, Wadhams LJ 1985 Absolute configuration of mosquito oviposition attractant pheromone, 6-acetoxy-5-hexadecanolide. J Chem Ecol 11:643–648

McCall PJ, Cameron MM 1995 Oviposition pheromones in insect vectors. Parasitol Today 11:352–355

Millar JG, Chaney JD, Mulla MS 1992 Identification of oviposition attractants for *Culex quinquefasciatus* from fermented Bermuda grass infusions. J Am Mosq Control Assoc 8:11–17

Millar JG, Chaney JD, Beehler JW, Mulla MS 1994 Interaction of the *Culex quinquefasciatus* egg raft pheromone with a natural chemical associated with oviposition sites. J Am Mosq Control Assoc 10:374–379

Mordue AJ, Blackwell A, Hansson BS, Wadhams LJ, Pickett JA 1992 Behavioural and electrophysiological evaluation of oviposition attractants for *Culex quinquefasciatus* Say (Diptera: Culicidae). Experientia 48:1109–1111

Mordue AJ, Blackwell A, Hansson BS, Wadhams LJ, Pickett JA 1993 Oviposition attractants for *Culex quinquefasciatus*. International Organisation for Biological and Integrated Control of Noxious Animals and Plants/West Palaearctic Regional Section Bulletin 16:335–340

Otieno WA, Onyango TO, Pile MM et al 1988 A field trial of the synthetic oviposition pheromone with *Culex quinquefasciatus* Say (Diptera: Culicidae) in Kenya. Bull Entomol Res 78:463–470

Pickett JA 1990 Gas chromatography–mass spectrometry in insect pheromone identification: three extreme case histories. In: McCaffery AR, Wilson ID (eds) Chromatography and isolation of insect hormones and pheromones. Plenum, New York, p 299–309

Prestwich GD 1993 Chemical studies of pheromone receptors in insects. Arch Insect Biochem Physiol 22:75–86

Siegel JP, Shadduck JA 1987 Safety of the entomopathogenic fungus *Lagenidium giganteum* (Oomycetes: Lagenidiales) to mammals. J Econ Entomol 80:994–997

Steelman CD, Gbur EE, Tolley G, Brown AH 1993 Individual variation within breeds of beef cattle in resistance to horn fly (Diptera: Muscidae). J Med Entomol 30:415–420

Wadhams LJ 1990 The use of coupled gas chromatography: electrophysiological techniques in the identification of insect pheromones. In: McCaffery AR, Wilson ID (eds) Chromatography and isolation of insect hormones and pheromones. Plenum, New York, p 289–298

DISCUSSION

Steinbrecht: Can you please clarify whether the 'unattractive cows' have a less attractive odour or whether there is a negative agent that is masking their attraction.

Pickett: Gethyn Thomas took various samples from attractive and unattractive cows, and he found that one of the fractions of these from the unattractive cows masked the attractiveness of the attractive material.

Curtis: Do you have any idea why *Culex* has a mechanism of attracting ovipositioning and hence larvae into one breeding site?

Pickett: At first this behaviour comes across as being maladaptive. The way I rationalize this is that food is not limiting because these mosquitoes employ polluted water oviposition sites where there is a plentiful supply of food. There is not even competition between species because all the *Culex* species that we have looked at, which have the same way of employing an oviposition pheromone, have the same compound. Even those that don't exhibit the same behavioural aspects still have the same pheromone component. If food isn't limiting, the only issue that the insect needs to establish is whether the site is mechanically safe. If they've evolved into urban insects to be near people, then they would obviously run the risk of trying to oviposit in sites that are mechanically disturbed. This pheromone denotes the safety of the site because it is present only after the eggs mature. The chemistry probably relates to the waterproofing of the egg raft to prevent hydrolytic damage, so it may be present originally as a by-product of some essential primary biochemistry.

Brady: But what is the benefit of the area to the eggs?

Pickett: They don't necessarily have to have a benefit. From a genetic point of view the pheromone may really be a kairomone because the beneficiaries are not genetically identical to those that have created the cue.

Steinbrecht: You showed that different insects prefer different types of pollution. Does this imply a universal pollution attractant or are different populations attracted to different components of polluted water?

Pickett: We don't yet know the answer to this question. When our colleagues left the London School of Hygiene and Tropical Medicine, we lost collaborators with whom we could do further field work. We have done one or two commercial trials but we have only looked at the pheromone in local pollution, and we have not done any field work. I would be interested if anyone else would like to collaborate with us.

Grant: What happens if the pheromone is placed in a less than optimum site?

Pickett: If we put the pheromone in unpolluted water, we do not observe a significant increase in ovipositing.

Klowden: You mentioned that this was attractive at a particular distance. What is that distance?

Pickett: We have provided chemicals for Mary Cameron and colleagues to do more work on that, but their experiments (Pile et al 1991) were critized from a statistical standpoint, so that although they claimed initially that olfaction was over a distance of several metres, they have had some reservations about the interpretation of the results (Pile et al 1993). From Brian Laurences' original experiments that employed the static air bioassay (Bruno & Laurence 1979, Laurence & Pickett 1985), the attractive distance is about 10 cm. However, one then has problems with determining how far insects are under the influence of the pheromone.

This work has not had any sophisticated behavioural studies applied to it. This is not in any way a criticism of Laurence or the people at Aberdeen, with whom we now collaborate, it simply does not fall into their sphere of scientific interest. If there are any people here who would be interested in doing some more behavioural work, then we will gladly provide the chemicals.

Klowden: I'm trying to imagine what's happening in the field. Is there a sort of pioneer mosquito that is attracted to a smelly place to lay eggs?

Pickett: Yes. There is a cue at the pollution site because polluted water gets more rafts than non-polluted water. This effect is amplified dramatically when the pheromone is present.

Klowden: Where does the pheromone come from?

Pickett: The apical droplet contains many unusual lipid materials, which is what Alvin Starratt was working on, when he was trying to identify the pheromone (Starratt & Osgood 1973). The droplet becomes apparent when the eggs mature, but it doesn't form properly until the eggs become melanized. Therefore, the pheromone is probably biosynthesized during the process of melanization.

Klowden: From where on the egg does the pheromone evaporate?

Pickett: The droplet forms at the top of the egg, which is the site of pheromone evaporation.

Klowden: Is there a structure that would allow that to emerge? Does it occur through the micropyle?

Pickett: I don't know how egg morphology relates to apical droplet production.

Gibson: Is it possible that this pheromone is actually a by-product of some aspect of the melanization process?

Pickett: No. Although the pheromone biosynthesis is known, the lipid component that acts as a precursor is *erythro*-5,6-dihydroxyhexadecanoic acid. The insect needs to saponify and then acetylate the compound to obtain the pheromone, in a similar way to our chemical synthesis of the pheromone.

Gibson: But is the pheromone something that the egg is trying to get rid of?

Pickett: No. The lipid precursor probably has a structural function. This adaptation must have been advantageous at some stage. It is possible that the original prehistoric *Culex* used it as a specific signal, but it then became advantageous as a kairomone to other genetic lines or species.

Galun: The literature is flooded with information on oviposition pheromones and oviposition attractants, either insect derived or environmentally derived from fermentation products. They all deal with different species of *Culex* and *Aedes*. Is anyone familiar with oviposition attractants for *Anopheles sp.*?

Pickett: We've looked at both the *Anopheles gambiae* complex and *Aedes aegypti* for this kind of chemistry. We were surprised to find it in *Ae. aegypti*. I'm not aware of any oviposition attractants in *An. gambiae*.

Galun: Are there any indications from its behaviour that they don't use any oviposition attractants?

Pickett: I am not aware of any.

Guerin: Are the indoles associated with faecal odour?

Pickett: Yes. Skatole, or 3-methylindole, is an example of this.

Bowen: What does the adaptation of a laboratory colony of *Culex* to London tap water involve? Is it a change in response to a specific odour, or is it a change in response threshold of receptors already present?

Pickett: I don't know for sure. *Culex quinquefasciatus* from different sites uses different pollution cues. It has an intrinsic neurophysiological response to specific indoles by different olfactory cells. It could have adapted to different pollution cues by a learning process. Alternatively, the strain used by Brian Laurence, which originally came from Lagos in Nigeria, may have been subject to selection for a subpopulation that was adapted positively to hypochlorous acid in London.

Hildebrand: When you talk about 'training' these insects to tap water, do you mean that in each generation you selected for mosquitoes that are attracted to the water's odour?

Pickett: That is essentially what happened. It is most likely that they associated being fed, or finding a profitable oviposition site, with the learned smell of the water. We know a lot about associative learning in hymenopterans. It's less well understood in dipterans, but it's certainly occurring.

Hildebrand: It would be interesting if insects that developed in tap water mature into insects that prefer tap water. This kind of 'imprinting' has been found in other insects; for example, in the case of induction of hostplant specificity in phytophagous Lepidoptera (Yamamoto 1974).

Guerin: One has to be careful because urban tap water has a high concentration of chlorine, which may play a part in selection process within cultures.

Pickett: We are discussing a behavioural phenomenon. However, it might be related to toxic effects from the water, which could generate the selection pressure for the London strain to be reduced down to a few individuals that show the ability to deal with the high oxidizing environment generated by the chlorine.

Knols: Concerning your field experiments in Kenya, would it be possible that certain parasitoids or predators use the oviposition pheromone of *Cx. quinquefasciatus* to locate a concentration of mosquito eggs? Are there any indications for the existence of such a phenomenon?

Pickett: This has not been studied. *Lagenidium giganteum* doesn't fly, but it would be interesting to see if it responded to kairomones.

Carlson: Are there chemosensory receptors around the ovipositor? And if so, does the pheromone stimulate them?

Pickett: Although we have identified a pheromone, we've done very little at the electrophysiological level. Mary Bowen has looked at this.

Bowen: There are receptors on the short, blunt sensilla trichodea of *Culex pipiens* that respond to egg raft pheromone by increasing spike frequency. It is possible that these receptors are also present on the ovipositor.

Davis: S. B. McIver (personal communication 1976) said that she could not find any chemoreceptors on the ovipositor. However, there are tarsal contact chemoreceptors that could provide this information.

Curtis: Have you considered the cost of your plant-derived products? In our work on treatment of pit latrines with polystyrene beads (Maxwell et al 1990), we considered using your tablets to trap oviposition in insecticide-treated pots and prevent its diversion to inaccessible sites, but the tablets would have cost more than the rest of the project. Are they now cheaper to buy?

Pickett: They could be made quite cheaply, especially if they were manufactured in a resource-poor country. Africa is looking for new cash crops to plant and *Kochia scoparia* grows well in Nigeria. There would also be an advantage in growing it here as an alternative to food cropping.

Brady: Is this kairomone attractive to other species?

Pickett: Yes, *Culex tarsalis*, *Culex molestus* and many others employ the same phenomena chemically. I am now working with Philip McCall on *Culex tigripes*. I have not been able to obtain a good sample of *Culex tritaeniorhynchus*.

Guerin: Could you comment on structure–activity relationships, particularly in terms of perfluorination of the side chain? Is this really the active moiety?

Pickett: This was a bit of a shock. Fortunately, Glen Prestwich was perfluorinating some lepidopteran pheromones at the same time, and he found a retention of activity in some cases. There is no doubt that you can't get far with structure–activity studies in terms of producing superpheromones. This is because these receptors have been exposed to a tremendous amount of selection pressure, so I don't recommend this as a real way forward, even though it's interesting chemically. In collaboration with Gordon Hamilton, we have identified the male-generated sex pheromones that attract the *Lutzomyia longipalpis* (sandflies) females. They are compounds, so we may have to construct analogues of these compounds to overcome this problem, unless we can use propheromones. I don't think that looking at analogues will be a profitable approach. The perfluorination is an anomaly. We have probably scored here because we've increased the vapour pressure so much in a compound where activity was limited by low vapour pressure.

References

Bruno DW, Laurence BR 1979 The influence of the apical droplet of *Culex* egg rafts on oviposition of *Culex pipiens fatigans* (Diptera: Culicidae). J Med Entomol 16:300–305

Laurence BR, Pickett JA 1985 An oviposition attractant pheromone in *Culex quinquefasciatus* Say (Diptera: Culicidae). Bull Entomol Res 75:283–290

Maxwell CA, Curtis CF, Haji H et al 1990 Control of Bancroftian filariasis by integrating therapy with vector control using polystyrene beads in wet pit latrines. Trans R Soc Trop Med Hyg 84:709–714

Pile MM, Simmonds MSJ, Blaney WM 1991 Odour-mediated upward flight of *Culex quinquefasciatus* mosquitoes elicited by a synthetic attractant. Physiol Entomol 16:77–85

Pile MM, Simmonds MSJ, Blaney WM 1993 Odour-mediated upward flight of *Culex quinquefasciatus* mosquitoes elicited by a synthetic attractant: a reappraisal. Physiol Entomol 18:219–221

Starratt AN, Osgood CE 1973 1,3-Diglycerides from eggs of *Culex pipiens quinquefasciatus* and *Culex pipiens pipiens*. Comp Biochem Physiol B 46:857–859

Yamamoto RT 1974 Induction of hostplant specificity in the tobacco hornworm *Manduca sexta*. J Insect Physiol 20:641–650

Introduction III: odours for host-finding mosquitoes

Carlo Costantini

Imperial College of Science, Technology & Medicine, Department of Biology, Silwood Park, Ascot, Berkshire SL5 7PY, UK

Host odours, together with visual and physical stimuli such as warm and moist convective currents, provide the necessary cues for mosquitoes to locate their hosts, although their relative importance is still debated (Kellogg & Wright 1962, Price et al 1979, Takken 1991). Cork (1996, this volume) has reviewed the relevance of several compounds as attractants (here termed kairomones, *sensu* Shorey 1977) for mosquitoes and other haematophagous Diptera. I will concentrate here on the effects of host odours on mosquito host finding and emphasize how a better knowledge of behaviour would help in the search for a synthetic bait to use in mosquito surveillance and control programmes (Kline 1994a), and in those studies aiming at the genetic manipulation of host preference of vectors of disease (Curtis 1994).

Behaviourally active compounds

One striking characteristic, when one reviews the current literature on odour-mediated mosquito host finding, is that, in spite of their theoretical interest and dramatic practical potential, very few active molecules have been discovered so far, and little is known about their actual mode of action at the physiological and behavioural level. Rather, some broad blends of volatiles obtained from sebum, sweat, blood, breath, or human arm, palm and foot emanations, plus some members of molecular families such as steroids, carboxylic acids and amino acids, have been recognized (with some conflicting conclusions) as potential kairomones (see Takken 1991, de Jong & Knols 1995). Moreover, their ability in eliciting behavioural responses when used without CO_2 is usually several times less potent than that of whole host odour, thus limiting their usefulness in field applications. CO_2, L-lactic acid and 1-octen-3-ol are the only compounds whose recognition as kairomones is at present less controversial. Their effects on behaviour, however, are still poorly known.

CO_2 is the molecule whose effects are best understood, although its specific role has been the subject of discussion (Bar-Zeev et al 1977), with some proposing that it both activates and 'attracts' mosquitoes (Van Thiel 1947, Brown 1951, Snow 1970), and others suggesting that it merely activates them (Willis 1947, Kellogg & Wright 1962, Daykin et al 1965, Khan & Maibach 1966). Gillies (1980) reviewed the subject and concluded that CO_2 has two distinct actions: first, as an activator eliciting kinesis and optomotor anemotaxis; and second as a 'synergist' in combination with warm moist convective currents at close range, and with odours at a distance from the host. In still air anemotaxis cannot take place, and only the kinetic effect is manifested.

L-Lactic acid has been occasionally reported as 'not attractive' or even as slightly repellent (Brown et al 1951), but observations in different experimental conditions have confirmed its role in host location at the doses normally released by human skin (Smith et al 1970). Alone, it neither activates nor elicits landing and probing behaviour. With CO_2, it increases take off and, in some concentration combinations, flight activity, landing and probing (Eiras & Jepson 1991).

No reports have been published on the behavioural effects of 1-octen-3-ol in mosquitoes. Its role has been mainly inferred from its enhancing effect on trap catches of certain culicid species, but only when used in conjunction with CO_2. Its efficacy on its own is less clear (Kline 1994b).

From a historical perspective, I suggest that part of the confusion and difference of opinion that is found in mosquito odour research (as outlined by Hocking [1963]), and the difficulty with which advances have been made compared to other areas of insect chemical ecology, are due in part to: (a) the loose use of the ambiguous term 'attraction'; (b) the variety of experimental conditions in which odours have been tested without proper recognition of the precise behavioural context examined; (c) the insufficient recognition of the different role that the same chemical may have at different concentrations, and in different behavioural contexts; (d) the underestimation of interspecific differences in behaviour other than the presence of certain general mechanisms; (e) the lack of links to the ecology of each species; and (f) the weak reciprocal feedback between laboratory and field studies.

Kairomones and attractants

Traditionally, debate has arisen as to whether to consider a certain substance an 'attractant' or a 'non-attractant', often based on the comparison of results obtained with different experimental techniques. Kennedy (1977a) warned us of how the term 'attractancy', when used to embrace all locomotory responses that bring a mosquito close to the source of odour, can be a vague and even misleading word. The problem goes beyond an academic dispute on terminology. This term has been used sometimes to describe the aggregative effect of a substance, whatever the underlying mechanisms of aggregation (e.g.

sugar, a non-volatile, is 'attractive' to houseflies). Thus, even compounds that cause activated mosquitoes to stop or slow down in front of the port of an olfactometer, or to increase their rate of turning if in flight, may confusingly be called 'attractants'.

Strictly speaking, attraction should involve only 'drawing from a distance', and it should describe only those substances whose behavioural effect is an oriented response towards the source, i.e. a taxis (Dethier et al 1960). However, the term 'attractant' has sometimes been attached to behavioural responses, such as probing and feeding, which do not involve any orientation towards the host. Such an approach underlies the counterproductive view of attraction as a unitary process (Kennedy 1977a), whereas a more precise description of the behavioural effects elicited by each substance would avoid ambiguities, and clear the field of potential teleological, anthropomorphic interpretations of host location.

Methodological approaches

Several models of olfactometers have been traditionally employed in studies of kairomones. Some of them use odour-laden air currents coming out of closely spaced ports, and repeated counts are made of the mosquitoes approaching the ports within a given distance or of those standing (and in certain instances of those probing) just in front of the port (Willis 1947, Van Thiel 1947). Bos & Laarman (1975) have pointed out that in a confined space, the mere presence of dead mosquitoes in front of one of the ports can bias the distribution of other responding mosquitoes: some sort of mass effect. An improvement to this basic design, as explained by Smith et al (1970), is to add some kind of trapping device at the ports where stimuli are presented; in this way responding mosquitoes are counted only once. Apart from the benefit of avoiding statistical pseudoreplication, this means that the mean response of the population cannot be unduly influenced by the behaviour of only a few individuals responding repeatedly and consistently. Moreover, kinetic effects alone are probably less likely to be effective, although this possibility cannot be ruled out. This design has been employed to demonstrate the 'attractive' effect of L-lactic acid (Acree et al 1968, Smith et al 1970). Nowadays, wind tunnels are more frequently used in studies of these kind of responses (Mayer & James 1969, Omer 1979, Eiras & Jepson 1991, Healy & Copland 1995). Even if they usually allow for several behaviours to be expressed and independently assessed, a detailed dissection of the in-flight behavioural mechanisms is possible only with video techniques (e.g. for tsetse flies, Colvin et al 1989).

Other authors have used either large cages (Brown & Carmichael 1961), rooms (Khan & Maibach 1972) or vertical towers (Khan et al 1966, 1967), with the stimuli presented from below in still air or carried along arising convective currents generated by heat and/or moisture. In these conditions, which simulate the final short-range approach of mosquitoes to the host, it is possible that those stimuli which intervene during previous phases of the host-finding process, either do not act at all, or may not do so in the same way. Directional cues are provided only by convective currents or, possibly, by odour concentration gradients (but see Kennedy [1977b] for a discussion on chemotaxis in flying insects).

Comparisons between different types of olfactometers are therefore meaningful only if it is kept in mind that, usually, completely different behaviours are measured, and sometimes different behavioural contexts are examined. Several key questions need to be addressed each time. What is being measured under the given experimental conditions? Which behavioural repertoires can be potentially expressed in these conditions, and which sequences are actually elicited? How are the stimuli presented, and how does this affect the specific behavioural responses?

Interspecific behavioural differences

Our knowledge of odour-mediated mosquito host location is even more incomplete than it appears because most of the laboratory studies have been on a single species, *Aedes aegypti*. The reasons for this are mainly the ease with which this species is reared in the laboratory, its diurnal activity and the traditional expertise of North American investigators (the most active in this particular field of medical entomology) on yellow fever and dengue vectors. Much less is known about the odour-mediated behaviour of other groups of Culicidae, such as the anophelines, although field studies usually show there is high interspecific variability in the response of mosquitoes to the chemicals tested (Kline et al 1990, Kline 1994b).

Given the importance of the malaria problem in the world (Curtis 1996, this volume), interest has been growing on the host-finding behaviour of the most powerful malaria vectors in the world, *Anopheles gambiae sensu stricto*, and its sibling, *Anopheles arabiensis*. In contrast to *Ae. aegypti*, these two species are typical night-biters and they actively disperse. Therefore, it would not be surprising to discover differences in the cues they use and how they use them compared to *Aedes* mosquitoes. Costantini et al (1996) have shown differences in the response to CO_2 between two anthropophilic members of the *An. gambiae* complex and more generalist sympatric species. Reeves (1953) long ago associated the CO_2 dose–response activities of three culicine species to their feeding habits. Gillies & Wilkes (1969) demonstrated interspecific differences in the active space of whole host odours. Omer (1979) highlighted differences between *An. arabiensis* and *Culex quinquefasciatus* in their upwind response to CO_2. It is likely that new insights in how mosquitoes use odour cues, and how these cues might modulate their host preferences, will come from such a comparative approach.

Feedback between laboratory and field studies

Relatively few studies have been carried out in the field on the odour-mediated host finding of mosquitoes. Even fewer studies have tried to verify laboratory results (Brown 1951), or to test laboratory-active molecules, and when this has been done, in some instances the attractiveness has not been confirmed (e.g. L-lactic acid, Stryker & Young 1970). Gillies & Wilkes (1969) concluded that *Anopheles melas* start responding to odours at greater distances than to CO_2. Unfortunately, this result did not lead to the search for non-CO_2 odours that might have an activating effect in this species.

For the most part, kairomones such as CO_2 have been employed for the practical scope of increasing the yield of trapping devices, such as light traps (see Service 1993). Unlike tsetse flies, where electric nets paved the way for field studies of kairomones, few specific trapping systems are available to collect mosquitoes mainly or exclusively by means of host odours. Unlighted light traps have been employed with the goal of studying the potential role of chemicals such as 1-octen-3-ol (see Kline 1994b). A more specific trapping device has been recently developed to catch mosquitoes using mainly odour cues, and therefore to study their odour-mediated 'host-seeking' behaviour in the field (Costantini et al 1993). This trap generates air currents against which mosquitoes must fly upwind and actively find their way into a cage, so that it catches negligible numbers without an appropriate odour bait. It is therefore sensitive even with odours that have a weak effect on the trap catch. Furthermore, it generates an odour plume that favours the entry of mosquitoes and, as the substances used to lure the trap are drawn from a tent under which there is a bait, whole odour from different kinds of hosts can be compared.

There are, of course, obvious important limitations with field studies. External, and to a certain extent even experimental, variables cannot be properly controlled and assessed by the experimenter. Responses are usually measured as the number of insects trapped, so they depend on the mode of action of the trap, and how that is perceived by the insect; video recording techniques may help in some of these interpretation difficulties (Griffiths & Brady 1994). Continuous feedback between laboratory and field studies has been certainly one of the reasons for the success of tsetse fly odour development (Colvin & Gibson 1992, Torr 1994).

Concluding remarks

A quantum leap in the discovery of attractants for tsetse flies (Hall et al 1984) and *Cochliomyia hominivorax* (Cork 1994) was the application of gas chromatography linked electrophysiology (Cork et al 1990). Even if their usefulness is still to be explored with mosquitoes (Cork 1996, this volume), it is likely that the employment of electroantennography will contribute to the identification of active molecules. Certainly, combined physiological–behavioural studies will considerably speed up our knowledge of behaviourally active compounds.

Clearly, there are still wide gaps in our knowledge of odour-mediated mosquito–host interactions. These will be filled only if there is a close interaction between chemists, physiologists, ethologists, geneticists, molecular biologists and entomologists, working both in the laboratory and in the field. The example of how such interactions have worked so successfully in achieving the surveillance and the control of tsetse flies through odour technology (Colvin & Gibson 1992, Torr 1994) should be a challenging model.

Acknowledgements

This work is part of the project 'Behavioural studies on malaria vectors' funded by the European Community under contract Nos. TS3-CT92-0101 and TS3-CT91-0032. Work cited in the text has received the support of the Centre National de Lutte contre le Paludisme, jointly sponsored by the Ministry of Health of Burkina Faso and the Programma di Assistenza Tecnica of the Direzione Generale per la Cooperazione allo Sviluppo of the Italian Ministry of Foreign Affairs. J. Brady, G. Gibson and M. Coluzzi substantially enriched the content of this contribution with their comments. S. Schofield is acknowledged for the many pleasant hours spent discussing insect behaviour.

References

Acree F Jr, Turner RB, Gouck HK, Beroza M, Smith N 1968 L-Lactic acid: a mosquito attractant isolated from humans. Science 161:1346–1347

Bar-Zeev M, Maibach HI, Khan AA 1977 Studies on the attraction of *Aedes aegypti* (Diptera: Culicidae) to man. J Med Entomol 14:113–120

Bos HJ, Laarman JJ 1975 Guinea pig, lysine, cadaverine and estradiol as attractants for the malaria mosquito *Anopheles stephensi*. Entomol Exp Appl 18:161–172

Brown AWA 1951 Studies of the responses of the female *Aedes* mosquito. IV. Field experiments on Canadian species. Bull Entomol Res 42:575–582

Brown AWA, Carmichael AG 1961 Lysine and alanine as mosquito attractants. J Econ Entomol 54:317–324

Brown AWA, Sarkaria DS, Thompson RP 1951 Studies on the responses of the female *Aedes* mosquito. I. The search for attractant vapours. Bull Entomol Res 42:105–114

Colvin J, Gibson G 1992 Host-seeking behavior and management of tsetse. Annu Rev Entomol 37:21–40

Colvin J, Brady J, Gibson G 1989 Visually-guided, upwind turning behaviour of free-flying tsetse flies in odour-laden wind: a wind-tunnel study. Physiol Entomol 14:31–39

Cork A 1994 Identification of electrophysiologically-active compounds for New World screwworm, *Cochliomyia hominivorax*, in larval wound fluid. Med Vet Entomol 8:151–159

Cork A 1996 Olfactory basis of host location by mosquitoes and other haematophagous Diptera. In: Olfaction and mosquito–host interactions. Wiley, Chichester (Ciba Found Symp 200) p 71–88

Cork A, Beevor PS, Gough JE, Hall DR 1990 Gas chromatography linked to electroantennography: a versatile technique for identifying insect semiochemicals. In: McCaffery AR, Wilson ID (eds) Chromatography and isolation of insect hormones and pheromones. Plenum, New York, p 271–280

Costantini C, Gibson G, Brady J, Merzagora L, Coluzzi M 1993 A new odour-baited trap to collect host-seeking mosquitoes. Parassitologia 35:5–9

Costantini C, Gibson G, Sagnon N'F, della Torre A, Brady J, Coluzzi M 1996 Mosquito responses to carbon dioxide in a West African Sudan savanna village. Med Vet Entomol 10, in press

Curtis CF 1994 The case for malaria control by genetic manipulation of its vectors. Parasitol Today 10:371–374

Curtis CF 1996 Introduction I: an overview of mosquito biology, behaviour and importance. In: Olfaction in mosquito–host interactions. Wiley, Chichester (Ciba Found Symp 200) p 3–7

de Jong R, Knols BGJ 1995 Selection of biting sites on man by two malaria mosquito species. Experientia 51:80–84

Daykin PN, Kellogg FE, Wright RH 1965 Host-finding and repulsion of *Aedes aegypti*. Can Entomol 97:239–263

Dethier VG, Barton Browne L, Smith CN 1960 The designation of chemicals in terms of the responses they elicit from insects. J Econ Entomol 53:134–136

Eiras AE, Jepson PC 1991 Host location by *Aedes aegypti* (Diptera: Culicidae): a wind tunnel study of chemical cues. Bull Entomol Res 81:151–160

Gillies MT 1980 The role of carbon dioxide in host-finding by mosquitoes (Diptera: Culicidae): a review. Bull Entomol Res 70:525–532

Gillies MT, Wilkes TJ 1969 A comparison of the range of attraction of animal baits and of carbon dioxide for some West African mosquitoes. Bull Entomol Res 59:441–456

Griffiths N, Brady J 1994 Analysis of the components of 'electric nets' that affect their sampling efficiency for tsetse flies (Diptera: Glossinidae). Bull Entomol Res 84:325–330

Hall DR, Beevor PS, Cork A, Nesbitt B, Vale GA 1984 1-Octen-3-ol: a potent olfactory stimulant and attractant for tsetse isolated from cattle odours. Insect Sci Appl 5:335–339

Healy TP, Copland MJW 1995 Activation of *Anopheles gambiae* mosquitoes by carbon dioxide and human breath. Med Vet Entomol 9:331–336

Hocking B 1963 The use of attractants and repellents in vector control. Bull WHO (suppl) 29:121–126

Kellogg FE, Wright RH 1962 The guidance of flying insects. V. Mosquito attraction. Can Entomol 94:1009–1016

Kennedy JS 1977a Behaviorally discriminating assays of attractants and repellents. In: Shorey HH, McKelvey JJ Jr (eds) Chemical control of insect behavior. Theory and application. Wiley, New York, p 215–229

Kennedy JS 1977b Olfactory responses to distant plants and other odor sources. In: Shorey HH, McKelvey JJ Jr (eds) Chemical control of insect behavior. Theory and application. Wiley, New York, p 67–91

Khan AA, Maibach HI 1966 Quantitation of effect of several stimuli on landing and probing by *Aedes aegypti*. J Econ Entomol 59:902–905

Khan AA, Maibach HI 1972 Effect of human breath on mosquito attraction to man. Mosq News 32:11–15

Khan AA, Maibach HI, Strauss WG, Fenley WR 1966 Quantitation of several stimuli on the approach of *Aedes aegypti*. J Econ Entomol 59:690–694

Khan AA, Strauss WG, Maibach HI, Fenley WR 1967 Comparison of the attractiveness of the human palm and other stimuli to the yellow-fever mosquito. J Econ Entomol 60:318–320

Kline DL 1994a Introduction to symposium on attractants for mosquito surveillance and control. J Am Mosq Control Assoc 10:253–257

Kline DL 1994b Olfactory attractants for mosquito surveillance and control: 1-octen-3-ol. J Am Mosq Control Assoc 10:280–287

Kline DL, Takken W, Wood JR, Carlson DA 1990 Field studies on the potential of butanone, carbon dioxide, honey extract, 1-octen-3-ol, L-lactic acid and phenols as attractants for mosquitoes. Med Vet Entomol 4:383–391

Mayer MS, James JD 1969 Attraction of *Aedes aegypti* (L.): responses to human arms, carbon dioxide and air currents in a new type of olfactometer. Bull Entomol Res 58:629–642

Omer SM 1979 Responses of females of *Anopheles arabiensis* and *Culex pipiens fatigans* to air currents, carbon dioxide and human hands in a flight-tunnel. Entomol Exp Appl 26:142–151

Price GD, Smith N, Carlson DA 1979 The attraction of female mosquitoes (*Anopheles quadrimaculatus* Say) to stored human emanations in conjunction with adjusted levels of relative humidity, temperature, and carbon dioxide. J Chem Ecol 5:383–395

Reeves WC 1953 Quantitative field studies on a carbon dioxide chemotropism of mosquitoes. Am J Trop Med Hyg 2:325–331

Service MW 1993 Mosquito ecology: field sampling methods, 2nd edn. Elsevier Science, New York

Shorey HH 1977 Interaction of insects with their chemical environment. In: Shorey HH, McKelvey JJ Jr (eds) Chemical control of insect behavior. Theory and application. Wiley, New York, p 1–5

Smith CN, Smith N, Gouck HK et al 1970 L-Lactic acid as a factor in the attraction of *Aedes aegypti* (Diptera: Culicidae) to human hosts. Ann Entomol Soc Am 63:760–770

Snow WF 1970 The effect of a reduction in expired carbon dioxide on the attractiveness of human subjects to mosquitoes. Bull Entomol Res 60:43–48

Stryker RG, Young WW 1970 Effectiveness of carbon dioxide and L(+) lactic acid in mosquito light traps with and without light. Mosq News 30:388–393

Takken W 1991 The role of olfaction in host-seeking of mosquitoes: a review. Insect Sci Appl 12:287–295

Torr SJ 1994 The tsetse (Diptera: Glossinidae) story: implications for mosquitoes. J Am Mosq Control Assoc 10:258–265

Van Thiel PH 1947 Attraction exercée sur *Anopheles maculipennis atroparvus* par l'acide carbonique dans un olfactomètre. Acta Trop 4:10–20

Willis ER 1947 The olfactory responses of female mosquitoes. J Econ Entomol 40:769–778

A search for components in human body odour that attract females of *Aedes aegypti*

Martin Geier, Hinrich Sass and Jürgen Boeckh

Universität Regensburg, Institut fürZoologie, Lehrstuhl Boeckh, Universitätsstrasse 31, D-93040 Regensburg, Germany

Abstract. In a new type of wind tunnel, mosquitoes fly upwind towards host odour sources and towards human skin wash extracts obtained by rubbing the skin with a pad soaked in ethanol. We used this behavioural response as a bioassay to identify attractants in liquid chromatography fractions of such extracts. L-Lactic acid is a major constituent of skin wash extracts and it is a necessary component for the extract's effectiveness. As a single stimulus, however, L-lactic acid is only slightly effective. This indicates that the extract's high degree of effectiveness is based on a synergism of L-lactic acid and other odour components. The separation of the extract by liquid chromatography revealed three distinct regions of active fractions, only one of which contained L-lactic acid. The components of the other two regions have not yet been determined. A combination of fractions in these two regions together with L-lactic acid is as attractive as the complete extract.

Olfaction in mosquito–host interactions. Wiley, Chichester (Ciba Foundation Symposium 200) p 132–148

In order to attain a blood meal, female mosquitoes use host odour for orientation. It has already been demonstrated that host odour attracts mosquitoes from a distance of several metres (Gillies & Wilkes 1972, summary in Takken 1991). However, all the relevant components of the odour have not yet been identified and the olfactory mechanisms for host recognition and orientation are not completely understood (Takken 1991). The olfaction of the yellow fever mosquito *Aedes aegypti* has been investigated extensively because this species is easy to breed and responds quickly to host stimuli. The abilities of CO_2 and L-lactic acid to act as attractants have been demonstrated in several behavioural studies (Acree et al 1968, Smith et al 1970, Eiras & Jepson 1991, summary in Davis & Bowen 1994). CO_2 increases flight activity and is an attractant (summary in Gillies 1980), and L-lactic acid alone is only a mild attractant, if at all. However, L-lactic acid does act as a synergistic attractant when combined with CO_2 (Acree et al 1968, Smith et al 1970). Nevertheless, the attractiveness of human host odour can only be partially explained by these two stimuli, suggesting that other components of host odour play a role in host location (Eiras & Jepson 1991). We are presently trying to identify these components using extracted odours from human skin, because humans are the preferred hosts for yellow fever mosquitoes (Freyvogel 1961, Clements 1992). We have compared the attractiveness of extracts to those of the natural host odour in a behavioural test using a Y-tube wind tunnel and used this as an assay to identify attractive components of the extract. L-Lactic acid was one of the main constituents of the extract, and we have determined the extent of this substance's attractiveness. Finally, we have fractionated the extract using liquid chromatography and used the behavioural response of the mosquitoes to detect the attractants.

Materials and methods

Animals
We used female *Aedes aegypti* aged 6–40 days raised from stems in the laboratory of BAYER (Leverkusen, Germany) for the behavioural tests. We maintained up to 300 females and males together in a container under the following conditions: 26–28 °C; 75% relative humidity; and a light:dark photoperiod of 12:12. We gave larvae the fish food Tetramin® (Tetra Werke GmbH, 4520 Melle, Germany) and adults a 10% glucose solution.

Bioassay
We conducted behavioural tests in an 80 cm long, 7 cm diameter transparent Y-shaped wind tunnel (Fig. 1). For each test, we lured 20 hungry females out of their container into a small mosquito chamber using the odour of a human hand. We fixed the chamber containing the mosquitoes at the stem of the Y-tube (start chamber, see Fig. 1) and we opened the traps at the branches. After 5 min, we applied the test odour in one branch and used the other branch as a control. We opened the start chamber with the mosquitoes, allowing the mosquitoes to fly through the wind tunnel. After 1 min, we trapped the mosquitoes at the ends of the branches by closing the screens. We noted the distribution of the mosquitoes and determined the flight activity and degree of attraction. After the test, we lured the mosquitoes back into the start chamber.

We evaluated the activity of the mosquitoes and the attractiveness of the odour source for each test. The per cent of mosquitoes that were found outside the start chamber after 1 min represents the flight activity. The per cent flight activity equals $A \times 100/G$, where A represents the number of mosquitoes outside the start chamber and G represents the total number of mosquitoes. The per cent of mosquitoes trapped in the test chamber after 1 min represents

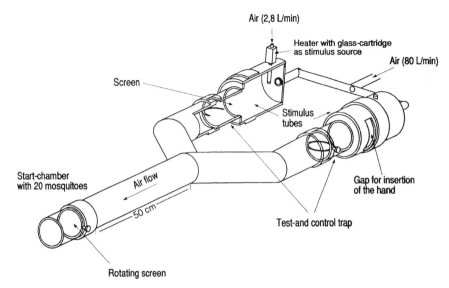

FIG. 1. Schematic drawing of the Y-tube wind tunnel and arrangement for application of the odour stimuli. Each end of the perspex Y-tube has a removable chamber. Both chambers on the branches of the Y-tube fit into stimulus tubes. A permanent airstream (cleaned by a charcoal filter, moistened to 70% ± 5% relative humidity and heated to 28 °C ± 1°C) flowed through the whole system at 80 l/min. The wind speed in the wind tunnel was 0.2–0.3 m/s in the branches and 0.4–0.6 m/s in the stem of the Y-tube.

the attractiveness of the stimulus. The per cent attraction equals T (or C) × 100/G, where T and C represent the number of mosquitoes in the test or control chamber, respectively. We compared the means of the test and control attractiveness values from several single experiments in order to estimate the attractiveness of an odour stimulus. We transformed the values by angle transformation (Sokal & Rohlf 1981) and calculated significant differences with the t-test for dependent random samples. After transforming the data, we performed an ANOVA procedure in order to determine significant differences of the means using the Tukey–Kramer test (HSD method) (Precht & Kraft 1993).

Odour stimuli

We used a human hand as the natural host odour because yellow fever mosquitoes find human hands particularly attractive (Rahm 1956, Freyvogel 1961). Human test subjects rubbed their hands, forearms, feet and calves with a pad soaked in ethanol, from which we extracted the attractants with methanol. We obtained the GEM 008 extract from 105 different samples. The volume of this extract was 90 ml and the concentration of rubbings in the extract was 1.2 rubbings/ml. For the application of olfactory stimuli, we filled a glass cartridge with the test solution. After evaporation of the solvent, we placed the glass cartridge into a heater ($60\,°C \pm 3\,°C$) to increase the volatility. In order to bring odour molecules into the wind tunnel, we blew air (2.8 l/min) through the cartridge. Simulation with smoke showed that the odour was completely mixed with the wind tunnel air 10 cm after the injection.

Removal of L-lactic acid from the extract
We removed L-lactic acid from the extract enzymatically with lactate oxidase from *Pediococcus* (Fluka Chemie AG, CH-9470 Buchs, Switzerland). Measurements showed that the amount of L-lactic acid in this extract had been reduced by 50-fold. We took a sample from this solution for the bioassay. We treated control samples in the same way, but did not add lactate oxidase.

Determination of L-lactic acid concentration
We determined the content of L-lactic acid using lactate dehydrogenase (Hohorst 1970) (from rabbit muscle, Boehringer Mannheim GmbH, Germany) and NAD (Boehringer Mannheim GmbH, Germany).

Chromatography
We fractionated 14 ml of GEM 008 extract using three different solvents of increasing polarity on a preparative silicagel column (Fig. 4a). We collected about 14 ml of each fraction for the behavioural assay and applied $5\,\mu$l of each fraction on the glass cartridge. This served as stimulus source in the bioassay.

Attractiveness of a human hand

We placed a human hand directly in the airstream of the Y-tube via a lateral opening in the stimulus tube, and used a clean piece of filter paper in the control side of the Y-tube to compensate for changes in airflow caused by the hand. We exposed the mosquitoes immediately to all of the stimuli (warmth, moisture and odours). Preliminary experiments showed that the hand is attractive in this test situation, so we tested the hands of three test persons over a half-year period. The hands of all three test persons activated and attracted 89–97% and 87–95%, respectively, of the mosquitoes (Table 1). The mosquitoes left the starting chamber a few seconds after the initial stimulation and flew upwind. They flew to the test side of the Y-tube and landed on the screen of the test chamber behind which the hand was situated.

In order to examine how mosquitoes behave in the Y-tube without host odour stimuli, we tested the following control conditions: (1) in the absence of stimuli, and test and control sides empty; and (2) with a piece of filter paper inserted into the test side, and the control side empty. Table 1 shows that only

TABLE 1 Responses of *Aedes aegypti* to human hands and to control stimuli in the Y-tube wind tunnel

Test stimulus chamber	Test attractiveness[a]	Stimulus in control chamber	Control attractiveness[a]	Flight activity[b]
Hand S	89 ± 6 (a)[c]	paper	1 ± 2	90 ± 6 (a)
Hand I	85 ± 5 (a)[c]	paper	2 ± 3	89 ± 4 (a)
Hand A	95 ± 3 (a)[c]	paper	1 ± 2	97 ± 5 (a)
None	3 ± 5 (b)	none	4 ± 5	34 ± 16 (b)
Paper	7 ± 6 (b)	none	5 ± 5	24 ± 12 (b)

[a]Percentage of mosquitoes in the test (control) chamber.
[b]Percentage of mosquitoes out of the start chamber.
[c]Significantly attractive compared to control (*t*-test, $P < 0.001$).
The means ± SD are given for 10 tests (20 mosquitoes in each test). Means followed by the same letter (a or b) do not differ significantly (ANOVA $P < 0.05$, Tukey's HSD). S, I and A are individual test persons.

30% of the mosquitoes were active and flew from the start chamber without additional stimuli. After 1 min, only 5–10% were trapped in the test chamber or in the control chamber. The filter paper did not elicit a higher response and there was no preference towards the test side.

Attractiveness of the GEM 008 extract

We tested different extraction procedures in preliminary experiments in order to extract active odour components. We found that the GEM 008 extract, which was obtained by rubbing the skin with an ethanol-soaked pad, was the most effective. We tested various doses of the GEM 008 extract, using heated glass tubes filled only with solvent as a control. Figure 2 shows the dose–response curves of this extract. The attractiveness increases from just over 5% (the control value) to over 90% within two magnitudes of the amount on the stimulus source. In the most effective dose, the extract is equally attractive to mosquitoes as the human hand.

In another experiment, we presented the GEM 008 extract and the hand simultaneously in order to determine whether they were equally attractive. We placed a human hand in the airstream on one side of the Y-tube and, at the same time, a glass cartridge with 50 µl GEM 008 extract on the other side. We found that the mosquitoes were evenly distributed in both chambers.

The role of L-lactic acid in the GEM 008 extract

Acree (1968) demonstrated that L-lactic acid is an attractive component of human skin acetone washes. The concentration of L-lactic acid in the GEM

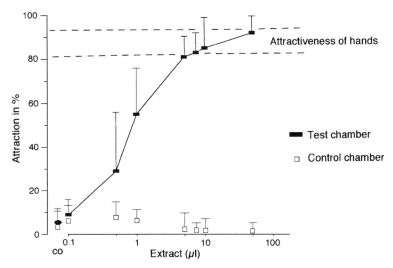

FIG. 2. Dose–response curve of the human skin wash extract GEM 008. The black (percentage of mosquitoes in the test chamber) and white (percentage of mosquitoes in the control chamber) symbols represent the means (±SD) of eight behavioural tests. Twenty mosquitoes were used for each test. The dotted lines show the level of atractiveness of human hands for comparison. The x-axis represents the amount of extract in the glass cartridge.

008 skin extract is 7 mg/ml (which is equivalent to 5.8 mg/rubbing), and the extract has a pH of 5. Based on this value, we calculated the extract's free L-lactic acid content to be 7% of the total lactate content (pK value of L-lactic acid = 3.862). This corresponds to an L-lactic acid concentration of 0.5 mg/ml in the extract, and 10 μl GEM 008 contains about 5 μg free L-lactic acid. We tested the attractiveness of pure L-lactic acid in four different doses ranging from 0.5 to 300 μg. L-Lactic acid in certain doses increased the flight activity of mosquitoes and was slightly attractive (Fig. 3a,b). Within 0.5 μg and 5 μg only both flight activity and the level of attractiveness increased from the control value to a saturation level. At most, 50–60% of the mosquitoes were activated and 20–30% were attracted to L-lactic acid. Increasing the dose of L-lactic acid did not increase the mosquitoes' responses (Fig. 3). However, the human skin rubbing extract containing 5 μg L-lactic acid attracted more than 80% of the mosquitoes.

To determine whether L-lactic acid is an essential component of the attractant, we removed L-lactic acid from the extract using the enzyme lactate oxidase. This enzymatic digestion resulted in a 50-fold decrease in the concentration of L-lactic acid.

We also tested the attractiveness of the following stimuli: (1) a sample with a reduced amount of L-lactic acid; (2) a sample that did not undergo the lactate

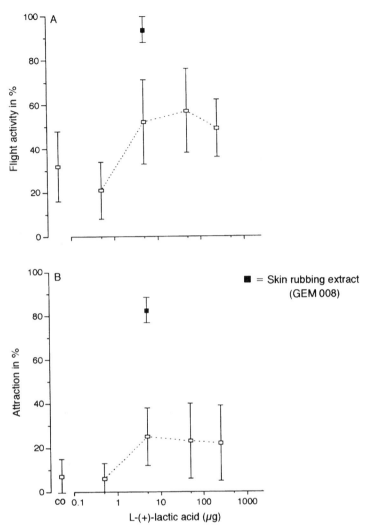

FIG. 3.   Dose–response curves of L-lactic acid. The behavioural responses of flight
activity (A) and attractiveness (B) to stimulation with L-lactic acid are shown. The
responses to the human skin wash extract GEM 008, which has a content of 5 μg L-
lactic acid, is also shown for comparison. The means (± SD) of 10 behavioural tests are
presented. Twenty mosquitoes were used for each test. The x-axis represents L-lactic
acid in the glass cartridge.

oxidase step; (3) a sample with a reduced amount of L-lactic acid to which we
added synthetic L-lactic acid; and (4) a sample containing only synthetic L-
lactic acid (Table 2). We found that the sample with a reduced amount of L-lactic
acid was not more attractive. When we added the original amount of synthetic

TABLE 2   Attractiveness of GEM 008 extract with a reduced content of L-lactic acid

Stimulus	L-lactic acid on test cartridge (μg)	Test[a]	Control[a]	Flight activity[b]
Reduced sample	0.15	7 ± 10 (c)	2 ± 2	33 ± 19 (c)
Control sample	5	72 ± 21 (a)[c]	2 ± 3	84 ± 12 (a)
Reduced sample plus L-lactic acid	5	66 ± 16 (a)[c]	1 ± 1	83 ± 11 (a)
L-lactic acid	5	21 ± 17 (b)[c]	1 ± 3	47 ± 19 (b)
Empty cartridge	0	6 ± 7 (c)	2 ± 2	35 ± 7 (c)

[a]Percentage of mosquitoes in the test (control) chamber.
[b]Percentage of mosquitoes out of the start chamber.
[c]Significantly attractive compared to control chamber (*t*-test, $P < 0.001$).
Means ± SD are given for 16 tests (20 mosquitoes in each test). Means followed by same letter (a, b or c) do not differ significantly (ANOVA $P < 0.05$, Tukey's HSD).

L-lactic acid to the reduced extract, the extract was just as attractive as the control sample (before we reduced the level of L-lactic acid). Synthetic L-lactic acid alone in the empty cartridge was only slightly attractive.

Liquid chromatography of the GEM 008 extract

Column chromatography on silica gel revealed that there were several active fractions (Fig. 4a–c). Enzymatic analyses for L-lactic acid, which cannot be detected by u.v. absorption, showed that L-lactic acid was eluted in a broad region between fraction 34 and 47 (Fig. 4b). The results mentioned above revealed that a behavioural response to the other components could not be expected without L-lactic acid. Therefore, in order to detect active fractions, we tested 5 μl of each fraction together with a standard dose of 5 μg L-lactic acid. Some fractions (11–13) which were eluted with the most lipophilic solvent increased the attractiveness of L-lactic acid significantly ($P < 0.05$). We also observed an increased attractiveness in another relatively broad region of fractions (25–39) that was eluted with a more polar solvent (Fig. 4c). These fractions overlap slightly with the fractions containing L-lactic acid (34–47). None of the active fractions were as attractive as 5 μl of the GEM 008 extract. None of the other fractions were effective.

We also tested combinations of active and non-active fractions in order to prove that important components were not lost in the separation procedure and to confirm the effectiveness of distinct fractions. We found that two combinations of fractions elicited a mosquito attractiveness response of 70–

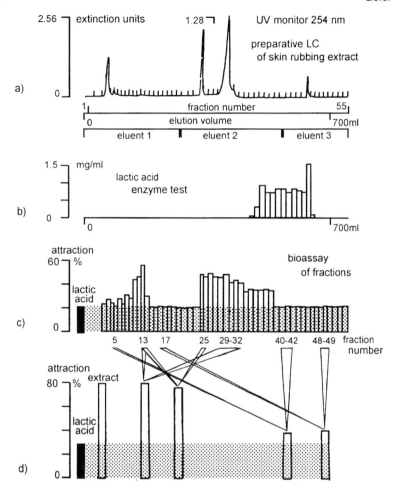

FIG. 4.   Preparative liquid chromatography (LC) of the human skin wash extract GEM 008
using three different solvents with increasing polarity and effectiveness of single fractions. (a)
Separation of 14 ml extract. Recorder trace of the u.v. detector (254 nm). Vertical lines mark
the fractions. Column (20 mm × 230 mm) made from silicagel (Lichroprep Si60® Merck,
KGaA D-64271 Darmstadt, Germany). Flow rate 6 ml/min. (b) Evaluation of the L-lactic
acid concentration of individual fractions using lactate dehydrogenase and NAD
(Boehringer GmbH, Mannhein, Germany). (c) Attractiveness of individual fractions (5 μl)
in combination with an L-lactic acid standard (5 μg). (d) Attractiveness of combined fractions
in comparison to the GEM 008 extract and L-lactic acid.

80% in the behavioural test (Fig. 4d). These combinations (fraction 13,
fraction 25 and L-lactic acid; and fraction 13, fractions 29–32 and L-lactic
acid) were almost as effective as the GEM 008 extract (Fig. 4d). Fractions 5, 41,
42 and L-lactic acid combined or fractions 17, 50, 51 and L-lactic acid

combined were not significantly more attractive ($P > 0.05$ Tukey's HSD) than L-lactic acid alone (Fig. 4d).

Discussion

The bioassay developed for this investigation enabled us to test the attractiveness of a human hand as a standard stimulus for natural host odours. *Aedes aegypti* found this natural scent source attractive (Rahm 1957, Freyvogel 1961, Eiras & Jepson 1994, Willis 1948, summary in Takken 1991). The clear behavioural responses to a human hand in our experimental set-up indicate that the mosquitoes can orient themselves well towards odour sources in the Y-tube wind tunnel. According to Miller & Strickler (1984), the response to the natural scent source represents the standard for which the effects of single odour compounds can be evaluated.

The collection of attractive material from human skin has been an important objective of this investigation. Many human scent sources have been tested previously. These include sweat (Parker 1948, Thompson & Brown 1955, Rössler 1961, Skinner et al 1968, Müller 1968, Eiras & Jepson 1991), urine (Rössler 1961) and blood (Rudolfs 1922, Schaerffenberg & Kupka 1959, Brown & Carmichael 1961). Odour extracts have also been obtained by washing the skin with different solvents (Acree et al 1968, Smith et al 1970, Schreck et al 1981, 1990) or by concentrating odours from hands and arms in cold traps (Bar-Zeev et al 1977). However, the results regarding the attractiveness of such odour sources are not conclusive. In our investigation, we obtained an attractive extract by rubbing human skin with a pad soaked in ethanol. As our behavioural tests indicate, such an extract is as attractive to mosquitoes as a human hand. Therefore, we assumed that the extract contains the essential components for host location. The attractiveness of this extract cannot be explained by its L-lactic acid content alone. The dose–response curves show that L-lactic acid attracts mosquitoes in the wind tunnel even without additional CO_2. However, L-lactic acid alone is only slightly effective and the extract is much more attractive than the most effective dose of L-lactic acid. After L-lactic acid is removed, the remaining extract does not induce flight activity and is no longer attractive. The initial level of attractiveness is restored by adding pure L-lactic acid. This demonstrates that L-lactic acid plays a key role in attracting yellow fever mosquitoes. In addition, the experiment indicates that this extract's high degree of effectiveness is based on a synergism of L-lactic acid and other unidentified odour components. We separated the extract into different active components by means of preparative liquid chromatography. We found that there were three active regions of fractions: fractions containing L-lactic acid and at least two other different fractions from different regions (fractions 11–13 and fractions 25–32). These fractions increase the effect of L-lactic acid. From the polarity of the different solvents used for elution of the fractions, we assume that fraction 13 and fractions 25–32 contain different compounds. The complete level of attractiveness can only be attained when these fractions are mixed and L-lactic acid is added. However, each fraction contains several components, so the activity of a given fraction may be based on more than one effective component. The relatively broad region of active fractions eluted by the second solvent can be attributed either to one component divided among some fractions (see L-lactic acid Fig. 4b) or to a few distinct components.

The separation by liquid chromatography is only the first step for isolating and identifying active compounds. Although the existing results clearly support an olfactory host recognition by a complex stimulus pattern, further chromatography is required to clarify this and to improve the separation of active compounds. It seems that some of the effective components cannot be detected by u.v. absorption when liquid chromatography is used for separation. In our behavioural test active compounds can be detected precisely. However, these tests are time consuming, so other bioassay methods should be taken into account, e.g. an electroantennogram test, which would allow many samples to be tested quickly.

A set of multiple odour components characterizes a host much better than only a single odour substance. Such a complex odour pattern not only enables the mosquitoes to recognize a suitable host, but also enables them to discriminate and select between different hosts. The ability to feed on various hosts would be limited by using only one very specific and characteristic host odour component. A stimulus pattern increases the flexibility of host selection. It is also possible that mosquitoes use more than one specific odour pattern for host location.

In order to study the mechanisms of olfactory host location in more detail, we need to identify all the components of the odour pattern. Once this has been done, we should be able to explain the difference in attractiveness of various hosts.

The results also have implications for the control of these vectors. For example, mosquito traps could be developed more effectively using attractive odour components. The effectiveness of a human skin wash extract, which is very attractive under laboratory conditions, could thus be investigated in field studies.

Acknowledgement

This study was performed in the course of a joint investigation of host odours for *Aedes aegypti* together with the research laboratories of Bayer AG, Leverkusen, Germany.

References

Acree F Jr, Turner RB, Gouck HK, Beroza M, Smith N 1968 L -Lactic acid: a mosquito attractant isolated from humans. Science 161:1346–1347

Bar-Zeev M, Maibach HI, Khan M 1977 Studies on the attraction *of Aedes aegypti* (Diptera: Culicidae) to man. J Med Entomol 14:113–120

Brown AWA, Carmichael AG 1961 Lysine and alanine as mosquito attractants. J Econ Entomol 54:317–324

Clements AN 1992 Mosquitoes, vol 1. Chapman & Hall, London

Davis EE, Bowen MF 1994 Sensory physiological basis for attraction in mosquitoes. J Am Mosq Control Assoc 10:316–325

Eiras AE, Jepson PC 1991 Host location by *Aedes aegypti* (Diptera: Culicidae): a wind tunnel study of chemical cues. Bull Entomol Res 81:151–160

Eiras AE, Jepson PC 1994 Responses of female *Aedes aegypti* (Diptera: Culicidae) to host odours and convection currents using an olfactometer bioassay. Bull Entomol Res 84:207–211

Freyvogel TA 1961 Ein Beitrag zu den Problemen um die Blutmahlzeit von Stechmücken. Acta Trop 18:201–251

Gillies MT 1980 The role of carbon dioxide in host-finding by mosquitoes (Diptera: Culicidae): a review. Bull Entomol Res 70:525–532

Gillies MT, Wilkes TJ 1972 The range of attraction of animal baits and carbon dioxide for mosquitoes. Studies of a freshwater area of West Africa. Bull Entomol Res 61:389–404

Hohorst HJ 1970 L-(+)-Laktat Bestimmung mit Laktat-Dehydrogenase und NAD. In: Bergemeyer HU (ed) Methoden der enzymatischen Analyse. Verlag Chemie, Germany

Lehane MJ 1991 Biology of blood-sucking insects. Harper Collins, London

Miller JR, Strickler KL 1984 Finding and accepting host plants. In: Bell WJ, Cardé RT (eds) Chemical ecology of insects. Chapman & Hall, London, p 127–157

Misaki H, Matsuura K, Horiuchi Y, Harada S 1980 Lactic oxidase and its use. Ger Offen 3:1–34

Müller W 1968 Die Distanz- und Kontaktorientierung der Stechmücken *Aedes aegypti* (Wirtsfindung, Stechverhalten und Blutmahlzeit). Z vergl Physiol 58:241–303

Parker AH 1948 Stimuli involved in the attraction of *Aedes aegypti* L., to man. Bull Entomol Res 39:387–397

Precht M, Kraft R 1993 Biostatistik 2. Oldenbourg Verlag, Munich

Rahm U 1956 Zum Problem der Attraktion von Stechmücken durch den Menschen. Acta Trop 13:319–344

Rahm U 1957 Wichtige Faktoren bei der Attraktion von Stechmücken durch den Menschen. Rev Suisse Zool 64:236–246

Rössler HP 1961 Versuche zur geruchlichen Anlockung weiblicher Stechmücken (Aedes aegypti L., Culicidae). Z vergl Physiol 44:184–231

Rudolfs W 1922 Chemotropism of mosquitoes. Bull N J Agr Exp Stn 367:4–23

Schaerffenberg B, Kupka E 1959 Der attraktive Faktor des Blutes für blutsaugende Insekten. Naturwiss 46:457–458

Schreck CE, Smith N, Carlson DA, Price GD, Haile D, Godwin DR 1981 A material isolated from human hands that attracts female mosquitoes. J Chem Ecol 8:429–438

Schreck CE, Kline DL, Carlson DA 1990 Mosquito attraction to substances from the skin of different humans. J Am Mosq Control Assoc 6:406–410

Skinner WA, Tong H, Johnson H, Maibach H, Skidmore D 1968 Human sweat components—attractancy and repellency to mosquitoes. Experientia 24:679–680

Smith CN, Smith N, Gouck HK et al 1970 L-Lactic acid as a factor in the attraction of *Aedes aegypti* (Diptera: Culicidae) to human hosts. Ann Entomol Soc Am 63:760–770

Sokal RR, Rohlf FJ 1981 Biometry. Freeman, New York

Sutcliffe JF 1987 Distance orientation of biting flies to their hosts. Insect Sci Appl 8:611–616

Takken W 1991 The role of olfaction in host-seeking of mosquitoes: a review. Insect Sci Appl 12:287–295

Thompson RP, Brown AWA 1955 The attractiveness of human sweat to mosquitoes and the role of carbon dioxide. Mosq News 15:80–84

Willis ER 1948 The olfactory responses of female mosquitoes. J Econ Entomol 40:769–778

DISCUSSION

Ziegelberger: You mentioned that your experimental set-up could discriminate between activators and attractants. However, as far as I understood, you only found compounds that were both attractants and activators, and you did not find any that were activators but not attractants.

Geier: Yes. All the components we tested were both activators and attractants. However, CO_2 seems to have a greater activating potential than attracting potential.

Costantini: In a completely control situation, i.e. clean air in both chambers, what was the per cent activation of the mosquitoes?

Geier: 30%.

Bowen: Were any of the fractions repellent in your bioassay?

Geier: No, because we used attractive people to obtain our extract. It is possible that if we used people who were less attractive, then we would find compounds that reduce the attractiveness of L-lactic acid.

Takken: Do you know why there were reports in the 1960s and 1970s (Acree et al 1968, Bar-Zeev et al 1977, Price et al 1979, Smith et al 1970) that L-lactic acid required CO_2 for it to become an attractant?

Geier: These results may be explained by the different biotests that were used. It could be that in my experimental set-up, using the Y-tube wind tunnel, the mosquitoes are more sensitive to weak attractants than in other biotests. We can exclude the possibility that the changes of the CO_2 concentration in wind tunnel air caused this response to L-lactic acid. This is proved by the fact that L-lactic acid was still attractive when we reduced the concentration of CO_2 to less than 10 ppm in the wind tunnel air.

Galun: In the presence of CO_2 these attractants may become more attractive. They may operate synergistically.

Boeckh: But in one experiment both low and high doses of CO_2 were added to the complete extract. Neither addition increased the attractiveness of the skin rubbing extract.

Costantini: What is the wind speed in your olfactometer?

Geier: The optimal wind speed was 20 cm/s in the branch and 40 cm/s in the stem of the Y-tube wind tunnel.

de Jong: You tested human skin rubbing extract versus human hands, and you found that the mosquitoes were equally attracted to both. However, did you compensate for temperature and humidity?

Geier: It has already been shown that warmth and humidity are attractive to *Aedes aegypti* (Bar Zeev et al 1977). In our experiments we minimize the influence of temperature and humidity by warming and humidifying the wind tunnel air so that the human hand does not cause an increase in temperature or humidity in the test situation. When I present the extract on a filter paper, I do not increase the temperature or the humidity. However, when I present the extract using the glass cartridge, I heat this cartridge to 60 °C. This causes an increased temperature of about 0.2 °C in the wind tunnel air. The mosquitoes did not respond to an empty, heated glass cartridge.

de Jong: Is it possible that the equal attraction of the mosquitoes to human hands and the extract is due to the high concentration of the extract?

Geier: This question is difficult to answer because we do not know the concentration of the attractive compounds in the wind tunnel air.

de Jong: Where did you take your skin rubbings from?

Geier: From hands.

de Jong: On the basis of the distribution of biting sites, it would be interesting to do it from shoulders as well.

Costantini: Did you notice any day-to-day variability in the activity levels of the mosquitoes?

Geier: At high concentrations of extract, the mosquitoes respond every day in nearly the same way. However, if the concentrations are lowered, the responses are more variable.

Brady: There were very few mosquitoes in the control side. Whereabouts in the wind tunnel do the mosquitoes choose between the clean airstream and the smelly airstream? What is the general pattern of their movement?

Geier: In the case of the attractive extract, the mosquitoes head straight towards the test chamber. However, the mosquitoes leave the start chamber and fly aimlessly around in the Y-tube when we use lower attractive stimuli e.g. L-lactic acid or CO_2.

Brady: If you put smoke down one side, what do you see at the junction?

Geier: At the junction there is not a complete mixing of both airstreams. If you use smoke you can see a weak gradient in the stem of the Y-tube.

Brady: Do the mosquitoes actually make a choice when they get to the junction of the Y-tube or do they sample both sides and then finally end up in one?

Geier: In most cases the mosquitoes make a choice at the junction. It is possible that this choice is facilitated by the gradient in the stem.

Davis: We have used a 1 m dual-port olfactometer, which has a flat face with two holes in it. We observed that the smoke gradually filled the whole cross-section near the end at which the mosquitoes were released, and that the mosquitoes flew directly up the tube. The mosquitoes that flew out of the odour-containing portion of the airstream came close to the clean air, and perhaps even entered it, but then they flew back down the tunnel towards the odour and reoriented to the odour.

Geier: In our experiments, some mosquitoes fly so quickly after they are released that they make a mistake and go up the wrong channel. After about 20 cm these mosquitoes turn, fly back towards the stem and enter the correct channel.

Guerin: Did you try using water or apolar solvents instead of ethanol?

Geier: We experimented with different solvents, and we found that the extracts were only attractive when ethanol was used. If we used water or acetone, the attraction decreased by 50%. The same was true for human sweat.

Guerin: It may be possible to infer something about the molecular mass or polarity of the missing attractant constituents from the types of solvents you used in the liquid chromatography separations. What solvents did you use?

Geier: I can't give you any further details about the components of the solvents because we obtained them from Bayer AC. The second eluent is more hydrophilic and in these more polar fractions we found L-lactic acid.

Pickett: How did you remove the ethanol before the bioassay?

Geier: By blowing air through the glass cartridge before I started the experiment.

Pickett: Presumably, you are using very purified ethanol, because higher homologues in the ethanol could remain.

Geier: Yes. I used purified ethanol as a control stimulus. After evaporation of the ethanol I did not observe any response to the control cartridge.

Klowden: Have you tried adding N,N-diethyl-m-toluamide (DEET) to the human stimuli or to the fractions? And if so, does it eliminate the response?

Geier: Yes. We have tested DEET plus a human hand or the skin extract, and most of the mosquitoes don't respond at all.

Curtis: Are you going to repeat these experiments with *Anopheles*?

Geier: No. I'm going to stick with *Ae. aegypti* because it responds to host stimuli quickly and is therefore the best species to use in this test. Another advantage is that the tests can be done at any time of the day.

Knols: We examined the possibility of using a Y-tube to study the behaviour of *Anopheles gambiae sensu stricto*, but the results were inconsistent. They did not fly as easily in this type of olfactometer as *Ae. aegypti*, which may explain the variability in responses to odours in this set-up.

Boeckh: We may have to invent other bioassays for *Anopheles*.

Galun: The optimal approach for developing a bioassay to identify human skin emanations would be to use *Ae. aegypti* for the assay because it is a 'better-behaved' mosquito. However, once the chemical nature of these attractants is recognized, their effect should be tested on *Anopheles*.

Dobrokhotov: In our experience, the genetic differences between *Aedes* and *Anopheles* mosquitoes are so great that we cannot extrapolate the results from one species to the other.

Knols: *Aedes albopictus* is becoming a problematic nuisance pest in southern Europe, and is much more closely related to *Ae. aegypti* than to anopheline mosquitoes. Therefore, it may be interesting to test behavioural responses of *Ae. albopictus* to the skin extract of Martin Geier. If effective, this extract may then be field tested as an odour bait for sampling this species in southern Europe or the USA.

Geier: It is important to investigate different species because it will be interesting to compare the behaviour of and the odours used by different mosquito species.

Takken: Another interesting aspect is that both *Ae. aegypti* and *An. gambiae* *s.s.* are anthropophilic, so one may expect that they share some of the chemical emanations produced by the human host.

Pickett: A comparable situation occurs in plant predation. Although the taxonomy of a group of pests predating one particular crop may be widely different, the actual compounds that they use are often similar. For example, the complex group of coleopterans and dipterans employ only about 20 key compounds and these are used by all the species. Therefore, although you are using the more convenient mosquito species, *Ae. aegypti*, you may still identify compounds that are used, possibly in different behavioural contexts, by *Anopheles* spp.

Geier: After removing L-lactic acid, the extract becomes completely unattractive, and the effect of removing other single components from the total extract has not yet been tested. However, it is also possible that small changes in the combinations of components in the mixture or in the concentration of distinct compounds can cause the different responses of different species.

Boeckh: A synergistic effect between these compounds raises problems for their isolation because they may only operate with other compounds. Indeed, some of these compounds may not have been detected if they did not have a synergistic effect with L-lactic acid, i.e. we already had a basic stimulus into which we could introduce other putative attractants. This synergism may be the key to understanding all the components of the host 'aroma'.

Kaissling: Can a similar synergism be obtained if L-lactic acid is replaced by CO_2?

Boeckh: That is a good question, and we are working on it. So far, we have found that, at lower CO_2 concentrations plus the extract, the attractiveness of the extract does not increase. However, we need to test this with lower concentrations of the extract.

Kaissling: Does a synergism exist between CO_2 and L-lactic acid?

Boeckh: Yes. A mixture of 4 mg L-lactic acid plus CO_2, such as is found in human breath, is highly attractive for mosquitoes. Indeed, the important point is that the compounds act in a synergistic way.

References

Acree F Jr, Turner RB, Gouck HK, Beroza M, Smith N 1968 L-Lactic acid: a mosquito attractant isolated from humans. Science 161:1346–1347

Bar-Zeev M, Maibach HI, Khan AA 1977 Studies on the attraction of *Aedes aegypti* (Diptera: Culicidae) to man. J Med Entomol 14:113–120

Price GD, Smith N, Carlson DA 1979 The attraction of female mosquitoes (*Anopheles quadrimaculatus* Say) to stored human emanations in conjunction with adjusted levels of relative humidity, temperature and carbon dioxide. J Chem Ecol 3:383–395

Smith CN, Smith N, Gouck HK et al 1970 L-Lactic acid as a factor in the attraction of *Aedes aegypti* (Diptera: Culicidae) to human hosts. Ann Entomol Soc Am 63:760–770

Introduction IV: coding mechanisms in insect olfaction

Hanna Mustaparta

Department of Zoology, University of Trondheim-AVH, N-7055 Dragvoll, Trondheim, Norway

The research on chemical communication in insects and insect olfaction has developed throughout the last 30 years and has been partly encouraged by the potential of using pheromones and other behaviourally modifying chemicals in insect pest control (Ridgeway et al 1990). The odours used by insects can be divided into pheromones and host odours, the latter being volatiles of plants or animal hosts. One central question in olfactory research is: which are the biological signals for the organism? This question has been best answered with respect to insect pheromones, and the research on olfactory coding mechanisms for pheromones is, therefore, ahead of research on host odour olfaction. Most studies on the identification of pheromones in insects are based on chemical analyses, combined with behavioural studies in the laboratory and in the field, as well as with electrophysiological recordings (Arn et al 1975). This has resulted in the identification of pheromone compounds in numerous insect species, including 4–500 lepidopterans (cf. Arn et al 1992), and the use of pheromones in insect pest control is now established. An important feature of insect pheromones is that they consist of several compounds, the ratio of which may be critical for the behavioural response (Cardé 1996, this volume). As well as acting intraspecifically as pheromones, the compounds may also have an interspecific effect. Interspecific interruption is common both in Lepidoptera and bark beetles, where one component interrupts the attraction of a related, sympatric species and thus contributes to reproductive isolation between them.

The olfactory system of insects

From the early studies on insect olfaction by Dietrich Schneider and his collaborators (Schneider 1992), we have learned the basic principles of the structure and function of insect olfactory organs (olfactory sensilla) that are located on the antennae (cf. Kaissling 1996, Steinbrecht 1996, this volume). The axons of the receptor neurons in the various sensilla form the antennal nerve, which enters the brain in the deutocerebrum. Here, the olfactory neurons terminate in the antennal lobe (cf. Masson & Mustaparta 1990, Boeckh & Tolbert 1993). The olfactory information is transmitted to central interneurons in the glomeruli containing the synapses (Anton 1996, this volume). These glomeruli can be individually identified in a species (Rospars 1983). Interestingly, the pheromone information is processed in one large, macroglomerular complex, which consists of two or more subcompartments and is male specific in some species. This complex is separated from the ordinary glomeruli in the antennal lobe, which are involved in the processing of plant odour information. Two main morphological types of antennal lobe neurons have been identified (cf. Boeckh & Tolbert 1993, Homberg et al 1989): local interneurons, which do not have axons and arborize in many if not all glomeruli; and projection neurons, or output neurons, which have one axon projecting to the protocerebrum. The projection neurons, which respond to pheromones, usually arborize in one glomerulus; whereas neurons responding to plant odours may be uniglomerular or multiglomerular (Kanzaki et al 1989, cf. Boeckh & Tolbert 1993). The local interneurons, which are γ-aminobutyric acid-(GABA)ergic, give inhibitory input to the projection neurons, suggesting that the excitatory responses of the projection neurons are due to disinhibition (Christensen & Hildebrand 1993). A direct input from the receptor neurons to projection neurons has also been demonstrated (Distler 1990). In the protocerebrum, the projection neurons terminate in two areas: the mushroom bodies associated with olfactory learning and the lateral lobe (cf. Homberg et al 1988, Hammer & Menzel 1995).

Receptor neuron responses

The conduction of the odour molecules to the receptor neuron membrane apparently consists of adsorption on the cuticular surface, diffusion through wall pores and transport by odorant-binding proteins (OBPs) in the receptor lymph (this volume: Steinbrecht 1996, Ziegelberger 1996). Here they interact with putative seven transmembrane helix membrane receptors, which activate G proteins (Breer et al 1988). Although the OBPs have a certain specificity (Ziegelberger 1996, this volume), the receptor proteins mainly determine the specificities of the receptor neuron. Via cascade reactions (Hekmat-Scafe & Carlson 1996, this volume) that involve inositol-1,4,5-trisphosphate as the second messenger, the influx of calcium and the subsequent opening of other cation channels elicits the receptor potential and nerve impulses (Stengl et al 1992, Stengl 1994, Breer et al 1990, Boeckhoff et al 1993). These potentials have been recorded extracellularly as summated receptor potentials (electroantennograms) from the whole antennae, as receptor potentials from single sensillum and as nerve impulses from single receptor neurons (cf. Kaissling 1996). In insects the responses to pheromones are always recorded as excitations. By screening the neurons for sensitivity to various relevant compounds at different concentrations, dose–response relationships are determined, which define the specificity of the olfactory receptor neuron. Studies of numerous insect species have shown that, in general, each pheromone receptor neuron responds best to one particular pheromone component (cf. Masson & Mustaparta 1990). It implies that information about pheromone mixtures are conveyed to the brain along labelled lines, and that a particular ratio of activity in the different labelled axons conveys the pheromone information to the brain.

Also of interest is that there seems to be no interactions between different pheromone components on the membrane receptors. By testing the receptor neurons for single compounds and mixtures, overlapping dose–response curves are found (Mustaparta 1990, Almaas & Mustaparta 1990). This also applies to compounds that interrupt pheromone attraction. Thus, an interspecific signal is not competitively blocking the pheromone receptors. Instead, the insects have specific receptor neurons that are activated by compounds which are not produced by their own species, but by sympatric species (Mustaparta et al 1977, 1980, Priesner 1979, Berg & Mustaparta 1996). This implies that the mechanism for interruption of pheromone attraction involves central nervous system integration of the two kinds of olfactory information.

Comparative studies of receptor neuron specificities in related species

Insect species of the same genus or subfamily have pheromone constituents or interspecific signal components in common, so it is of interest to compare the receptor neuron specificities in different species (Priesner 1979). In early studies of olfactory receptor neurons in bark beetles, it was shown that all receptor neurons tuned to the same compound showed similar dose–response relationships to various chemical analogues (Mustaparta 1990). This occurred independently of the message that the compound mediated, i.e. whether it either acted as a pheromone or food odour, or interrupted pheromone attraction. Similar results were obtained in another insect group of the subfamily heliothinae, with only one exception (Almaas & Mustaparta 1990, 1991, Almaas et al 1991, Berg et al 1995, Berg & Mustaparta 1996). In this group, all of five species studied possessed two major groups of receptor neurons, each tuned to one of two compounds (called A and B for simplicity) (Mustaparta 1996). Neurons from different species tuned to the major pheromone component A have similar specificities, as do those neurons that are tuned to compound B (in all but one species). This suggests that similar functional receptor neurons possess the same type of receptor proteins, a relationship which in general seems to have been preserved through evolution.

Projections of receptor neurons to the antennal lobe

Whether functionally different types of receptor neurons project to different glomeruli in the central nervous system has been of interest both in vertebrates and invertebrates (cf. Mustaparta 1996). This has been shown to be true for several insect species by functional characterization of receptor neurons and exposure to stain marking (Anton 1996, this volume). Projections of different types of pheromone receptor neurons in the turnip moth *Agrotis segetum* have mainly been found in separate macroglomerular complex lobes (Hansson et al 1992). In two species of heliothine moths, *Heliothis virescens* and *Helicoverpa zea*, similar studies have revealed that the receptor neurons mediating information about pheromones and interspecific signals project into different macroglomerular complex lobes (Hansson et al 1995, Almaas et al 1995), whereas those responding to the two essential pheromone components in one species (*H. virescens*) projected into the same macroglomerular complex compartment.

Antennal lobe output

The functional subdivision of the macroglomerular complex has been previously shown by different patterns of dendritic arborizations of antennal lobe neurons in *Manduca sexta*, *H. zea* and *H. virescens*. Intracellular recordings from projection neurons combined with staining showed that in *M. sexta* each of the neuron types responding to stimulation with one of the essential pheromone components arborized in one particular macroglomerular complex compartment, whereas the neurons responding to both components seemed to integrate the information via local interneurons (Christensen & Hildebrand 1987, Hansson et al 1991). In the two heliothine moths, information about pheromones and interspecific signals is to a large extent kept in separate pathways throughout the antennal lobe (cf. Mustaparta 1996). In *H. virescens* the projection neurons either respond to antennal stimulation with the major pheromone component A or to both essential components A and B, whereas a few projection neurons are activated exclusively by stimulation with the interspecific signal (Christensen et al 1991). In *H. zea*, where pheromone component B is the interspecific signal, almost all the projection neurons either respond to the major pheromone component A or to the interspecific signal B (Christensen et al 1995). However, in this species, the system is more complicated, since the B-responding neurons seem to mediate information also about the second principal pheromone component (Almaas et al 1991, Christensen et al 1991, Vickers et al 1991). Arborizations of the projection neurons in the macroglomerular complex of the two species correlate with the projections to the macroglomerular complex of the different receptor neuron types. Thus, the studies of pheromone receptor neurons and antennal lobe projection neurons, representing input and output of the antennal lobe, have demonstrated that the macroglomerular complex is functionally subdivided, and that each glomerular compartment is specified either for one or more pheromone components, or for interspecific signals.

Plant odours of hosts and non-hosts

Although the importance of host odours has been recognized for a long time in numerous herbivorous insect species, little is known about which compounds are used. Most electrophysiological studies of responses to host odours have been carried out using synthetic compounds, and have suggested that host odour receptor neurons are broadly tuned (cf. Masson & Mustaparta 1990). However, an olfactory receptor neuron may respond to many compounds at high concentrations that may not necessarily be of biological significance. Nevertheless, in some investigations biologically relevant compounds have been tested. In the cockroach different classes of receptor neurons each have a particular response spectrum, suggesting an across-fibre patterning mechanism for host odour reception (Sass 1978). The integration of this information has been studied by recordings from antennal lobe projection neurons, showing wide variations of the response spectra (cf. Boeckh et al 1987). It has been suggested that each of the ordinary glomeruli receive a complex, but unique, input from a subset of receptor neurons. Recently, progress has also been made on the identification of odours released from mammals that influence the behaviour of mosquitoes (this volume: Cork 1996, Pappenberger et al 1996, Takken 1996).

Gas chromatography linked to electrophysiology can be used to identify the host chemical signals for which receptor neurons have evolved. This technique was first employed for pheromone identification by recording electroantenno-graph responses to components of the pheromone gland content separated by gas chromatography (cf. Arn et al 1975), and it has subsequently been used for identifying host plant odours (Guerin et al 1983). Recordings from single receptor neurons have been performed in combination with the gas chromatography separation of pheromones (Wadhams 1990), and they are now being used for the identification of plant odour components that influence receptor neuron responses in several insect species (A. Wibe, A.-K. Borg-Karlson, T. Norin & H. Mustaparta, unpublished results 1995, E. N. Barata J. A. Pickett, H. Mustaparta & J. Araújo, unpublished results 1995). In these studies, the odours released from host and non-host plants are collected by drawing air around the plant material through an adsorbant. The odours trapped in the adsorbant are then used for stimulating the receptor neurons. When a single receptor neuron responds to a volatile mixture, a sample is injected into the gas chromatography column for separation. At the end of the column, half of the effluent proceeds to the gas chromatography detector and the other half exits the apparatus into an airstream blowing over the insect antennae. In this manner, the gas chromatograms and the stimulatory effect of each component on a single receptor neuron can be recorded simultaneously (Wibe & Mustaparta 1996).

In the pine weevil (*Hylobius abietis*) these experiments have revealed 20 different types of receptor neurons that respond to plant odours and can be classified according to the gas chromatography component which elicits the strongest stimulatory effect. Responses have been recorded both to mono-terpenes and sesquiterpenes, but a larger number of neurons respond to the more volatile monoterpenes. Selective or best (i.e. highest firing frequency) responses to single components were usually found, although some neurons also respond to several other components. All compounds that elicit responses in the same neuron are structurally related (e.g. they are either bicyclic, monocyclic or acyclic monoterpenes or sesquiterpenes [Wibe et al 1996]), implying that they may stimulate the same membrane receptors. It has been suggested that each receptor neuron has one type of membrane receptor, which has different affinities for the various compounds. The receptor specificities for optical isomers are also important, and the plant materials contain different ratios of, for example, monocyclic and bicyclic monoterpene enantiomers. The pine weevil seems to have receptor neurons tuned to one or to each of the optical configurations (A. Wibe, A.-K. Borg-Karlson, T. Norin & H. Mustaparta, unpublished results 1995), as does the pheromone receptor neurons in bark beetles (Mustaparta et al 1980). Another challenging problem that may be resolved by this technique is the importance of minor components in the plant volatile mixtures used by the insects. Selective responses to such components, sometimes hardly detected by gas chromatography, are common in the various species studied. When identified, they may provide important information as to which components influence the behaviour of the insects.

These analyses of plant odours have revealed that insects possess a large but restricted number of receptor neuron types which receive plant odour information. With some exceptions, most compounds are present in both host and non-host plants, suggesting that the ratio between them mediates the odour message of a host. Furthermore, the neurons seem to be specialized for either one or a few compounds of structural relatedness, indicating that each neuron may possess one or a few related types of membrane receptors. However, in order to draw definite conclusions about the mechanisms by which the plant odour information is mediated to the brain, it is necessary to determine the dose–response relationships for each compound, including optical isomers and mixtures. These analyses of compounds activating receptor neurons have to be followed up by behavioural studies in order to prove their importance as biological signals.

References

Almaas TJ, Mustaparta H 1990 Pheromone reception in the tobacco budworm moth *Heliothis virescens*. J Chem Ecol 4:1331–1347

Almaas TJ, Mustaparta H 1991 *Heliothis virescens*: response characteristics of receptor neurons in *sensilla trichodea* type 1 and type 2. J Chem Ecol 17:953–972

Almaas TJ, Christensen TA, Mustaparta H 1991 Chemical communication in heliothine moths. I. Antennal receptor neurons encode several features of intra- and interspecific odorants in the corn earworm moth *Helicoverpa zea*. J Comp Physiol A 169:249–258

Almaas TJ, Amundgaard AK, Berg BG, Mustaparta H 1995 Functional subdivision of the macroglomerular complex in different heliothine moth species. Nervous systems and behaviour. Proceedings of the International Congress of Neuroethology, Thieme Medical Publishers, New York, p 397

Anton S 1996 Central olfactory pathways in mosquitoes and other insects. In: Olfaction in mosquito–host interactions. Wiley, Chichester (Ciba Found Symp 200) p 184–196

Arn H, Städler E, Rauscher S 1975 The electroantennographic detector—a selective and sensitive tool in the gas chromatographic analysis of insect pheromones. Z Naturforsch 30:722–725

Arn H, Tóth M, Priesner E 1992 List of sex pheromones of Lepidoptera and related attractants, 2nd edn. International Organization of Biological Control, Montfavet

Berg BG, Mustaparta H 1996 The significance of major pheromone components and interspecific signals as expressed by receptor neurons in the oriental budworm moth, *Helicoverpa assulta*. J Comp Physiol A 177:683–694

Berg BG, Tumlinson JH, Mustaparta H 1995 Chemical communication in heliothine moths. IV. Receptor neuron responses to pheromone compounds and formate analogues in the male tobacco budworm moth *Heliothis virescens*. J Comp Physiol A 177:527–534

Boeckh J, Tolbert LP 1993 Synaptic organization and development of the antennal lobe in insects. In: Johnson JE (ed) Microscopy research and technique, part II: Olfactory centers in the brain. p 260–280

Boeckh J, Ernst KD, Selsam P 1987 Neurophysiology and neuroanatomy of the olfactory pathway in the cockroach. Ann N Y Acad Sci 510:39–43

Boeckhoff I, Seifert E, Göggerle S, Lindeman M, Krüger BW, Breer H 1993 Pheromone-induced second messenger signaling in insect antennae. Insect Biochem Mol Biol 23:757–762

Breer H, Raming K, Boeckhoff I 1988 G-proteins in the antennae of insects. Naturwiss 75:627

Breer H, Boeckhoff I, Tareilus E 1990 Rapid kinetics of second messenger formation in olfactory transduction. Nature 345:65–68

Cardé RT 1996 Odour plumes and odour-mediated flight in insects. In: Olfaction in mosquito–host interactions. Wiley, Chichester (Ciba Found Symp 200) p 54–70

Christensen TA, Hildebrand JG 1987 Male-specific, sex pheromone-selective projection neurons in the antennal lobes of the moth *Manduca sexta*. J Comp Physiol A 160:553–569

Christensen TA, Hildebrand JG 1993 Local interneurons and information processing in the olfactory glomeruli of the moth *Manduca sexta*. J Comp Physiol A 173:385–399

Christensen TA, Mustaparta H, Hildebrand JG 1991 Chemical communication in heliothine moths. II. Central processing of intra- and interspecific olfactory messages in the male corn earworm moth *Helicoverpa zea*. J Comp Physiol A 169:259–274

Christensen TA, Mustaparta H, Hildebrand JG 1995 Chemical communication in heliothine moths. VI. Parallel pathways for information processing in the macro-glomerular complex of the male tobacco budworm moth *Heliothis virescens*. J Comp Physiol A 177:545–557

Cork A 1996 Olfactory basis of host-location by mosquitoes and other haematophagous Diptera. In: Olfaction in mosquito–host interactions. Wiley, Chichester (Ciba Found Symp 200) p 71–88

Distler P 1990 GABA-immunohistochemistry as a label for identifying types of local interneurons and their synaptic contacts in the antennal lobes of the American cockroach. Histochemistry 93:617–626

Guerin PM, Städler E, Buser HR 1983 Identification of host plant attractants for the carrot fly *Psila rosae*. J Chem Ecol 9:843–861

Hammer M, Menzel R 1995 Learning and memory in the honeybee. J Neurosci 15:1617–1630

Hansson BS, Christensen TA, Hildebrand JG 1991 Functionally distinct subdivisions of the macroglomerular complex in the antennal lobe of the male sphinx moth *Manduca sexta*. J Comp Neurol 312:264–278

Hansson BS, Ljungberg H, Hallberg E, Löfstedt C 1992 Functional specialization of olfactory glomeruli in a moth. Science 256:1313–1315

Hansson BS, Almaas TJ, Anton S 1995 Chemical communication of heliothine moths. V. Antennal lobe projection patterns of pheromone-detecting olfactory receptor neurons in the male *Heliothis virescens* (Lepidoptera: Noctuidae). J Comp Physiol A 177:535–543

Hekmat-Scafe DS, Carlson JR 1996 Genetic and molecular studies of olfaction in *Drosophila*. In: Olfaction in mosquito–host interactions. Wiley, Chichester (Ciba Found Symp 200) p 285–301

Homberg U, Montague RA, Hildebrand JG 1988 Anatomy of antenno-cerebral pathways in the brain of the sphinx moth *Manduca sexta*. Cell Tissue Res 254:255–281

Homberg U, Christensen TA, Hildebrand JG 1989 Structure and function of the deutocerebrum in insects. Ann Rev Entomol 34:477–501

Kaissling K-E 1996 Peripheral mechanisms of pheromone reception. Chem Senses 20:132

Kanzaki R, Arbas EA, Strausfeld NJ, Hildebrand JG 1989 Physiology and morphology of projection neurons in the antennal lobe of the male moth *Manduca sexta*. J Comp Physiol A 165:427–453

Masson C, Mustaparta H 1990 Chemical information processing in the olfactory system of insects. Physiol Rev 70:199–245

Mustaparta H 1990 Evolutionary aspects of chemical communication in insects: pheromone perception. In: Døving KBD (ed) Proceedings from the Xth International symposium on olfaction and taste, Oslo, p 164–174

Mustaparta H 1996 Central mechanisms of pheromone information processing. Chem Senses 20:209

Mustaparta H, Angst ME, Lanier GN 1977 Responses of single receptor cells in the pine engraver beetle *Ips pini* Say (Coleoptera: Scolytidaeto) to its aggregation pheromone, ipsdienol, and the aggregation inhibitor, ipsenol. J Comp Physiol A 121:343–347

Mustaparta H, Angst ME, Lanier GN 1980 Receptor discrimination of enantiomers of the aggregation pheromone, ipsdienol, in two species of *Ips*. J Chem Ecol 6:689–701

Pappenberger B, Geier M, Boeckh J 1996 Responses of antennal olfactory receptors in the yellow fever mosquito *Aedes aegypti* to human body odours. In: Olfaction in mosquito–host interactions. Wiley, Chichester (Ciba Found Symp 200) p 254–266

Priesner E 1979 Specificity studies on pheromone receptors of noctuid and tortricid lepidoptera. In: Ritter FJ (ed) Chemical ecology: odour communication in animals. Elsevier, New York, p 57–71

Ridgeway RL, Silverstein RM, Inscoe May N 1990 Behavior-modifying chemicals for insect management. Applications of pheromones and other attractants. Marcel Dekker, New York

Rospars JP 1983 Invariance and sex specific variations of the glomerular organization in the antennal lobes of a moths. J Comp Neurol 220:80–96

Sass H 1978 Olfactory receptors on the antennae of Periplantea: response constellations that encode food odours. J Comp Physiol A 128:227–233

Schneider D 1992 100 years of pheromone research. An essay on Lepidoptera. Naturwissenschaften 79:241–250

Steinbrecht RA 1996 Structure and function of insect olfactory sensilla. In: Olfaction in mosquito–host interactions. Wiley, Chichester (Ciba Found Symp 200) p 158–177

Stengl M 1994 Inositol-triphosphate-dependent calcium currents precede cation currents in insect olfactory receptor neurons *in vitro*. J Comp Physiol A 174:187–194

Stengl M, Hatt H, Breer H 1992 Peripheral processes in insect olfaction. Ann Rev Physiol 54:665–681

Takken W 1996 Synthesis and future challenges: the response of mosquitoes to host odours. In: Olfaction in mosquito–host interactions. Wiley, Chichester (Ciba Found Symp 200) p 302–320

Vickers, NJ, Christensen TA, Mustaparta H, Baker TC 1991 Chemical communication in heliothine moths. III. Flight behavior of male *Helicoverpa zea* and *Heliothis virescens* in response to varying ratios of intraspecific and interspecific sex pheromone compounds. J Comp Physiol A 169:275–280

Wadhams LJ 1990 The use of coupled gas chromatography: electrophysiological techniques in the identification of insect pheromones. In: McCaffery AR, Wilson ID (eds) Chromatography and isolation of insect hormones and pheromones. Plenum Press, New York, p 289–298

Wibe A, Mustaparta H 1996 Encoding of plant odours by receptor neurons in the pine weevil *Hylobius abietis* studied by linked gas chromatography–electrophysiology. J Comp Physiol, in press

Wibe A, Borg-Karlson A-K, Norin T, Mustaparta H 1996 Identification of plant volatiles which activate the same receptor neurons in the pine weevil, *Hylobius abietis*. J Entomol Appl, in press

Ziegelberger G 1996 The multiple role of the pheromone-binding protein in olfactory transduction. In: Olfaction in mosquito–host interactions. Wiley, Chichester (Ciba Found Symp 200) p 267–280

Structure and function of insect olfactory sensilla

Rudolf Alexander Steinbrecht

Max-Planck-Institut für Verhaltensphysiologie, D-82319 Seewiesen, Germany

Abstract. Olfactory sensilla show a large diversification of sensillum types even in the same species. Thus, double-walled and single-walled sensilla with highly different wall pores are usually found on the same antenna, and these may appear in the form of long slender hairs, pore plates or pit pegs. The selective constraints leading to this diversification are evident only in a few cases, e.g. the demand for extreme sensitivity in moth pheromone communication supported the evolution of long sensilla trichodea with high efficiency of capturing odour molecules. The structural diversity continues with the odorant-binding proteins (OBPs) in the sensillum lymph surrounding the sensory dendrites. These proteins may be subdivided into pheromone-binding proteins and two classes of general odorant-binding proteins according to their primary sequence. Different sensilla of the same morphological type may contain different OBPs of the same or of different subclasses. However, OBPs of different subclasses are not co-localized in the same individual sensory hair. The presence of a given OBP is related more to the functional specificity of the receptor cells than to the morphological type of the sensillum, suggesting a role of OBPs in stimulus recognition.

1996 Olfaction in mosquito–host interactions. Wiley, Chichester (Ciba Foundation Symposium 200) p 158–177

The complex diversity of sensillar structures on insect antennae has puzzled scientists since the last century. The advent of electrophysiological recording methods enabled this diversity to be explained in terms of function. The grouping of receptor cells in separate entities, i.e. the sensilla (Fig. 1), enables a reproducible recording to be made from defined receptor cells. These recordings are more difficult with sensory epithelia, such as the vertebrate olfactory mucosa.

Extensive literature exists on the types of sensilla encountered on the antennae of nearly every insect taxon. The scanning electron microscope is useful for obtaining the first survey of external structure and distribution, but important aspects of innervation and modality-specific structures require transmission electron microscopy for clarification. Since the heroic days of Slifer et al (1959), considerable progress in specimen preparation and structural

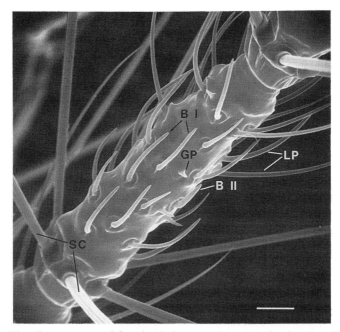

FIG. 1. Flagellar segment of female *Aedes aegypti* as seen in the scanning electron microscope. bI, blunt-tipped I trichodea; bII, blunt-tipped II trichodea; GP, grooved peg; LP, base of long pointed-tipped trichodea; SC, sensilla chaetica. Bar = 10 μm (from McIver 1982).

resolution has been made, e.g. by the introduction of cryofixation techniques (Steinbrecht 1980).

In this paper I will not deal extensively with the well-known general principles of sensillar structure. Rather I will emphasize the structural, functional and chemical diversity of sensilla, using moths as an example because of the ample data in this area obtained from electron microscopy, electrophysiology and protein biochemistry. The remarkable ability of male moths to find their mates from a distance by using airborne pheromone signals has fascinated biologists for over a hundred years, has led to the characterization of the first insect pheromone and has brought about the first electrophysiological recordings of insect olfactory sense organs (reviewed in Schneider 1992). The discovery of odorant-binding proteins (OBPs) in moth antennae (Vogt & Riddiford 1981) has opened up another promising line of research.

Sensillar structure in relation to olfactory function

The uniform bauplan of all insect sensilla—one or several primary receptor cells and usually three auxiliary cells in combination with a special cuticular apparatus (seta)—is well known and can be explained by a similar development from epidermal mother cells (reviewed in Keil 1992). Modality-specific structures in the sensory dendrites and the cuticular apparatus are remarkably conserved throughout the insect orders. Consequently, predictions of sensillar function can be made from the morphological data (Fig. 2). (All modalities are reviewed in Altner & Prillinger [1980] and Zacharuk [1985]; mechanosensitive and olfactory sensilla are reviewed in Keil & Steinbrecht [1984]; olfactory, gustatory, hygro/thermosensitive sensilla are reviewed in Steinbrecht [1984];

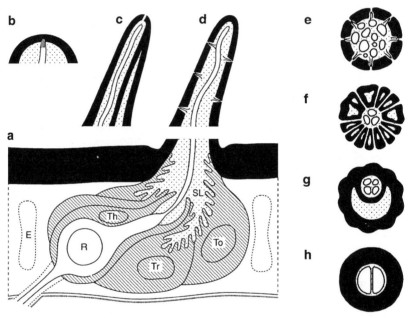

FIG. 2. Schematic representation of insect sensilla. (a) The cellular organization is rather uniform regardless of the specific receptor modality. One or several bipolar receptor cells (R) send an axon to the brain and a dendrite (white profiles in b–h) to the peripheral region of stimulus uptake which displays a specific cuticular apparatus. Three auxiliary cells (Th, Tr, To) surround the receptor cell(s) and border the sensillum lymph cavity (SL). The cuticle is black and undifferentiated epidermal cells (E) are shown as white shapes. Modality-specific specializations of the cuticular apparatus are shown in longitudinal section (b–d) and in cross-section (e–h). (b) Mechanosensitive campaniform sensillum. The dendrite displays a tubular body where it is compressed by deformations of the cuticle. (c, g) Gustatory sensillum. The dendrites of usually four taste receptor cells are exposed via a terminal pore. (d, e) Olfactory sensillum (single walled). The dendrites of several receptor cells responding to different odour qualities are accessible through wall pores (compare Figs 3 and 4). (f) Double-walled olfactory sensillum with different structure of wall pores (compare Fig. 5). (h) Poreless, thermo/ hygrosensitive sensillum with two hygroreceptive dendrites (ending of thermoreceptor not shown at this level) (modified from Steinbrecht 1992).

thermo/hygrosensitive sensilla are reviewed in Altner & Loftus [1985]; and pheromone-sensitive olfactory receptors are reviewed in Steinbrecht 1987.)

Olfactory sensilla, for example, are characterized by numerous pores in the sensillar wall, which are the supposed site of entry of odour molecules. The wall pores may be of the pore–tubule type, such as in single-walled sensilla, or of the spoke–channel type, such as the pores of the complex double-walled, wall pore sensilla (Figs 3–5). Olfactory sensilla may be hair shaped and up to $500\,\mu m$ long, such as the pheromone-sensitive sensilla trichodea of the moth *Manduca sexta* (Keil 1989), or they may be short pegs less than $10\,\mu m$ in length, such as the grooved pegs (A_3-type sensilla) of the yellow fever mosquito *Aedes aegypti* (McIver 1974, Cribb & Jones 1995) (Fig. 1). In honey bees and other Hymenoptera the hairs are represented by pore plates that flush with the antennal surface. Olfactory sensilla may also be hidden in pits that are connected to the environment only by small apertures. Examples of these include the sensilla in the sacculus on dipteran antennae or the sensilla coeloconica of the locust (reviewed in Kaissling 1971, Altner & Prillinger 1980). As a rule, the functional diversity of insect olfactory sensilla is much greater than the diversity of morphological types, and sensilla with receptors of a uniform specificity are exceptions (e.g. the pheromone-sensitive long sensilla trichodea of male silkmoths, see below).

There are usually several types of olfactory sensilla on the antenna of a given species. These are more frequent and more diverse than the sensilla serving the other modalities. For example, moths have the sensilla trichodea, which are long, thick-walled hairs with unbranched dendrites (Fig. 3); the sensilla basiconica, which are short, thin-walled pegs with branched dendrites (Fig. 4); and the sensilla coeloconica, which are grooved pit pegs with a double wall and spoke channels (Fig. 5) (Keil & Steinbrecht 1984). These types are found with species-specific distribution patterns. In males of the large saturniid silk moth, *Antheraea polyphemus,* an 'odour sieve' is formed by rows of 50 000 long sensilla trichodea on large feathered antennae. These sensilla are responsible for the detection of the female sex pheromone, and they are absent on the female antenna (reviewed in Steinbrecht 1987 and Kaissling 1987). Sensilla basiconica occur in both sexes and their receptor cells respond to plant and other general odours (Schneider et al 1964, Kafka 1987). The function of sensilla coeloconica was unknown in moths, but recently receptor cells responding to mulberry leaf headspace and to short-chain fatty acids have been found in sensilla coeloconica of *Bombyx mori* (B. Pophof, unpublished results 1995). Morphological subtypes can be distinguished on the basis of the size and position of the setae and also on the basis of receptor cell numbers, e.g. long and medium-sized sensilla trichodea, and large and small sensilla basiconica (Steinbrecht 1973, Meng et al 1989). Functional verification of these subtypes by electrophysiology is, however, lacking in most instances.

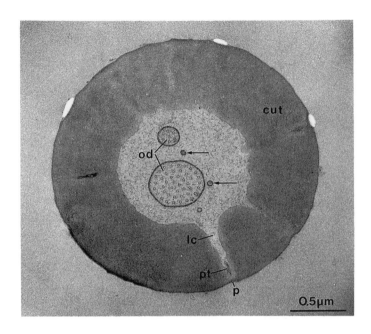

FIG. 3. Cross-section of a thick-walled olfactory sensillum trichodeum of male *Bombyx mori*. The cuticular wall (cut) is perforated by narrow pores (p) each leading into several pore tubules (pt) which are surrounded by a wider liquor channel (lc) filled with sensillum lymph such as in the hair lumen. The outer dendritic segments (od) of the two pheromone-sensitive receptor cells have different calibres and only occasionally show some branching (arrows) (from Keil & Steinbrecht 1984).

In mosquitoes some 90% of the antennal sensilla have an olfactory function (reviewed in McIver 1982, Sutcliffe 1994) (Fig. 1). There are sensilla trichodea of various length and form (McIver 1978, Muir & Cribb 1994) which contain receptor cells for compounds related to oviposition sites (Bentley et al 1982), nectar sources (Davis 1977) and certain repellents (Lacher 1971, Davis & Rebert 1972). Receptor cells for behaviourally active host attractants, such as L-lactic acid, have been found only with the short A_3-type sensilla, which belong to the double-walled type of multiporous sensilla (McIver 1974, Bowen 1995, Cribb & Jones 1995, see also Pappenberger et al 1996, this volume). The sensilla coeloconica of *A. aegypti*, on the other hand, are aporous pegs with an ultrastructure similar to known thermo/hygroreceptive units (McIver 1973). This example shows that the old nomenclature may be misleading with regard to modality-specific structures and sensillar function. In addition to those on the antennae, olfactory sensilla are also found on the maxillary palps: the 'capitate pegs' are a uniform population of single-walled multiporous sensilla that respond mainly to CO_2. One of the dendrites displays extensive

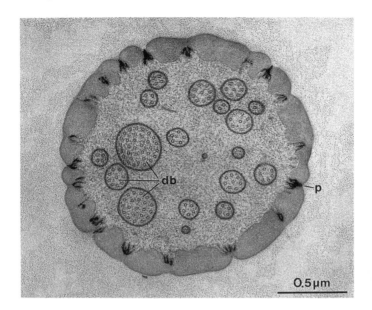

FIG. 4. Cross-section of thin-walled sensillum basiconicum of *Bombyx mori* displaying a high density of pores (p) each leading into numerous pore tubules. The outer segments of three olfactory receptor cells form many dendritic branches (db) (from Keil & Steinbrecht 1984).

lamellation as do those of other CO_2-sensitive sensilla (McIver 1972). The different sensillum types show characteristic differences in sensillum numbers and distribution patterns between the sexes, between culicine and anopheline mosquitoes, and even more drastic differences between blood and nectar feeders (McIver 1982).

Distribution of odorant-binding proteins and olfactory function

OBPs were first discovered in moths (Vogt & Riddiford 1981), but they now appear to be a consistent feature of insect chemosensory sensilla and are being found in other orders, such as Diptera (McKenna et al 1994, Pikielny et al 1994), Phasmidoptera (Tuccini et al 1996) and Heteroptera (Dickens et al 1995). Similar proteins are also present in taste sensilla (Ozaki et al 1995). OBPs are soluble proteins of low molecular weight (about 17 kDa), they have an acidic isoelectric point (pH = 4.7) and they can be divided into three subclasses according to their amino acid sequence: the pheromone-binding proteins (PBPs), which are particularly abundant in male moths; and two classes of general odorant-binding proteins (GOBP1 and GOBP2), which are found in equal amounts in both sexes (reviewed in Pelosi & Maida 1995).

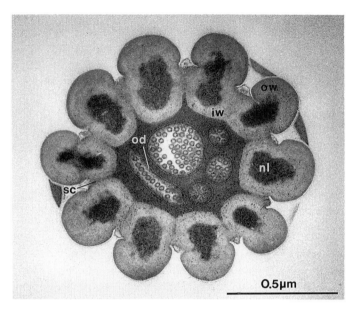

FIG. 5. Cross-section of double-walled sensillum coeloconicum of *Bombyx mori*. Five dendrites (od) are surrounded by an inner (iw) and outer (ow) cuticular wall, the latter being longitudinally grooved. Radial spoke channels (sc) lead from the inner lumen into the longitudinal grooves. The inner lumen as well as the non-innervated lumina (nl) between the cuticular walls are filled with extremely dense sensillum lymph and extracellular material (from Keil & Steinbrecht 1984).

Immunolabelling studies have shown that OBPs are biosynthesized in the sensillar auxiliary cells and secreted into the sensillum lymph (Fig. 6) (Steinbrecht et al 1992). Biosynthesis continues through adult life and is balanced by simultaneous endocytosis and breakdown (Vogt et al 1989, Steinbrecht et al 1992). The functional significance of OBPs was unclear for many years, but it now appears that they may serve as transporters and deactivators of the stimulus molecules (Ziegelberger 1995, see also Ziegelberger 1996, this volume). In addition, they may play a role in stimulus recognition (see below).

The number of fully sequenced OBPs is increasing rapidly. In particular, PBPs are very diverse and in moths some have as little as 29% amino acid identity (Pelosi & Maida 1995). It has already been noted by Vogt et al (1991) that those species which use the most different pheromones also have the most different PBPs (e.g. *A. polyphemus* and *Lymantria dispar*). Comparative immunolabelling studies from my laboratory confirm this notion. Antiserum raised against *A. polyphemus* PBP also cross-reacts strongly with the sensillum lymph of long sensilla trichodea in *Antheraea pernyi*, *B. mori* and *M. sexta*, i.e. all species that use C-16 alkyl compounds for pheromone communication. This

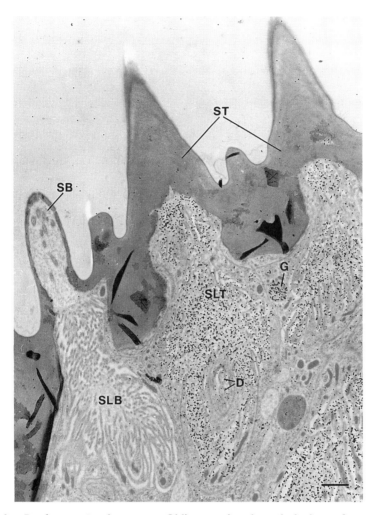

FIG. 6. *Bombyx mori* male antenna. Oblique section through the base of two sensilla trichodea (ST) and one sensillum basiconicum (SB). The sensillum lymph of the pheromone-sensitive sensilla trichodea (SLT) is labelled by anti-PBP antiserum, whereas that of the sensillum basiconicum (SLB) is not. D, dendritic outer segments surrounded by labelled inner sensillum lymph cavity; G, heavily labelled granule in auxiliary cell. Bar = 1 μm (from Steinbrecht et al 1992).

antiserum cross-reacts only weakly with PBP of sensilla trichodea of *L. dispar* and *Dendrolimus kikuchii* (Table 1). Taxonomically, the lasiocampid *Dendrolimus* is closer to *Bombyx* than is the sphingid *Manduca*. However, the pheromone components of *Dendrolimus* are less similar to those of *Bombyx* than are those of *Manduca* (all *Dendrolimus* species studied so far use C-12

TABLE 1 Cross-reactivity of immunolabelling with antiserum against *Antheraea polyphemus* pheromone binding protein (a-PBP[Apol]) in pheromone-sensitive long sensilla trichodea of male moths

Species[a]	Family	Pheromone components[b]		a-PBP(Apo) labelling[c]
Antheraea polyphemus	Saturniidae	E6, Z11-**16**:Ac	E6, Z11-**16**:Ald	+ + + +
Antheraea pernyi	Saturniidae	E6, Z11-**16**:Ald	E6, Z11-**16**:Ac	+ + + +
Bombyx mori	Bombycidae	E10, Z12-**16**:OH	E10, Z12-**16**:Ald	+ + + +
Dendrolimus kikuchii	Lasiocampidae	Z5, E7-**12**:Ac[d]	Z5, E7-**12**:OH[d]	+
Manduca sexta	Sphingidae	E10, E12, Z14-**16**:Ald	E10, Z12-**16**:Ald	+ + + +
Lymantria dispar	Lymantriidae	cis-7,8-epo-2Me-**18**:Hy		+

[a]Species are arranged according to taxonomic proximity.
[b]Main pheromone components are indicated in abbreviated formulation indicating chain length of alkyl group (bold face), location and type of double bonds and terminal functional group.
[c] + + + +, strong cross-reactivity; +, weak cross-reactivity; i.e. close to background.
[d]Pheromone components of *Dendrolimus punctatus*. Those of *D. kikuchii* are not yet known, but all *Dendrolimus* species identified so far use C-12 alkyl compounds.

alkyl compounds as pheromones). Likewise, the pheromone of *Lymantria* is a methylated C-18 compound with an epoxy group, which differs from the C-16 acetates and alcohols used by *Antheraea*, *Bombyx* and *Manduca* (Arn et al 1992). These results corroborate the notion that PBPs are tuned to their specific ligands and thus may contribute to stimulus recognition, an idea that gets further support from recent binding studies (reviewed in Prestwich et al 1995).

The term GOBP had been proposed, based on the amino acid sequence homology and equal occurrence of these proteins in males and females, in anticipation of their association with a different set of olfactory sensilla (Vogt et al 1991). This has now been confirmed by immunolabelling experiments: antiserum raised against *A. polyphemus* GOBP2 labels predominantly sensilla basiconica that contain receptor cells for plant volatiles and other 'general odours' (Laue et al 1994, Steinbrecht et al 1995) (Figs 7 and 8).

To date, PBP and GOBP2 have not been found to be co-localized in the same sensillum (Fig. 8). However, the distribution of PBP and GOBPs in

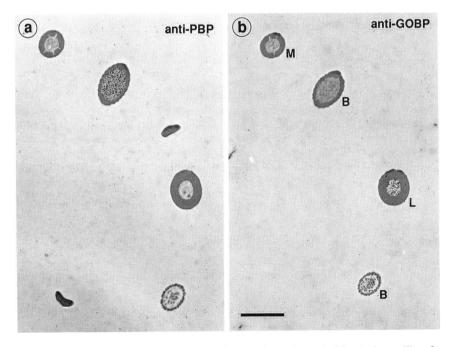

FIG. 7. *Bombyx mori* female. Consecutive sections through identical sensilla after labelling with anti-PBP (a) and anti-GOBP2 (b) antisera. Except for the medium-sized sensillum trichodeum (M), which is not labelled by either serum, all other sensilla show complementary labelling with anti-PBP antiserum (such as the upper sensillum basiconicum [B]) or with anti-GOBP2 antiserum (such as the sensillum trichodeum [L] and the lower sensillum basiconicum [B]). Bar = 3µm (from Steinbrecht et al 1995).

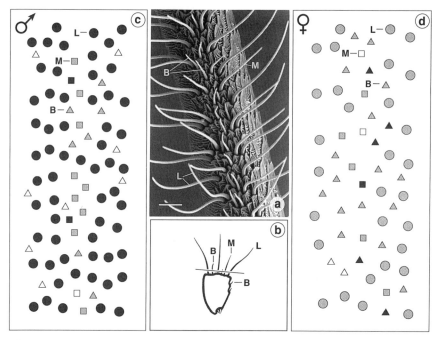

FIG. 8. *Bombyx mori.* (a) Scanning electron micrograph of a male antennal branch. Bar = 20 μm. (b) Schematic drawing of branch cross-section showing sectioning plane of maps. (c, d) Semischematic maps showing the distribution, type and labelling pattern of all sensilla encountered in a large section along a male (c) and female (d) antennal branch. B, sensillum basiconicum; L, long sensillum trichodeum; M, medium-sized sensillum trichodeum. Black symbols, sensilla labelled by anti-PBP antiserum; hatched symbols, sensilla labelled by anti-GOBP2 antiserum; open symbols, same sensillum types but not labelled by either serum (from Steinbrecht et al 1995).

moths is not correlated strictly with the morphological types of sensilla trichodea and sensilla basiconica, respectively. In several species PBP rather than GOBP is found in some sensilla basiconica, and in *B. mori* GOBP rather than PBP is found in some medium-sized sensilla trichodea (Fig. 8). Moreover, this was found in both sexes, suggesting that silkmoth females are also sensitive to their pheromone (Laue et al 1994). Female moths, in general, are believed to be incapable of smelling their own pheromone. Only a few exceptions to this have been reported in noctuids and tortricids (for references see Steinbrecht et al 1995).

For most of these cases of atypical localization of an OBP, electrophysiological verification of the sensillar function is still pending. In one case, however, it has been established that the presence of a particular class of OBP is related directly to the functional specificity of a sensillum irrespective of its morphological type. This is in the long sensilla trichodea of the silk moth *B.*

mori, which in males respond uniformly to pheromone (one cell in each sensillum to bombykol and the other one to bombykal) but in females have a different specificity: in each sensillum one cell responds to linalool and the other one to benzoic acid (these compounds are not pheromones but plant odours that are present in mulberry leaves). Concomitantly with these physiological results, our immunocytochemical results show that all male long sensilla trichodea contain PBP, whereas all female long sensilla trichodea contain GOBP2 (Figs 7 and 8) (Laue et al 1994, Steinbrecht et al 1995). There are no significant morphological differences between male and female long sensilla trichodea, and they are found in the same position on the antennae (Steinbrecht 1973).

Antisera against GOBP1 are not yet available, so the localization of this subclass of binding proteins is unknown. Possible candidates in *Bombyx* are the sensilla coeloconica and some sensilla basiconica that are neither labelled by anti-PBP nor by anti-GOBP2 antisera (Steinbrecht et al 1995).

Sensillum types and evolution

The main features of modality-specific structures are well conserved throughout all insect orders, suggesting that they must have evolved prior to the radiation of the present orders. Results and hypotheses on the evolution of the various forms of pore plates are available for the strongly radiating group of the lamellicorn beetles (Meinecke 1975). It is particularly puzzling why two different types of wall pore sensilla are used for olfaction in all insect groups (except in Collembola): (1) the single-walled sensilla, which have pores with pore tubules for stimulus transport; and (2) the double-walled sensilla, which use spoke channels devoid of pore tubules (Steinbrecht 1969, Altner & Prillinger 1980). Studies of *Periplaneta* have suggested that spoke channels filled with sensillum lymph may be advantageous for the transport of more polar stimuli, such as short-chain fatty acids, whereas pore tubules may have been adapted for the transport of apolar stimuli, such as long-chain aliphatic esters and alcohols (Altner et al 1977). The physiological data, however, are still too scarce and contradictory to allow generalizations to be made.

Almost nothing is known about the physico-chemical nature of pore tubules and spoke channels, nor is there any direct evidence of how the pores are involved in stimulus transport. For a detailed discussion of the possible stimulus transport pathways and mechanisms see the review of Steinbrecht (1987). It suffices to mention that the presence of OBPs in the sensillum lymph has solved at least one problem, i.e. the question of how can a highly apolar molecule, such as a pheromone compound, pass through an aqueous fluid to reach the dendritic membrane (Ziegelberger 1996, this volume). In this context it would be extremely interesting to know whether the double-walled sensilla also contain OBPs, for example GOBP1, or whether these sensilla employ another stimulus transport mechanism.

As far as the single-walled multiporous sensilla are concerned, the classification into sensilla trichodea and sensilla basiconica has probably been overemphasized in the past. In order to guide our ideas in a different direction, I would like to propose the concept of a continuum of single-walled multiporous sensilla with the typical sensilla trichodea and sensilla basiconica just emerging as peaks at both ends of this continuum. It is likely that most criteria used for the discrimination of sensilla trichodea and sensilla basiconica are actually interdependent. This is easy to understand for sensillum length and wall thickness: longer hairs need a thicker wall. The inverse correlation of pore numbers per surface area with sensillum size can be explained, because the total numbers of pores in sensilla trichodea and sensilla basiconica in *Bombyx* is roughly the same (Steinbrecht 1973). The dendritic branching that is present in sensilla basiconica is also found in the tip region of many sensilla trichodea. It is possible that a minimum outer dendritic membrane area is required to obtain an outer membrane resistance compatible with electrophysiological function. Shorter sensilla basiconica might have employed branching of the dendrite in order to compensate for a deficiency in length. This hypothesis could be checked by using voltage-clamp and current-clamp techniques, as were recently used for measuring the electrical parameters of long sensilla trichodea in *Bombyx* (Redkozubov 1995).

Selection must have favoured long sensilla for maximizing the sensitivity of pheromone detection at threshold levels. Both theoretical considerations of convective diffusion and experiments with radiolabelled pheromones have shown convincingly that it is the surface of the evenly spaced long sensilla trichodea which attracts the majority of stimulus molecules passing the antenna, despite their surface being only a minor fraction of the total antennal surface (less than 13% in *B. mori*) (Adam & Delbrück 1968, Steinbrecht & Kasang 1972, Kanaujia & Kaissling 1985). Thus, long sensilla trichodea are necessary to build an efficient odour sieve. However, even in moths, the morphological differences between sensilla trichodea and sensilla basiconica are gradual rather than clear cut. The existence of intermediate types has been known for a long time (Steinbrecht 1973).

There are also constraints that favour short sensilla, e.g. sensitivity to damage and contamination. In bark beetles, for example, the thin-walled sensilla basiconica protrude only slightly from the surface of the antennal club, and the whole antenna can be folded back into a groove on the head when the animal is boring into a tree (Dickens & Payne 1978). In hymenopterans there is a tendency to reduce the sensory hair to a pore plate, which makes cleaning easier (Walther 1981) but at the expense of molecule catch (see Table 5 in Kaissling 1971).

Our knowledge of factors that influence the evolution of sensillar diversity is still meagre. Eventually, however, we may learn how insects as tiny as a mosquito or an ant, which live under conditions where humans may perish with thirst, can successfully conserve their few milligrams of body water whilst exposing sensory processes to the environment and detecting the stimuli that guide them to their food, prey, host or mate.

Acknowledgements

I thank S. McIver and J Med Entomol for the permission to use Fig. 1. I also thank my colleagues and friends K.-E. Kaissling, T. A. Keil, M. Laue and G. Ziegelberger for critical reading of the manuscript, B. Müller for technical help and R. Alton for preparing the manuscript.

References

Adam G, Delbrück M 1968 Reduction of dimensionality in biological diffusion processes. In: Rich A, Davidson N (eds) Structural chemistry and molecular biology: a volume dedicated to Linus Pauling by his students, colleagues, and friends. Freeman, San Francisco, p 198–215

Altner H, Loftus R 1985 Ultrastructure and function of insect thermo- and hygroreceptors. Ann Rev Entomol 30:273–295

Altner H, Prillinger L 1980 Ultrastructure of invertebrate chemo-, thermo-, and hygroreceptors and its functional significance. Int Rev Cytol 67:69–139

Altner H, Sass H, Altner I 1977 Relationship between structure and function of antennal chemoreceptive, hygroreceptive, and thermoreceptive sensilla in *Periplaneta americana*. Cell Tissue Res 176:389–405

Arn H, Tóth M, Priesner E 1992 List of sex pheromones of Lepidoptera and related attractants, 2nd edn. International Organization for Biological Control, Montfavet

Bentley MD, McDaniel IN, Davis EE 1982 Studies of 4-methylcyclohexanol: an *Aedes triseriatus* (Diptera: Culicidae) oviposition attractant. J Med Entomol 19:589–592

Bowen MF 1995 Sensilla basiconica (grooved pegs) on the antennae of female mosquitoes: electrophysiology and morphology. Entomol Exp Appl 77:233–238

Cribb BW, Jones MK 1995 Reappraisal of the pore channel system in the grooved pegs of *Aedes aegypti*. Tissue & Cell 27:47–53

Davis EE 1977 Response of the antennal receptors of the male *Aedes aegypti* mosquito. J Insect Physiol 23:613–617

Davis EE, Rebert CS 1972 Elements of olfactory receptor coding in the yellow fever mosquito. J Econ Entomol 65:1058–1061

Dickens JC, Payne TL 1978 Structure and function of the sensilla on the antennal club of the southern pine beetle, *Dendroctonus frontalis* (Zimmerman) (Coleoptera: Scolytidae). Int J Insect Morph Embryol 7:251–265

Dickens JC, Callahan FE, Wergin WP, Erbe EF 1995 Olfaction in a hemimetabolous insect: antennal-specific protein in adult *Lygus lineolaris* (Heteroptera, Miridae). J Insect Physiol 41:857–867

Kafka WA 1987 Similarity of reaction spectra and odor discrimination: single receptor cell recordings in *Antheraea polyphemus* (Saturniidae). J Comp Physiol A 161:867–880

Kaissling K-E 1971 Insect olfaction. In: Beidler LM (ed) Handbook of sensory physiology, vol IV: Chemical senses: olfaction. Springer-Verlag, Berlin, p 351–431

Kaissling K-E 1987 In: Colbow K (ed) R. H. Wright lectures on insect olfaction. Simon Fraser University Press, Burnaby

Kanaujia S, Kaissling K-E 1985 Interactions of pheromone with moth antennae: adsorption, desorption and transport. J Insect Physiol 31:71–81

Keil TA 1989 Fine structure of the pheromone-sensitive sensilla on the antenna of the hawkmoth, *Manduca sexta*. Tissue & Cell 21:139–151

Keil TA 1992 Fine structure of a developing insect olfactory organ: morphogenesis of the silkmoth antenna. Microsc Res Techn 22:351–371

Keil TA, Steinbrecht RA 1984 Mechanosensitive and olfactory sensilla of insects. In: King RC, Akai H (eds) Insect ultrastructure, vol 2. Plenum, New York, p 477–516

Lacher V 1971 Arbeitsbereiche von Geruchsrezeptoren auf der Moskitoantenne (*Aedes aegypti L.*). J Insect Physiol 17:507–517

Laue M, Steinbrecht RA, Ziegelberger G 1994 Immunocytochemical localization of general odorant-binding protein in olfactory sensilla of the silkmoth *Antheraea polyphemus*. Naturwissenschaften 81:178–180

McIver SB 1972 Fine structure of pegs on the palps of female culicine mosquitoes. Can J Zool 50:571–576

McIver SB 1973 Fine structure of antennal sensilla coeloconica of culicine mosquitoes. Tissue Cell 5:105–112

McIver SB 1974 Fine structure of antennal grooved pegs of the mosquito, *Aedes aegypti*. Cell Tissue Res 153:327–337

McIver SB 1978 Structure of sensilla trichodea of female *Aedes aegypti* with comments on innervation of antennal sensilla. J Insect Physiol 24:383–390

McIver SB 1982 Sensilla of mosquitoes (Diptera: Culicidae). J Med Entomol 19:489–535

McKenna MP, Hekmat-Scafe DS, Gaines P, Carlson JR 1994 Putative *Drosophila* pheromone-binding proteins expressed in a subregion of the olfactory system. J Biol Chem 269:16340–16347

Meinecke C-C 1975 Riechsensillen und Systematik der Lamellicornia (Insecta, Coleoptera). Zoomorphologie 82:1–42

Meng LZ, Wu CH, Wicklein M, Kaissling K-E, Bestmann HJ 1989 Number and sensitivity of three types of pheromone receptor cells in *Antheraea pernyi* and *Antheraea polyphemus*. J Comp Physiol A 165:139–146

Muir LE, Cribb BW 1994 *Aedes aegypti*: sensilla trichodea and stimulus-conducting structures. J Insect Physiol 40:1017–1023

Ozaki M, Morisaki K, Idei W, Ozaki K, Tokunaga F 1995 A putative lipophilic stimulant carrier protein commonly found in the taste and olfactory systems: a unique member of the pheromone-binding protein superfamily. Eur J Biochem 230:298–308

Pappenberger B, Geier M, Boeckh J 1996 Responses of antennal olfactory receptors in the yellow fever mosquito *Aedes aegypti* to human body odours. In: Olfaction in mosquito–host interactions. Wiley, Chichester (Ciba Found Symp 200) p 254–266

Pelosi P, Maida R 1995 Odorant-binding proteins in insects. Comp Biochem Physiol B 111:503–514

Pikielny CW, Hasan G, Rouyer F, Rosbash M 1994 Members of a family of *Drosophila* putative odorant-binding proteins are expressed in different subsets of olfactory hairs. Neuron 12:35–49

Prestwich GD, Du G, LaForest S 1995 How is pheromone specificity encoded in proteins? Chem Senses 20:461–469

Redkozubov A 1995 High electrical resistance of the bombykol cell in an olfactory sensillum of *Bombyx mori*: voltage-clamp and current-clamp analysis. J Insect Physiol 41:451–455

Schneider D 1992 100 years of pheromone research—an essay on Lepidoptera. Naturwissenschaften 79:241–250

Schneider D, Lacher V, Kaissling K-E 1964 Die Reaktionsweise und das Reaktionsspektrum von Riechzellen bei *Antheraea pernyi* (Lepidoptera, Saturniidae). Z vergl Physiol 48:632–662

Slifer EH, Prestage JJ, Beams HW 1959 The chemoreceptors and other sense organs on the antennal flagellum of the grasshopper (Orthoptera; Acrididae). J Morph 105:145–191

Steinbrecht RA 1969 Comparative morphology of olfactory receptors. In: Pfaffmann C (ed) Olfaction and taste, vol III. Rockefeller University Press, New York, p 3–21

Steinbrecht RA 1973 Der Feinbau olfaktorischer Sensillen des Seidenspinners (Insecta, Lepidoptera): Rezeptorfortsätze und reizleitender Apparat. Z Zellforsch Mikrosk Anat 139:533–565

Steinbrecht RA 1980 Cryofixation without cryoprotectants. Freeze substitution and freeze etching of an insect olfactory receptor. Tissue & Cell 12:73–100

Steinbrecht RA 1984 Chemo-, hygro-, and thermoreceptors. In: Bereiter-Hahn J, Matoltsy AG, Richards KS (eds) Biology of the integument—invertebrates. Springer-Verlag, New York, p 523–553

Steinbrecht RA 1987 Functional morphology of pheromone-sensitive sensilla. In: Prestwich GD, Blomquist GJ (eds) Pheromone biochemistry. Academic Press, San Diego, CA, p 353–384

Steinbrecht RA 1992 Cryotechniques with sensory organs. Microscopy & Analysis A Sept, p 21–23

Steinbrecht RA, Kasang G 1972 Capture and conveyance of odour molecules in an insect olfactory receptor. In: Schneider D (ed) Olfaction and taste, vol IV. Wiss Verlagsges, Stuttgart, p 193–199

Steinbrecht RA, Ozaki M, Ziegelberger G 1992 Immunocytochemical localization of pheromone-binding protein in moth antennae. Cell Tissue Res 270:287–302

Steinbrecht RA, Laue M, Ziegelberger G 1995 Immunolocalization of pheromone-binding protein and general odorant-binding protein in olfactory sensilla of the silk moths *Antheraea* and *Bombyx*. Cell Tissue Res 282:203–217

Sutcliffe JF 1994 Sensory bases of attractancy: morphology of mosquito olfactory sensilla—a review. J Am Mosq Control Assoc 10:309–315

Tuccini A, Maida R, Rovero P, Mazza M, Pelosi P 1996 Putative odorant-binding protein in antennae and legs of *Carausius morosus*. Insect Biochem Mol Biol 26:19–24

Vogt RG, Riddiford LM 1981 Pheromone binding and inactivation by moth antennae. Nature 293:161–163

Vogt RG, Köhne AC, Dubnau JT, Prestwich GD 1989 Expression of pheromone binding proteins during antennal development in the gypsy moth *Lymantria dispar*. J Neurosci 9:3332–3346

Vogt RG, Prestwich GD, Lerner MR 1991 Odorant-binding protein subfamilies associate with distinct classes of olfactory receptor neurons in insects. J Neurobiol 22:74–84

Walther JR 1981 Die Morphologie und Feinstruktur der Sinnesorgane auf den Antennengeisseln der Männchen, Weibchen und Arbeiterinnen der roten Waldameise *Formica rufa* Linne 1758 mit einem Vergleich der antennalen Sensillenmuster weiterer Formicoidea (Hymenoptera). PhD thesis, Freie Universität, Berlin

Zacharuk RY 1985 Antennae and sensilla. In: Kerkut GA, Gilbert LI (eds) Comprehensive insect physiology, biochemistry and pharmacology, vol VI. Pergamon, Oxford, p 1–69

Ziegelberger G 1995 Redox-shift of the pheromone-binding protein in the silkmoth *Antheraea polyphe mus*. Eur J Biochem 232:706–711

Ziegelberger G 1996 The multiple role of the pheromone-binding protein in olfactory transduction. In: Olfaction in mosquito–host interactions. Wiley, Chichester (Ciba Found Symp 200) p 267–280

DISCUSSION

Boeckh: Could you speculate on why there are so many different types of sensilla?

Steinbrecht: Sensilla involved in pheromone detection usually have long hairs, but it is not clear why some other sensilla are hidden in pits. These sensilla are usually of the double-walled type with spoke channels, and it is possible that these sensilla are more prone to water loss. We have never observed very large hairs with spoke channels. It is possible that the evolution of pore tubules enabled the development of very long sensory hairs. To verify these speculations, we need data on the rate of water loss through the different types of sensilla.

Mustaparta: Some of the older data suggest that there is a difference between the sensitivity of pheromone receptors and host odour receptors. However, our results from gas chromatography linked to electrophysiology show that host odour receptor neurons may be as sensitive as the pheromone receptor neurons.

Steinbrecht: Yes, that is correct. The work of Kafka (1970) suggests that even sensilla coeloconica may respond to single molecules. However, the chance to catch a stimulus molecule is definitely greater with the long hairs.

Brady: People always talk about the feathery moth antennae as being molecular sieves. You showed some beautiful stereoscan pictures of these antennae, but I cannot understand how anything can pass through them.

Steinbrecht: The whole process is called convective diffusion. It has been examined in detail by Adam & Delbrück (1968). An airstream is present, which is convective, but diffusion is more important within the antennae.

Brady: You showed a picture of a pore tubule going into the cuticle, but there was also a wide gap before the dendrite. Therefore, what's the point of having pore tubules?

Steinbrecht: We used to argue that contact between the pore tubules and the dendrites (which we sometimes observed) was essential because we couldn't imagine how a small hydrophobic molecule could leave the cuticle and diffuse to the dendrites through the aqueous sensillum lymph. But we now know that there are carrier proteins or transport proteins, so this is no longer as difficult to understand. We don't actually know why there are pore tubules, but the pores are not simply holes; they are covered by an electronlucid epicuticular layer, which possibly continues into the lumen of the pore tubules.

Guerin: It is worth mentioning the work of Thomas Keil here, which shows that the sensillum is not a static milieu, but that the dendrites are relatively dynamic structures (Keil 1993). They move up and down inside the sensillum, so the pore tubules may not contact the same part of the dendritic membrane all of the time.

Kaissling: We do not know whether the dendrites move up and down in the closed hair. Williams (1988) observed dendrites protruding from the distal ends of hairs when cut under Ringer's solution. By exerting positive or negative pressure on the haemolymph side the dendrite can be partially moved out or back in. Also, as you mentioned, Keil (1993) observed slow active bending of dendrites outside the hair. It seems possible that dendrites perform 'stirring' movements in the closed hair which could lead to formation and disruption of contacts between pore tubules and the dendritic membrane.

Pickett: We have had serious problems with getting good recordings from midge antennae. In collaboration with Mario Solinas, we have recently found the reason for this problem: when we cut the top of the antenna, we lose all the dendritic extensions, which are immediately drawn down from the sensillum. Why should that happen? I don't understand the basis of this and neither does Mario, in spite of his eminence in insect morphological studies.

Kaissling: I would like to comment on that. We can do this experimentally by opening the hair and slipping the recording electrode over the open tip of the hair. Then, by exerting a slight negative pressure on the haemolymph side, we can withdraw the dendrite from the hair lumen. After this operation, the cell does not respond to local application of pheromone to the hair. However, if we push the dendrite back into the hair by exerting positive pressure on the haemolymph side, we again observe a response to pheromone (Kaissling et al 1991).

Carlson: How flexible are the sensilla? Are they more like iron rods or like rubber hoses?

Steinbrecht: They are relatively elastic. The long hairs have thicker walls than the shorter ones, probably to ensure mechanical support. They can be deflected with an electrode but not with an airstream. After withdrawing from the electrode they snap back into their original position.

Hildebrand: Has anyone done thin-section studies or looked at the sexual dimorphism of mosquito antennae?

Steinbrecht: Yes. Susan McIver has done several studies of this kind. She has shown that there are differences in sensillum type and numbers both between the sexes and between blood suckers and plant feeders. I can recommend her review in *Medical Entomology* in 1982 and Sutcliffe's (1994) recent review.

Costantini: It is interesting that, within the *Anopheles gambiae* complex, there are quantitative differences in the total number of sensilla basiconica and sensilla coeloconica on the antennal flagellum.

Steinbrecht: Yes, Anophelini also have large grooved sensilla coeloconica, which are not present in Culicini.

Costantini: But it's even more complicated than that, because by looking at the distribution and total number of sensilla coeloconica on the female antennal segments, one could, in principle, distinguish freshwater-breeding *An. gambiae sensu stricto* from saltwater-breeding *Anopheles merus* (Coluzzi 1964). Moreover, there is a general trend for freshwater species (*An. gambiae* and *Anopheles arabiensis*) to have lower numbers of sensilla basiconica than saltwater species (*Anopheles melas* and *An. merus*). Given the different feeding and oviposition habits of these two groups, with the former representing the two most anthropophilic species of the complex, one wonders what is the adaptive significance, if any, of this difference.

Grant: You mentioned that pheromone-binding proteins (PBPs) were not very abundant in the female moth antenna. Are there correlates between PBP abundance and sensillar morphology?

Steinbrecht: Those sensilla trichodea or basiconica that are labelled with anti-PBP antibody have exactly the same morphology as their siblings which are labelled with anti-GOBP antibody. On the other hand, in some noctuid species the females have sensilla trichodea that are as sensitive to their own pheromone as those of males (Ljungberg et al 1993). We did some labelling of these antennae and we found that our anti-PBP antibody reacts with a greater number of hairs than in *Bombyx*, for example.

Grant: Is the density of the stain related to the relative sensitivity of the receptor neurons in the sensillum?

Steinbrecht: We are not able to determine this. The density of the labelling can only tell us how much PBP is there or how well the antibody cross-reacts.

Klowden: Is it possible to inject your anti-PBP antibody into an intact animal to inhibit olfaction?

Steinbrecht: That's a wonderful suggestion, and we have thought about doing this experiment. However, the concentration of PBP in the sensillum lymph is as high as 10 mM (about 100-fold higher than the maximal concentration we can achieve with purified PBP in the test tube). The sensory hairs of *Antheraea* are 300 μm long and only 3 μm wide. Therefore, it would be almost impossible to get sufficient antibody into the sensillum lymph to block all the binding protein.

Boeckh: But preparations with isolated dendrites have been provided by M. Stengl in Regensburg for patch-clamp experiments. She obtained good receptor responses *in vitro*.

Ziegelberger: But she neither adds nor removes any binding protein in her tissue cultures, which not only contain receptor cells but also auxiliary cells.

We know that these auxiliary cells continually synthesize PBP (Vogt et al 1989, Steinbrecht et al 1992). Therefore, it is possible that a low concentration of binding protein is present. Western blots should be performed to prove the presence or absence of PBP in Monika Stengel's preparation.

Steinbrecht: Moreover, if Monika adds PBP to the preparation then the responses of the receptor cells are more like those *in vivo*.

Carlson: The other way of addressing the function of binding proteins, besides through antibody neutralizing experiments, is genetically through mutation analysis.

References

Coluzzi M 1964 Morphological divergences in the *Anopheles gambiae* complex. Riv Malariol 43:197–232

Kafka WA 1970 Molekulare Wechselwirkungen bei der Erregung einzelner Riechzellen. Z vergl Physiol 70:105–143

Kaissling K-E, Keil TA, Williams L 1991 Pheromone stimulation in perfused sensory hairs of the moth *Antheraea polyphemus*. J Insect Physiol 37:71–78

Keil TA 1993 Dynamics of 'immotile' olfactory cilia in the silkmoth *Antheraea*. Tissue & Cell 25:573–587

Ljungberg H, Anderson P, Hansson BS 1993 Physiology and morphology of pheromone-specific sensilla on the antennae of male and female *Spodoptera littoralis* (Lepidoptera: Noctuidae). J Insect Physiol 39:253–260

McIver SB 1982 Sensilla of mosquitoes (Diptera: Culicidae). J Med Entomol 19:489–535

Steinbrecht RA, Ozaki M, Ziegelberger G 1992 Immunocytochemical localization of pheromone-binding protein in moth antennae. Cell Tissue Res 270:287–302

Sutcliffe JF 1994 Sensory bases of attractancy: morphology of mosquito olfactory sensilla—a review. J Am Mosq Control Assoc 10:309–315

Vogt RG, Köhne AC, Dubnau JT, Prestwich GD 1989 Expression of pheromone binding proteins during antennal development in the gypsy moth *Lymantria dispar*. J Neurosci 9:3332–3346

Williams JLD 1988 Nodes on the large pheromone-sensitive dendrites of olfactory hairs of the male silkmoth, *Antheraea polyphemus* (Cramer) (Lepidoptera: Saturniidae). Int J Insect Morphol Embryol 17:145–151

General discussion III

Guerin: I would like to ask John Pickett a question about recordings from single sensilla with the oviposition pheromone as a stimulant. Which sensilla did you record from, and is it easy to do this?

Pickett: We have had numerous problems with linking our recordings with the morphology. Our approach has been to search for olfactory cells using pharmacological approaches rather than morphological approaches. We have studied aphids extensively, and we have mapped out the function of the antenna and linked the neurophysiology to the actual morphological appearance. However, when we've been working with dipterans and coeleopterans, we often don't know exactly which type of sensillum we're studying.

Guerin: Did you use glass or tungsten electrodes?

Pickett: We used the type of electrode that Jürgen Boeckh developed, i.e. electrolytically sharpened tungsten.

Guerin: What exactly do you mean by a pharmacological approach?

Pickett: We search for a cell that's exhibiting spontaneous activity and then investigate its behaviour by testing with whole extracts or pure compounds. This serves as a preliminary experiment to doing coupled gas chromatography analysis or investigating its dose–response characteristics.

Guerin: Does this require a lot of searching?

Pickett: Yes. Our major impetus is for the identification of behaviourally active compounds for a wide range of insects, and we don't see ourselves as doing basic neurophysiological investigations. We often find cells paired together morphologically that are responding to two different compounds from different biosynthetic pathways. It is possible that this pairing allows the insect to determine the relative concentrations of different components, which may be necessary for the location of specific plant host species. These ideas have been developed in collaboration with Jan Pettersson, Tom Christianson and Bill Hansen.

Steinbrecht: We need a good, simple labelling technique that allows us to do electron microscopy after making the recordings, so that we can identify the sensillum type exactly.

Pickett: We prefer using light microscopy. We've been trying to explore better ways of finding out this information.

Davis: In mosquito antennae we have found that it's relatively easy to mark the base of the sensillum from which we're recording and make a map of its location. We then look at this sketch under a scanning microscope and can identify the sensillum type.

Pickett: We ought to do something along those lines.

Grant: It isn't that difficult to identify individual sensilla from which recordings have been made, if the optics are good, by removing adjacent sensilla and making a sketch. With these 'landmarks' it is relatively easy to find them later during scanning microscopy. The problem with any direct photographic method is that the depth of focus is limited so that it is difficult to resolve all the necessary landmarks.

Kaissling: There are pheromone components of one species which inhibit the attraction of another species. Also, Priesner (1986) found up to three types of receptor cells in one species of moth that respond only to compounds produced by other species of moths. What is the advantage of producing these inhibitory compounds?

Mustaparta: The advantage is to provide reproductive isolation between sympatric species, which has been demonstrated for bark beetles by Lanier & Wood (1975). Related species that are geographically isolated are cross-attractive; whereas sympatric, related species are not. This is because a single pheromone component inhibits the attraction of the sympatric species—it's called interspecific interruption. In heliothine moths all but one species produce the same compound as the major pheromone component, and the addition of a second component provides the species specificities of the pheromone blend. In addition, the second compound often inhibits the attraction of the sympatric males, so that the reproductive isolation becomes more efficient. This is of interest for both the producing and the receiving species. In terms of physiological mechanisms for the behavioural inhibition, it seems more efficient to activate specific receptor neurons that may block the pheromone responses centrally rather than to competitively block the pheromone receptors. We have tested whether competitive blocking takes place in addition to the specific activation of separate neurons. We found that the pheromone receptor neurons responded equally to the pheromone and to the mixture when the 'inhibitor' was added, which ruled out a competitive blocking mechanism. We then thought that pheromone responses might be inhibited in the antennal lobe, where there is a strong convergence of receptor neurons onto central neurons. However, stimulation with the interspecific signal does not block the pheromone responses. The pheromone- and interspecific-signalling pathways are separate in the antennal lobe. Therefore, the eventual interaction between the two information pathways may either be integrated at a higher level or another behavioural motor programme has to be activated. The presence of interspecific interruption in many insects suggests that it is important for reproductive isolation between species.

Pickett: We have found a tremendous investment in negative behavioural responses. In the case of host location, we have observed cells responding to specific non-host compounds. Examples of this include the black bean aphid *Aphis fabae* and the cabbage aphid *Brevicoryne brassicae*, which have exactly the same cells for isothiocyanate reception. Therefore, although *A. fabae* is not feeding on members of the Brassicaceae and is specializing on Fabaceae, it has invested in isothiocyanate cells and uses these compounds as repellents. This is a lovely issue to work on, because there is a total parsimony in receptor physiology—similar receptors with the same structure–activity relationships—but a totally different behaviour, i.e. a compound such as an isothiocyanate is a repellent for one species and for another it is an attractant.

Mustaparta: There are also some compounds that mediate negative messages to the insect. This has been demonstrated in contact chemosensory signal studies in Spodoptera, where receptor neurons of monophagous insects are more sensitive to toxic compounds not present in their host plants than to toxic compounds present in non-host plants. In contrast, the polyphagous insects are not sensitive to many toxic compounds that are present in their host plants. Similar cases may be apparent in the field of olfaction. It is important that the insects know where to go, but it is also important that they know where not to go.

de Jong: On the other hand, if the insects are specialized, then one wouldn't expect them to respond to non-host plants anyway.

Pickett: I would like to relate this point to mosquitoes. We would like to investigate mosquito–host interactions as we have done for flies that attack cattle, i.e. to find if there are similarly specialized non-host cells in the other haematophagous insects. This would then represent a logical approach for identifying host-masking agents and repellents.

Guerin: I have difficulties understanding the term 'masking', as against behavioural preferences. How can one conclude that certain hosts are masked? Is it possible that the masked individuals just release less of the attractants?

Pickett: No. This is a behavioural phenomenon. The underlying neurophysiology, as we understand it, involves attractant-responding cells and a different population of masking agent-responding cells. The behavioural phenomenon that allows us to use the masking terminology is that in, for example, a linear track olfactometer, we can eliminate an attractant behavioural effect by adding a masking agent.

Guerin: But you don't yet know what the attractants are.

Pickett: In some aspects I'm extrapolating from the plant area, in terms of the basic knowledge, because we know this area very well: we know what the attractant chemicals are, we know that they elicit responses from specific cells and we know the masking agents. For example, in *A. fabae*, an isomer of myrtenal stops the attractiveness of compounds from the host, field bean *Vicia faba*.

Guerin: Are you sure that it's not a concentration effect?

Pickett: There may be a concentration effect as well, but there is definitely evidence for a masking effect related to the different neurophysiological perception of the compounds.

Guerin: This is relevant to the case of sick animals being more attractive to mosquitoes than healthy ones, which may be simply related to the higher amount of a metabolite, such as CO_2, released.

Pickett: I am not denying that a quantitative effect may be involved, but this masking effect is a very real effect.

Brady: Masking could imply all kinds of behavioural things. What is actually going on with regard to the cows and the odours? Is this masking agent a repellent in the sense that the flies turn away? Or do the flies just arrive less often?

Pickett: Most of that work has been done by Gethyn Thomas. He has shown that the activity of the attractive extract is eliminated by a particular fraction, i.e. there is a masking effect. He hasn't tested the masking material on its own, but we have done this in the plant experiments, and we find that these also behave as repellents in, for example, the linear track or T-wire olfactometer.

Brady: A crucial point when using a T-wire olfactometer is to get the insect to walk up the wire in the first place. When the masking agent is present in the stem of the tube, do they still walk up the wire?

Pickett: Yes. And we get the same results when we use the Pettersson four-way olfactometer.

Boeckh: There are also problems with our definition of a 'repellent'. Repellents can behave in different ways. For example, they may counteract the activity in the receptors themselves or they may switch on other input channels that relay different messages to the CNS. In the behaviour experiments with the repellents, it looks as though the host information is not wiped out completely, even with a high concentration of N,N-diethyl-m-toluamide (DEET). The mosquitoes fly close to the source but then they seem to smell something different. It's not a complete inhibition. If DEET alone is given in the control side, the mosquitoes do not turn around and fly away, i.e. they are not 'repelled'.

Pickett: The experiments with DEET are different from ours.

Boeckh: That's exactly my point. Even though DEET is an artificial substance, there are different types of repellent. No one really knows which type of channel is switched on by DEET, or how this channel might work under natural circumstances and when host stimuli are present.

Guerin: But the interaction of DEET with the olfactory receptor for L-lactic acid is known.

Boeckh: Yes, but it does not wipe it out. DEET itself elicits large responses from other types of sensilla that do not house receptors for host odours. If DEET is mixed with skin extract, the latter is still smelled by the mosquito.

Guerin: That's why I raised the point about the nature of these so-called masking agents. We need to know what they are and then we need to understand what might be happening. Do they interact with the host odour receptors or do they activate another receptor family?

Pickett: In the case I talked about, the masking agents activated a different family of receptors.

Boeckh: But in other cases they may interact with the host odour receptors.

Pickett: Yes, that's possible.

Bowen: In order to see the masking effect electrophysiologically you have to present DEET concurrently with a host attractant, such as L-lactic acid. These experiments show that DEET depresses the sensitivity of the L-lactic acid receptor because more L-lactic acid is required to drive the cell. One can also observe behaviourally that DEET is not acting as a repellent. We've used a dual-chamber olfactometer, consisting of an upper and a lower chamber, to look at the repellent effect of DEET on four different mosquito species. If *Aedes aegypti* are placed in the upper chamber and the host (i.e. a hand) at the bottom of the chamber, they take about 5–7 min to move from the upper to the lower chamber. If 10% DEET is applied to the hand, or on a piece of gauze placed directly above the hand, the mosquitoes behave very differently. They begin to fly, so they're obviously activated, almost as though they have detected CO_2. However, after a short while they alight on the sides of the upper chamber; they do not move into the lower chamber. This suggests that DEET inhibits the approach to the host, and it's clearly a masking effect because they're not making movements oriented away from the DEET source.

Ed Davis (personal communication 1986) has also done some wind tunnel studies of the effect of DEET on mosquito behaviour and they show basically the same thing.

Geier: We have done similar experiments with CO_2 and the human skin rubbing extract (M. Geier & I. Mutzbauer, unpublished work 1995). If we add DEET to CO_2 or to the extract, we observe different effects. The increase of flight activation by stimulation with CO_2 is only slightly reduced by adding DEET. However, DEET strongly reduces the flight activity normally elicited by the extract.

Guerin: Are you releasing DEET and the extract from the same or different substrates?

Geier: From different substrates. We placed a net containing DEET behind the human hand or the odour source with the attractive extract.

Bowen: Another important point to make about DEET is that its effect on the olfactory receptor cells is a general one because it masks the detection of oviposition site attractants and also inhibits oviposition site approach by the gravid female (Kuthiala et al 1992).

References

Kuthiala A, Gupta R, Davis EE 1992 The effect of repellents on the responsiveness of the antennal chemoreceptors for oviposition of the mosquito *Aedes aegypti*. Med Vet Entomol 29:639–643

Lanier GN, Wood DL 1975 Specificity of response to pheromones in the genus Ips (Coleoptera: Scolytidae). J Chem Ecol 1:9–23

Priesner E 1986 Correlating sensory and behavioural responses in multichemical pheromone systems of Lepidoptera. In: Payne TL, Birch MC, Kennedy CEJ (eds) Mechanisms in insect olfaction. Oxford University Press, Oxford, p 225–233

Central olfactory pathways in mosquitoes and other insects

Sylvia Anton

Department of Ecology, Ecology Building, Feromongruppen, Lund University, S223 62 Lund, Sweden

Abstract. Studies of CNS processing of olfactory information have contributed significantly to understanding olfactory-guided behaviour in insects. Evidence in moths suggests that each glomerulus in the antennal lobes has a unique property: receptor–axon projections and dendritic arborizations of uniglomerular output neurons can relate to anatomically and functionally distinct glomeruli. Similar correlations are not typical of the concentrically organized locust antennal lobes. Insights about odour processing in the CNS of female mosquitoes should help us to understand how sensory information can lead to host-seeking behaviour. It will be interesting to learn how inputs from CO_2 receptors on the maxillary palps and inputs from antennal olfactory receptors that respond to host odours are integrated centrally, so that pharmacological manipulation of olfactory neuron activity might provide a tool for the control of mosquitoes as important vector insects. The antennal lobe of male and female *Aedes aegypti* contains 20–25 glomeruli. Primary afferent projections from the antennae and maxillary palps target specific glomeruli of the ipsilateral antennal lobe. Maxillary palp projections are restricted to two posteromedial glomeruli, which do not receive antennal afferents. The latter arborize in the remaining glomeruli.

1996 Olfaction in mosquito–host interactions. Wiley, Chichester (Ciba Foundation Symposium 200) p 184–196

Odours play an important role in guiding the behaviour of insects. Sex pheromones help insects to find a mate. Plant volatiles, flower odours or host odours lead them to oviposition sites or food sources. Aggregation pheromones let them gather in groups. In most insect species olfactory stimuli of different origins are responsible for the different behaviours. Studying the structure and function of olfactory receptor neurons and antennal lobe interneurons gives insight into central nervous processing mechanisms, which are the basis for eliciting different types of olfactory-guided behaviour. The structure of the primary olfactory neuropil, the antennal lobe, consisting of varying numbers of glomeruli in different insect species, seems to have functional significance. Information on olfactory cues involved in a certain type of behaviour are often transmitted by labelled lines, represented by neurons connected within the same glomerulus, that carry specific information.

The present knowledge on central nervous processing of olfactory information is restricted to a few insect species. Although extensive studies have also been performed on cockroaches and bees, only three examples (moths, locusts and flies) will be given here for comparison with the relatively recent studies on mosquitoes.

Moths

In moths the sex pheromone processing system of the male is well studied in a number of species (for example, *Manduca sexta* and some noctuid species), using tip recordings from olfactory sensilla, selective backfills with cobalt lysine, intracellular recording and staining of antennal lobe interneurons. The sexual dimorphism of the antennal structure and function in moths is reflected by a structural specialization of the male antennal lobe. A cluster of enlarged glomeruli, the macroglomerular complex, situated at the entrance of the antennal nerve, is the processing site of female-produced sex pheromone stimuli (Christensen & Hildebrand 1987, Koontz & Schneider 1987). Each olfactory receptor axon arborizes in one glomerulus, and it has been shown that specific glomeruli in the male macroglomerular complex are targeted by functionally distinct receptor neurons. Receptor neurons responding specifically to one sex pheromone component always arborize in one specific glomerulus within the macroglomerular complex in different noctuid species (Fig. 1a) (Hansson et al 1992, Ochieng' et al 1995, Todd et al 1995). Local interneurons in the antennal lobe connect a large number of glomeruli independently of their functional specificity (Christensen et al 1993, Anton & Hansson 1994, 1995). Most projection neurons, which relay information from the antennal lobe to the protocerebrum, have uniglomerular arborizations (Fig. 1b) that correlate with specific response characteristics. Projection neurons responding specifically to one sex pheromone component in *M. sexta* always arborize in one specific glomerulus within the macroglomerular complex (Hansson et al 1991). Physiological characteristics of multiglomerular projection neurons (Homberg et al 1988) in moths have not yet been described.

In female *Spodoptera littoralis* (Lepidoptera: Noctuidae), receptor axons responding to the sex pheromone produced by females have been shown to project to glomeruli close to the entrance of the antennal nerve (Fig. 1c), at a similar position as the macroglomerular complex in males (Ochieng' et al 1995). Uniglomerular projection neurons in *S. littoralis* females, arborizing in glomeruli close to the antennal nerve entrance (Fig. 1d), always responded to the sex pheromone produced by females, whereas projection neurons responding specifically to single plant-related odours had no specific antennal lobe arborization patterns (Anton & Hansson 1994).

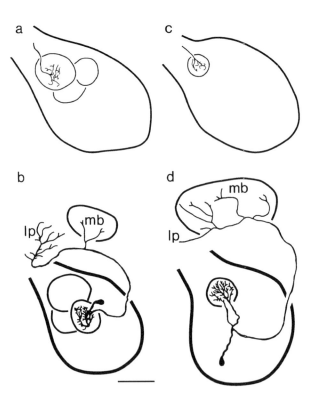

FIG. 1. Structure of sex pheromone sensitive receptor and projection neurons in *Spodoptera littoralis* males (a, b) and females (c, d). All neurons shown responded to the major sex pheromone component Z,E-9,12-tetradecadien-1-yl acetate. (a) Receptor neurons arborizing in the biggest glomerulus of the macroglomerular complex in the male. (b) A projection neuron in a male arborizing in one macroglomerular complex glomerulus and projecting to the calyces of the mushroom bodies (mb) and the lateral protocerebrum (lp). (c) Receptor neurons arborizing in a glomerulus close to the entrance of the antennal nerve in the female. (d) A projection neuron in a female arborizing in a glomerulus close to the entrance of the antennal nerve and projecting to the calyces of the mushroom bodies and the lateral protocerebrum. Reconstructions from frontal sections. Medial is right. Bar for a, b, c and d = 100 μm.

Locusts

In locusts different aspects of CNS processing of odours have been studied using neuroanatomical and electrophysiological methods. In the desert locust, *Schistocerca gregaria,* the central nervous basis for olfactory guided aggregation behaviour is under investigation. In the American locust, *Schistocerca americana,* membrane potential oscillations of groups of antennal lobe interneurons are discussed as a means of odour coding (Laurent & Davidowitz 1994).

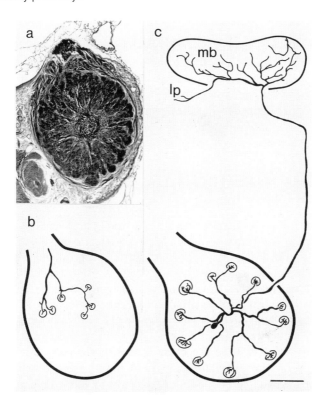

FIG. 2. Neuroanatomy of the antennal lobe in *Schistocerca gregaria*. (a) Reduced silver staining showing small glomeruli arranged around a central fibre core in a frontal section. (b) Receptor neuron arborizing in several glomeruli within the lobe. (c) Projection neuron with dendritic arborizations in several glomeruli and projecting to the calyces of the mushroom bodies (mb) and the lateral protocerebrum (lp). b and c are reconstructions from frontal sections. Medial is on the right. Bar for a, b, c and d = 100 μm.

Generally, a different anatomical organization of the antennal lobe in locusts implies a different functional organization as compared to other insects. A large number of small glomeruli is arranged concentrically around a central fibre core (Fig. 2a). Olfactory receptor axons branch within the antennal lobe and innervate several glomeruli in a sector of the lobe (Fig. 2b). Projection neurons, responding to aggregation pheromone components, arborize in 10–25 glomeruli, equidistantly from the central fibre core (Fig. 2c). Correlations between the arborization patterns and physiological response characteristics of projection neurons have yet to be found.

Flies

Odours guide *Drosophila melanogaster* to oviposition sites and food sources. Deoxyglucose studies showed that stimulation of the antennae with different odours elicits different activity patterns within the glomeruli of the antennal lobe (Rodrigues 1988). Neuroanatomical investigations have revealed bilateral receptor axon projections, which innervate one glomerulus in each antennal lobe, as well as unilateral projections. Unilateral and bilateral antennal lobe interneurons have been described anatomically (Stocker et al 1990).

Mosquitoes

In female mosquitoes, olfactory cues are involved in host-seeking behaviour. In both sexes, olfaction plays a role in finding nectar as an energy source (for review see McIver 1982). In *Aedes aegypti* females, antennal receptors are involved in the detection of host odours, such as L-lactic acid (for review see Bowen 1991); whereas sensilla on the maxillary palps detect CO_2 stimuli. The latter process is essential for eliciting host-seeking behaviour (Kellogg 1970). Investigation of CNS odour processing in female mosquitoes should help to gain a better understanding of how sensory information can lead to host-seeking behaviour. It will be especially interesting to learn how inputs from CO_2 receptors on the maxillary palps and inputs from antennal olfactory receptors that respond to host odours are centrally integrated.

CNS investigations in *Ae. aegypti* initially involved neuroanatomical studies of the antennal lobes of both sexes. Reduced silver stainings revealed some general neuroanatomical features of the antennal lobes, and they have also added some information to the previous studies by Christophers (1960) and Childress & McIver (1984). Golgi staining experiments allowed a description of receptor and projection neurons. Neurobiotin mass fills visualized with Lucifer yellow-coupled avidin have been used to study afferent projections from the antennae and maxillary palps.

In common with other Diptera, it is difficult to determine the exact number of glomeruli in *Ae. aegypti* within the antennal lobe because no distinct glial sheet separates the glomeruli from each other (as revealed by propidium iodide staining). This is probably the reason why the number of glomeruli counted varied between individual mosquitoes. In four males and four females, I counted between 20 and 25 glomeruli, independent of sex. I observed that the cell bodies of antennal lobe neurons were mainly situated in a lateral cell group (Fig. 3), and that few cell bodies were found medially and dorsally of the glomerular neuropil. In addition, I found that the average size of the antennal lobe was slightly larger in females than in males, and that antennal commissures were present as described earlier by Childress & McIver (1984).

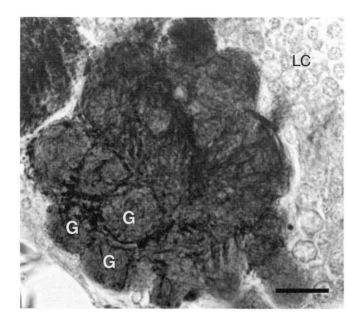

FIG. 3.   Neuroanatomy of the antennal lobe in *Aedes aegypti*. Frontal section, reduced silver staining. Note the poor separation of the glomeruli (G). The majority of the cell bodies from antennal lobe neurons are situated in a lateral cell group (LC). Bar = 15 μm.

Mass fills from the maxillary palps revealed massive unilateral projections in two posteromedial glomeruli in males and females (Fig. 4a). Mass fills from the antennae showed projections in the remaining glomeruli in both sexes (Fig. 4b). Contralateral projections of olfactory receptor neurons have not yet been found.

From Golgi preparations, randomly stained single receptor axons could be reconstructed that showed uniglomerular arborizations with few, strongly blebbed branches (Fig. 5a). A projection neuron stained in a Golgi preparation arborized in one glomerulus and sent its axon to the ipsilateral protocerebrum through the antennal glomerular tract (Childress & McIver 1984) with axonal branches in the calyces of the mushroom body (Fig. 5b).

Conclusions

Although there is a clear sexual dimorphism in olfactory-guided behaviour and in the antennal sensilla between male and female mosquitoes, no obvious dimorphism was found in the deutocerebrum, apart from a size difference in the antennal lobe. More detailed studies of the central olfactory pathways are

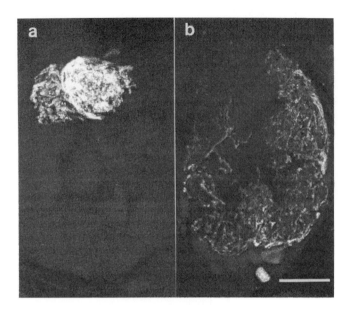

FIG. 4. Confocal microscope reconstructions of receptor neuron projections from a female *Aedes aegypti*. (a) Maxillary palps. The palpal projections innervate two posteromedial glomeruli. (b) Antennae. The antennal projection innervates the remaining glomeruli. Medial is on the left. Bar for a and b = 25 μm.

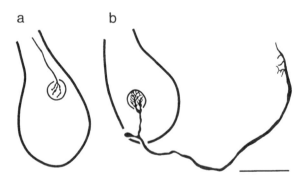

FIG. 5. (a) Structure of a single receptor neuron projection. A few blebbed branches innervate one glomerulus. (b) Structure of a single projection neuron. The cell body of the projection neuron is situated in the small medial cell group. The dendritic arborizations are restricted to one glomerulus, and the axon arborizes in the calyces of the mushroom bodies. Reconstructions from horizontal sections. Medial is on the left. Bar for a and b = 50 μm.

necessary to find if there are any other differences in olfactory processing between the sexes. Investigation of single receptor neuron projections as well as interneuron structure and function in females will also help to understand olfactory integration in the CNS. This knowledge might lead to the possibility of pharmacological manipulation of olfactory neuron activity and thus, in the future, provide a tool for the control of mosquitoes as important vector insects.

Acknowledgements

I wish to thank E. E. Davis and M. F. Bowen for encouraging me to work on mosquitoes, and B. S. Hansson (Lund, Sweden) and N. J. Strausfeld (Tucson, Arizona) for their support. I would also like to thank B. S. Hansson for comments on the manuscript and S. Ochieng for providing Figs 1a and 1c. This study was supported by fellowships from the Deutscher Akademisches Austauschdienst and Deutsche Forschungs gemeinschaft, Germany, the Swedish Institute and the Kungl. Fysiogr. Sällskapet in Lund, Sweden, and by grants from Swedish Research Councils to B. S. Hansson (Naturvetenskapliga Forskningsrådet, Swedish Agency for Research Co-operation with Developing Countries and Forskningsrådsnämden).

References

Anton S, Hansson BS 1994 Central processing of sex pheromone, host odour, and oviposition deterrent information by interneurons in the antennal lobe of female *Spodoptera littoralis* (Lepidoptera: Noctuidae). J Comp Neurol 350:199–214
Anton S, Hansson BS 1995 Sex pheromone and plant-associated odour processing in antennal lobe interneurons of male *Spodoptera littoralis* (Lepidoptera: Noctuidae). J Comp Physiol A 176:773–789
Bowen MF 1991 The sensory physiology of host-seeking behavior of mosquitoes. Ann Rev Entomol 36:139–158
Childress SA, McIver SB 1984 Morphology of the deutocerebrum of female *Aedes aegypti* (Diptera: Culicidae). Can J Zool 62:1320–1328
Christensen TA, Hildebrand JG 1987 Male-specific, sex pheromone-selective projection neurons in the antennal lobes of the moth *Manduca sexta*. J Comp Physiol A 160:553–569
Christensen TA, Waldrop B, Harrow ID, Hildebrand JG 1993 Local interneurons and information processing in the olfactory glomeruli of the moth *Manduca sexta*. J Comp Physiol A 173:385–399
Christophers SR 1960 *Aedes aegypti* (L.): the yellow fever mosquito. Cambridge University Press, Cambridge, MA
Hansson BS, Christensen TA, Hildebrand JG 1991 Functionally distinct subdivisions of the macroglomerular complex in the antennal lobe of the male sphinx moth *Manduca sexta*. J Comp Neurol 312:264–278
Hansson BS, Ljungberg H, Hallberg E, Löfstedt C 1992 Functional specialization of olfactory glomeruli in a moth. Science 256:1313–1315
Homberg U, Montague RA, Hildebrand JG 1988 Anatomy of antenno-cerebral pathways in the brain of the sphinx moth *Manduca sexta*. Cell Tissue Res 254:255–281
Kellogg FE 1970 Water vapor and carbon dioxide receptors in *Aedes aegypti*. J Insect Physiol 16:99–108

Koontz MA, Schneider D 1987 Sexual dimorphism in neuronal projections from the antennae of silk moths (*Bombyx mori, Antherea polyphemus*) and the gypsy moth (*Lymantria dispar*). Cell Tissue Res 249:39–50

Laurent G, Davidowitz H 1994 Encoding of olfactory information with oscillating neural assemblies. Science 265:1872–1875

McIver SB 1982 Sensilla of mosquitoes (Diptera: Culicidae). J Med Entomol 19:489–535

Ochieng' SA, Anderson P, Hansson BS 1995 Antennal lobe projection patterns of olfactory receptor neurons involved in sex pheromone detection in *Spodoptera littoralis* (Lepidoptera: Noctuidae). Tissue & Cell 27:221–232

Rodrigues V 1988 Spatial coding of olfactory information in the antennal lobe of *Drosophila melanogaster*. Brain Res 453:299–307

Stocker RF, Lienhard MC, Borst A, Fischbach KF 1990 Neuronal architecture of the antennal lobe in *Drosophila melanogaster*. Cell Tissue Res 262:9–34

Todd JL, Anton S, Hansson BS, Baker TC 1995 Functional organization of the macroglomerular complex related to behaviorally expressed olfactory redundancy in male cabbage looper moths. Physiol Entomol 20:349–361

DISCUSSION

Mustaparta: How did you stain the single receptor neurons of the olfactory sensilla in the locust? Did you insert a capilliary tube into the sensillum base?

Anton: We don't know the physiological characteristics of the stained locust receptor neurons. We first recorded with tungsten electrodes and then tried to insert a glass capillary in the same place.

Steinbrecht: Did you say that the locust sensilla coeloconica respond to pheromones but not to plant odours?

Anton: In the locust sensilla coeloconica, there are receptor neurons that are both inhibited by pheromone components and excited by plant odours.

Boeckh: From the antenna the central projections remain ipsilateral, but what is the situation with the maxillary palps?

Anton: In all preparations that I have they remain ipsilateral.

Boeckh: It may not be a general rule but, with a few exceptions, in all the specimens we have looked at the antennae project only ipsilaterally. This situation is different in insects with very short antennae that are inserted closely together in the head capsule, such as in flies like *Drosophila* or *Calliphora*. There, many receptor axons project directly into the contralateral antennal lobe (but this is not the case for mosquitoes). However, the maxillary (or labial) palps of all the investigated insects (except locusts) have bilateral projections.

Anton: So far I have not observed any contralateral projections from maxillary palps, but I probably don't have enough preparations to be able to say this definitively.

Geier: In the sensilla of the maxillary palps there are two other receptors in addition to the CO_2 receptor. Therefore, can you be sure that the projections from this region are only from the CO_2 receptor?

Anton: No. I can't be sure.

Hildebrand: You showed that the female antennal lobe is larger than that of the male. Are there more receptor cells in female antennae than in male antennae?

Anton: Yes.

Hildebrand: Is this the extent of the sexual dimorphism that you see? Did you observe any recognizably distinct female specific structures?

Anton: We have not seen any yet, but more detailed studies might reveal such structures.

Davis: With respect to the sexual dimorphism of the antenna itself, in males only the two terminal flagellar segments have chemosensory neurons, whereas the remaining 11 possess only mechanoreceptors. This contrasts with the female, where all 13 flagellar segments have olfactory inputs. Therefore, the greater number of chemosensory neurons in the female could account for the greater size of the antennal lobes of females.

Guerin: The brain acts as an integrating centre. Considering the size of the periphery and how difficult it is to record from it, is it possible to obtain more information on the various olfactory stimuli by recording centrally?

Anton: I'm not sure that this will be easier than recording from the periphery. It's not that easy to do. I started to develop a preparation in which I could get contact with central neurons, but I don't know how easy it will be to obtain good contacts and record from olfactory neurons in the antennal lobe, which is rather small. It is not possible to open the head capsule of a mosquito far enough to see the antennal lobes. Therefore, the only way to locate the antennal lobe whilst inserting an electrode is stereotactically.

Guerin: But the antennal lobe is certainly bigger than single sensilla at the periphery.

Anton: Yes, that's true. Sometimes the screening of odour responses might be easier because there are more widely responding neurons in the CNS than in the periphery. Once we have developed a good preparation in an animal species, it might be quicker to test a large number of odours than to screen individual sensilla, especially for components to which only few receptor neurons respond.

Guerin: And what about recording directly from the antennal nerve?

Boeckh: The fibres only have a diameter of about $0.1\,\mu m$. Therefore, you would need special microelectrodes, for example like the ones that have been developed for recording from dorsal root fibres or the fila olfactoria. The CNS would be an interesting target for interference with the animals' behaviour, but we need to develop an approach. If, for example, the synergism of different host odour components occurs at the level of the CNS, one could think of manipulating neurons that are responsible for this synergism by means of drugs. I would go as far as to say that blocking or 'confusing' just one major individual neuron in the mosquito brain will be more effective for controlling the insect than knocking out or altering half the receptor population. This is because there is a lot of redundancy at the level of the receptor neurons in the co-operative coding of olfactory stimuli. However, this redundancy is not so apparent in the insect brain.

Carlson: There was a good paper from Martin Heisenberg's lab a couple of years ago in Science, in which he used a drug to ablate the mushroom bodies, and he showed that this had a severe effect on olfactory-based learning (de Belle & Heisenberg 1994).

Hildebrand: We have discussed peripheral receptors and the central pathway, but how can we relate this basic work on the olfactory system to the practical problems of controlling disease vectors?

Galun: At this stage we should not worry how synergism between CO_2 and L-lactic acid affects vector-borne diseases. I think that Sylvia Anton asked the right question, i.e. how does the synergism between CO_2 (palpal receptors) and L-lactic acid (antennal receptors) take place in the CNS? To date, work of this type has been conducted with moths and *Drosophila*, and she is harnessing this information to the study of mosquitoes.

Hildebrand: I also believe that these experiments are worth doing. However, we have to explain the connections between what seems like remote basic research and the very pressing problem of controlling vector-borne diseases.

Galun: It has taken a long time to bring both neurobiology and molecular biology into the field of mosquito research, and now we are beginning to see the benefits.

Hildebrand: If we could surmount the technical difficulties presented by the mosquito and do the kinds of recordings that have been possible in moths and cockroaches, we might be able to make predictions about odour-regulated behaviour from recordings in the CNS. This has been possible with heliothine moths (Vickers et al 1991).

Grant: This is true. However, it's also true that the peripheral nervous system places an upper limit on the array of odorants that the mosquito can actually smell. The problem remains that we need a fundamental outline of this information as a natural starting point.

Mustaparta: This is an important point. Receptor neuron responses may be a good starting point to find out what the mosquitoes smell. Also, in Lepidoptera electrophysiological recordings of receptor neuron responses have been used to identify the biological signals. I agree that there is a wealth of information to be gained from studying the CNS in mosquitoes, but it is difficult to record electrophysiologically from such small brains. Perhaps, in the future, molecular biology may be helpful by determining the expression patterns of receptor proteins in receptor neuron terminals. However, in the short-term, the combination of chemical analyses and receptor neuron responses seems more promising for gaining information on chemical signals that modify behaviour in mosquitoes.

Boeckh: Everybody who wants to know how the brain works stumbles over the problem of pattern recognition. In the mosquito we have such a complex stimulus pattern. Here, we know at least several components and their corresponding receptor classes. We can artificially change the composition of the blend and look at changes in behaviour. From this we can obtain some idea of the relevant input pattern. This approach has been done for moth pheromones. However, we are still far from understanding how pattern recognition functions.

Hildebrand: I am reminded of an earlier discussion about *N,N*-diethyl-*m*-toluamide (DEET). If we can understand mechanistically exactly what DEET does, this would be an important guide for the design of new strategies to interfere with host recognition.

Boeckh: It may be possible to manipulate certain channels with a single agent. If one could exclude a certain input by introducing a counter agent, it would be a very useful tool.

Brady: It depends on the questions you're asking. If the goal is to control malaria, then the question is much more pragmatic than those proposed in this discussion. If we go back to the tsetse fly as an example, apart from the electroantennograms and single-cell recordings, the entire control technology was developed without using a single electrode.

Dobrokhotov: Our goal is to develop new tools for malaria control within 15–20 years. Let us forget about what is available now, but instead think about how we will implement disease prevention strategies in the future. We should not limit our discussion to vector control, because it is becoming apparent that we cannot eradicate mosquitoes. It may be even ecologically harmful to do so, because mosquitoes are an important component of food chains in nature. Therefore, we can investigate the possibilities for the interruption of malarial parasite development in mosquitoes without harming them. Modern advances in genetics and biotechnology provide a way to create such malarial non-competent mosquitoes. It will not be long before the genes responsible for *Plasmodium* development in the mosquito midgut, haemocoele and salivary glands are discovered and manipulated. Can we change mosquito olfaction mechanisms for host-seeking behaviour in such a way as to divert them from human baits? These are very interesting subjects for research.

Boeckh: But we have to study the ecology and behaviour of the mosquito first. We can develop molecular tools to interfere with processes at the molecular genetic level, but we need to study the behaviour of the mosquito in order to understand what these manipulations are doing. It would be miserable if the classical entomologists became extinct because they are responsible for laying out the plans upon which the rest of the research is based.

Dobrokhotov: Of course we are not going to stop research into these areas. That's why we encourage symposia like this one—so that entomologists, biochemists and geneticists, for example, have the opportunity to communicate with each other and develop new ideas for future research.

References

de Belle SJ, Heisenberg M 1994 Associative odor learning in *Drosophila* abolished by chemical ablation of mushroom bodies. Science 163:692–695

Vickers NJ, Christensen TA, Mustaparta H, Baker TC 1991 Chemical communication in heliothine moths. III. Flight behavior of male *Heliocoverpa zea* and *Heliothis virescens* in response to varying ratios of intra- and interspecific sex pheromone components. J Comp Physiol A 169:275–280

Sensory aspects of host location in mosquitoes

M. F. Bowen

SRI International, Life Sciences Division, 333 Ravenswood Avenue, Menlo Park, CA 94025–3493, USA

Abstract. Visual, thermal and olfactory stimuli all contribute to blood meal host location in mosquitoes but olfaction is probably the dominant sensory modality used for this purpose. Much attention has been devoted to the L-lactic acid receptor because it is well characterized and because its sensitivity is a major determinant of host responsiveness in the anautogenous species *Aedes aegypti*. Studies employing statistical analysis of close to 500 single unit recordings and the scanning electron microscope have demonstrated that L-lactic acid-excited neurons are associated with the shortest sensilla basiconica (grooved pegs) in *Ae. aegypti, Culex pipiens, Aedes atropalpus* and *Aedes epactius*. L-Lactic acid-inhibited neurons are found in either short or long grooved pegs, depending on mosquito species. Video recording analysis, a vertical dual-chamber olfactometer and a horizontal dual-port wind tunnel olfactometer have been used to study host location behaviour and nutritional preferences in the obligately autogenous *Ae. atropalpus*, the facultatively autogenous *Aedes bahamensis* and the adult diapausing species *Cx. pipiens*. These behavioural studies, in conjunction with electrophysiological analysis, illustrate the ways in which the interactions between reproductive condition, developmental stage and L-lactic acid receptor sensitivity determine the nutritional choice made between blood and sugar by mosquitoes and demonstrate the role that olfactory sensitivity plays in this process.

1996 Olfaction in mosquito–host interactions. Wiley, Chichester (Ciba Foundation Symposium 200) p 197–211

Visual, thermal and olfactory stimuli all contribute to a mosquito's ability to locate a blood meal host[1]. *Aedes aegypti* females visually identify potential blood meal hosts as solidly coloured stationary objects of low reflectance; they are not attracted to specific colours (Muir et al 1992). There are arguably many objects in a mosquito's environment that fit this description, so even day-active mosquitoes must rely on non-visual cues to discriminate between objects in their environment

---

[1]It has been suggested that the term 'host location' is more appropriate and less teleological than 'host seeking' (Clements 1993). The terms are used interchangeably here in order to facilitate reference to previous publications.

which are viable resources and those which are not. Non-visual cues for host location include CO_2, thermal convection currents arising from warm-blooded hosts and volatile chemicals emitted as components of host effluence.

Mosquitoes are well equipped to exploit olfactory information in their environment. There are about 2000 antennal neurons in female *Ae. aegypti* mosquitoes, over 90% of which have an olfactory function, the rest being mechanoreceptive and thermoreceptive (McIver 1978). Mosquitoes use olfactory cues to guide them to resources as diverse as blood meal hosts, oviposition sites and plants, so it is not surprising that the range of olfactory stimuli which they can detect is quite broad, from simple carboxylic acids (Davis 1988) to complex terpenes (Bowen 1992a) and pheromones (Pile et al 1991, M. F. Bowen, unpublished observations 1990). Mosquitoes possess five morphologically distinct olfactory sensilla types, each of which contains more than one neuron. Each of these neurons may respond to different stimuli. Only electrophysiological studies of single cells can establish function and stimulus identity.

Clearly identified single compounds that have been shown to be behaviourally active when present in the vapour phase at physiological concentrations are an essential tool for identifying and characterizing olfactory receptors. The identification of host attractants is not a trivial pursuit, as there are over 102 volatile compounds found in human breath (Krotoszynski et al 1977) and over 300 in whole-body effluence (Ellin et al 1974). A few host-related volatiles have been identified as mosquito host attractants (Takken 1991, Kline 1994). These include 1-octen-3-ol, L-lactic acid and certain phenols. They are not all effective in all species. Trap catches of most *Culex* species (except for *Culex salinarius*) are not enhanced by 1-octen-3-ol, for example, and some species are actually repelled by this compound (see Kline 1994). It is unknown to what extent this interspecific variation is attributable to interspecific differences in sensory physiology.

Much attention has been paid to the L-lactic acid receptor, not because L-lactic acid is the only host attractant for mosquitoes, but because it is a particularly well-characterized host attractant receptor. This paper addresses the location, developmental patterns and physiology of the L-lactic acid receptor in a variety of mosquito species with different gonotrophic and behavioural strategies. Temperature receptors have been reviewed elsewhere (Bowen 1991) and CO_2 reception is treated as a separate topic in this symposium (Grant & O'Connell 1996, this volume). Recent developments in host attractant identification are also discussed separately (this volume: Cork 1996, Geier et al 1996).

Methods

We obtained electrophysiological recordings using standard techniques described in detail in previous publications (Bowen 1992a, Bowen et al

1994a, 1995), and a video camera to record sugar-feeding behaviour (Bowen & Romo 1995a). We used a vertical dual-chamber olfactometer based on the olfactometer design of Feinsod & Spielman (1979) to assess responsiveness to the odours of a blood meal host, and a horizontal dual-port olfactometer wind tunnel of our own design (Bowen et al 1995) to assess odour preferences.

General characteristics of the L-lactic acid receptor

Cells can be excited or inhibited by L-lactic acid (Fig. 1). The dynamic range of these cells is well within the concentration range of L-lactic acid given off by a human hand (9–25 × 10^{-11} mol/s, calculated from the data of Smith et al 1970). The receptor is relatively specific, its optimal stimulus configuration being a 3-carbon, α-hydroxy, monocarboxylic acid: L-lactic acid (Davis 1988). The α side

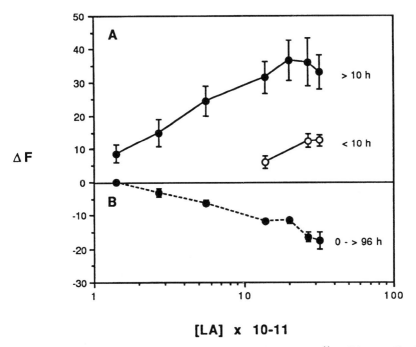

[LA] x 10-11

FIG. 1. Effect of L-lactic acid (LA, concentration given as 10^{-11} mol/s) on cells. ΔF represents the difference in nerve spike frequency between spontaneous activity and the peak phasic response to stimulus presentation. (A) Average dose responses of all (low and high sensitivity) L-lactic acid-excited cells from adult *Aedes atropalpus* females over 10 h old ($\bullet$) ($n = 24$ cells) and under 10 h old ($\circ$) ($n = 8$ cells). (B) Average dose responses of L-lactic acid-inhibited cells ($n = 11$) from females that ranged in age from newly emerged to over 96 h (from Bowen et al 1994a).

chain requirement is not rigidly specific but alterations in chain length, the monocarboxylic acid group, the number of side chains or the number of saturated sites significantly decrease neuronal responsiveness (Davis 1988).

L-Lactic acid receptors are associated with the sensilla basiconica or grooved pegs (McIver 1974). There are at least two types of sensilla basiconica: those which contain L-lactic acid-excited cells and those which do not. The presence of the cells is correlated with sensillar length. L-Lactic acid-excited cells are found only in the shortest grooved pegs in *Ae. aegypti*, *Aedes atropalpus*, *Aedes epactius* and *Culex pipiens* (Table 1). L-Lactic acid-inhibited cells are found in either short or long grooved pegs, depending on the species. These two different types of grooved pegs are readily distinguished on electron micrographs of antennae of each of these species (Bowen 1995).

High L-lactic acid sensitivity is essential for the expression of host seeking: when sensitivity is low, host seeking is not expressed. The control of receptor sensitivity is one way that mosquitoes control the expression of this behaviour. Receptor sensitivity is governed to a large extent by endogenous physiological events. A number of different reproductive strategies and blood-feeding patterns have evolved in mosquitoes, and the role of the L-lactic acid receptor in the modulation of host location behaviour is species specific and adapted to fit specific reproductive strategies.

Anautogeny

The first species in which this phenomenon was explored was *Ae. aegypti*. This species is anautogenous, i.e. it has an absolute requirement for a blood meal for egg maturation. Newly-emerged *Ae. aegypti* females do not express host location immediately after adult emergence, rather the behaviour takes several days to appear. During this time the receptors undergo an apparent maturation at the end of which receptor sensitivity is high and host seeking is manifest (Davis 1984a). Host seeking is inhibited in *Ae. aegypti* after a vitellogenic blood meal in two stages. After a threshold volume of blood is ingested, host seeking is inhibited by abdominal distension, a process mediated by neural signals originating in abdominal stretch receptors (Klowden & Lea 1979a). L-Lactic acid receptor sensitivity remains high during this period (Davis 1984b). After distension is alleviated, a brain decapeptide appears in the haemolymph which inhibits host seeking until the eggs are laid (Klowden & Lea 1979b, Brown et al 1994). Hormonal inhibition of host seeking involves the down-regulation (decrease in sensitivity) of the L-lactic acid-excited receptors, which subsequently maintains host seeking inhibition for the duration of the gonotrophic cycle (Davis 1984b).

TABLE 1   Relationship of length to electrophysiological response in the grooved peg type sensilla of four mosquito species [a]

Species	Response to L-lactic acid								
	Excited	Range	n	Inhibited	Range	n	None	Range	n
Aedes aegypti	7.4 ± 0.1 (a)	5.6–11.4	106	7.6 ± 0.2 (a)	5.6–8.6	46	8.5 ± 0.3 (b)	5.7–11.4	21
Aedes atropalpus	7.8 ± 0.2 (a)	5.0–10.0	50	8.9 ± 0.3 (b)	6.0–10.9	24	9.7 ± 0.3 (b)	6.0–13.6	41
Culex pipiens	9.7 ± 0.2 (a)	7.0–13.6	62	9.9 ± 0.4 (a)	7.1–11.8	10	10.5 ± 0.2 (b)	8.0–13.6	33
Aedes epactius	7.0 ± 0.3 (a)	5.0–10.9	42	ND	ND	ND	8.8 ± 0.2 (b)	6.0–11.8	58

[a]Mean sensillar length (μm) ± SEM. Average lengths within a species that are followed by a different number (a, b) are significantly different ($P \leq 0.01$, analysis of variance).
ND, not determined. Insufficient data for analysis.
Data from Bowen (1995).

Obligate autogeny

Ae. atropalpus undergoes obligate autogeny during its first gonotrophic cycle, i.e. a blood meal is not required for egg maturation nor is one taken (O'Meara 1985). Adults emerge with extensive reserves accumulated during larval development (Van Handel 1976). Egg maturation proceeds without interruption after adult emergence, and a full complement of eggs (about 170/female) is produced autogenously. Females are not host responsive during the first gonotrophic cycle (Bowen et al 1994b).

In this species L-lactic acid receptor sensitivity develops rapidly after emergence, and receptor sensitivity is high both during and after egg development (Table 2). In contrast to *Ae. aegypti*, haemolymph transfers from females with developing or fully developed eggs into either nulliparous *Ae. aegypti* or parous non-gravid *Ae. atropalpus* did not induce inhibition of host seeking in recipients, and ovariectomy did not release host seeking in operated females (Bowen et al 1994b). Distension inhibition due to large numbers of developing eggs is apparently responsible for the absence of host seeking during the first gonotrophic cycle. A hormone that inhibits host seeking by down-regulating the antennal host attractant receptors (as is found in blood-fed vitellogenic *Ae. aegypti*) is not present in *Ae. atropalpus* either during or after autogenous egg development.

Facultative autogeny

Aedes bahamensis can emerge as either autogenous or anautogenous adults, depending on larval nutrition (O'Meara et al 1993). Food-deprived larvae develop into small, anautogenous females. Larvae fed to repletion develop into large autogenous adults.

TABLE 2 L-Lactic acid-excited cells in *Aedes atropalpus*

Age (h)	Behaviour[a]	Gonotrophic status	High (%)	Low (%)	Other[b]	Number of sensilla	Number of females
0–12	NR	nulliparous[c]	0	23	77	31	11
12–24	NR	gravid	50	11	39	18	7
24–96	NR	gravid	31	25	44	36	14
> 96	R	parous	22	44	33	9	4

[a]NR, non-host responsive, non-host seeking; R, host responsive, host seeking.
[b]Includes cells that had no spike activity, cells that had no response to any stimulus tested and cells that displayed non-specific responses to more than one stimulus.
[c]Nulliparous *Ae. atropalpus* females were defined as those between the ages of 0 and 10–12 h of age, i.e. the time during which previtellogenic development is completed at 27°C.
Data from Bowen et al (1994).

In autogenous females L-lactic acid receptor development is rapid and complete (Table 3). In newly emerged females spikes were present in all neurons examined and 23% were already of high sensitivity. High sensitivity receptors were found in 100% of the females examined 48–72 h after emergence. The percentage of high sensitivity neurons increased steadily to 83% by 96 h after emergence. The number of low sensitivity, non-specific or non-responsive neurons decreased from 77% at emergence to 17% by the time the mosquitoes were over 96 h old.

Anautogenous females showed a different pattern of receptor development (Table 3). Half of the neurons examined in newly emerged females had no spike activity and only 2% were of high sensitivity. High sensitivity receptors were found in only 50% of females 96 h old or older. The percentage of low sensitivity, non-specific or non-responsive neurons showed no appreciable change as the mosquitoes aged from 0 to over 96 h old.

These developmental patterns in the peripheral sensory system are consistent with the observed olfactory mediated behaviour of both anautogenous and autogenous females. Host seeking is expressed by both but the response is more robust in autogenous females (Bowen & Romo 1995a). In fact, anautogenous females were reluctant to seek hosts unless deprived of sugar (Bowen & Romo 1995b). Anautogenous females fed on sugar more frequently than autogenous females (Bowen & Romo 1995a) and, when given a choice between the odours of a blood meal host and those of a plant sugar source (fruit), they were equally likely to approach either one even when mature (Table 4).

TABLE 3   Summary of electrophysiological responses of L-lactic acid-excited neurons in autogenous and anautogenous *Aedes bahamensis* females

	Age (h)	% High sensitivity	% Other[a]	% Spikes absent	Number of sensilla	% Females with high sensitivity	Number of females
Autogenous	0–24	23	77	0	22	60	5
	24–48	15	85	0	20	60	5
	48–72	40	60	0	20	100	5
	72–96	68	32	0	22	100	5
	⩾96	83	17	0	6	100	2
Anautogenous	0–24	2	47	51	45	11	9
	24–48	18	70	12	57	58	12
	48–72	20	80	0	20	60	5
	⩾96	35	65	0	34	50	8

[a]Includes neurons that displayed low sensitivity, non-specific responses, or no responses to any stimulus tested. Data from Bowen et al (1995).

TABLE 4   Responses of autogenous, parous and anautogenous *Aedes bahamensis* **females to a fruit–host choice in a two-choice olfactometer**

	Age (days)	n	Choice Host (%)	Fruit (%)	P
Autogenous	1–2	69	31.9	68.1	***
	8–10	63	66.7	33.3	***
Parous	1–2	ND	ND	ND	ND
	8–10	49	67.3	32.7	***
Anautogenous	1–2	55	36.4	63.6	**
	8–10	60	53.3	46.7	NS

n, number of females; ND, not determined; NS, not significant.
***$P < 0.001$.
**$P < 0.01$.
Data from Bowen et al (1995).

Although these results were initially surprising, they are understandable when one considers the reproductive advantages which accrue from blood feeding in autogenous and anautogenous *Ae. bahamensis*. The autogenous egg clutch of *Ae. bahamensis* females is relatively small: approximately 60 eggs/female are produced in this way. Blood feeding almost doubles egg production from 60 to over 100 eggs/female (Bowen et al 1995). Although anautogenous females have an absolute requirement for blood feeding in order to mature eggs, the number of eggs that they produce upon ingestion of a replete blood meal increases as they become older (Bowen et al 1995). Presumably, the more extensive the reserves accumulated as a result of adult sugar feeding are (i.e. the longer adults feed on sugar after emergence), the more reserves are available for reproductive purposes and, consequently, the more eggs are matured.

The physiological explanation for the lack of high sensitivity receptors in anautogenous females is unknown. Two obvious possibilities are: (1) the anautogenous sensory system matures at a much slower rate in anautogenous compared to autogenous females; and/or (2) intense sugar feeding by anautogenous females affects haemolymphatic changes which, in turn, affect L-lactic acid receptor function. The second hypothesis offers a potential explanation for the correlation between sugar feeding and the absence of host seeking observed in other mosquitoes (Foster 1995).

Diapause

Adult diapause in mosquitoes, as in other insects, is characterized by delayed reproduction, fat body hypertrophy, low metabolic rate and other behavioural and metabolic changes that enhance survival during inclement conditions brought about by seasonal change. *Cx. pipiens* females enter adult diapause in response to short daylengths experienced during larval and pupal development. The follicles of diapausing adult females fail to undergo previtellogenic development and remain teneral until diapause is terminated. During the first several weeks of adult life diapausing females refused to seek hosts (Bowen et al 1988), but they fed frequently on sugar and displayed clear olfactory preferences for plant-related odours instead of blood meal host odours (Bowen 1992b). Females forced to take a blood meal retained only a small proportion of the blood ingested and were incapable of either digesting it completely or utilizing it for either reproduction or augmentation of fat body reserves (Mitchell & Briegel 1989a,b). Non-diapausing females became host responsive after feeding on sugar for the first few days of adult life (Bowen 1992b, Bowen et al 1988). These mosquitoes are anautogenous and require a blood meal for egg maturation.

L-Lactic acid receptor sensitivity parallels closely the developmental features of diapause and non-diapause development in *Cx. pipiens*. Non-diapausing females almost invariably possess high sensitivity L-lactic acid receptors; such receptors are never found on diapausing females (Bowen et al 1988). The termination of diapause is associated not only with the initiation of ovarian follicle development but also with the appearance of high sensitivity receptors (Bowen 1990).

Table 5 summarizes L-lactic acid receptor development and its regulation during egg maturation in mosquitoes.

TABLE 5   Reproductive strategy, L-lactic acid receptor development and host-seeking behaviour in five mosquito species

Reproductive strategy	Anautogeny			Autogeny	
	Obligate	Facultative	Diapause	Obligate	Facultative
Host-seeking behaviour	present	delayed (if sugar present)	delayed	absent	optional
L-lactic acid receptor development	complete	incomplete (or modulated)	incomplete	complete	complete
Down-regulation of receptors during egg development	yes	ND	yes[a]	no	ND
Species	*Aedes aegypti*	*Aedes bahamensis*	*Culex pipiens*	*Aedes atropalpus*	*Aedes bahamensis*

[a]M. F. Bowen, unpublished observations 1987. ND, not determined.
Data from Bowen et al (1995).

Summary

Insects not only regulate the amount of food they ingest but also express nutritional preferences that depend on physiological state. How this is accomplished is not well understood and is a topic of much current interest (Simpson & Raubenheimer 1993). Mosquitoes express nutritional preferences for plant sugar or blood depending on physiological state and developmental age. The modulation of host attractant receptor sensitivity is one way by which this preference is implemented.

Acknowledgement

This work was supported by National Institutes of Allergy and Infectious Diseases grants AI 23336 and AI 21267.

References

Bowen MF 1990 Post-diapause sensory responsiveness in *Culex pipiens*. J Insect Physiol 6:923–929
Bowen MF 1991 The sensory physiology of host-seeking behavior of mosquitoes. Ann Rev Entomol 36:139–158
Bowen MF 1992a Terpene-sensitive receptors in female *Culex pipiens* mosquitoes: electrophysiology and behavior. J Insect Physiol 38:759–764
Bowen MF 1992b Patterns of sugar feeding in diapausing and nondiapausing *Culex pipiens* (Diptera: Culicidae) females. J Med Entomol 29:843–849
Bowen MF 1995 Sensilla basiconica (grooved pegs) on the antennae of female mosquitoes: electrophysiology and morphology. Entomol Exp Appl 77:233–238
Bowen MF, Romo J 1995a Host-seeking and sugar-feeding in the autogenous mosquito *Aedes bahamensis* (Diptera: Culicidae). J Vector Ecol 20:195–202
Bowen MF, Romo J 1995b Sugar deprivation in *Aedes bahamensis* females and its effects on host seeking and longevity. J Vector Ecol 20:211–215
Bowen MF, Davis EE, Haggart DA 1988 A behavioral and sensory analysis of host-seeking behaviour in the diapausing mosquito *Culex pipiens*. J Insect Physiol 34:805–813
Bowen MF, Davis EE, Romo J, Haggart D 1994a Lactic acid sensitive receptors in the autogenous mosquito *Aedes atropalpus*. J Insect Physiol 40:611–615
Bowen MF, Davis EE, Haggart D, Romo J 1994b Host-seeking behaviour in the autogenous mosquito *Aedes atropalpus*. J Insect Physiol 40:511–517
Bowen MF, Haggart D, Romo J 1995 Long distance orientation, nutritional preference, and electrophysiological responsiveness in the mosquito *Aedes bahamensis*. J Vector Ecol 20:203–210
Brown MR, Klowden MJ, Crim JW, Young L, Shrouder LA, Lea AO 1994 Endogenous regulation of mosquito host-seeking behaviour by a neuropeptide. J Insect Physiol 40:399–406
Clements A 1993 Behavioral causation: a new look at mosquito activities. In: Borovsky V, Spielman A (eds) Host regulated developmental mechanisms in vector arthropods: proceedings of a symposium, Vero Beach, Florida. Entomological Society of America, Lanham, MD, p 302–308

Cork A 1996 Olfactory basis of host location by mosquitoes and other haematophagous Diptera. In: Olfaction in mosquito–host interactions. Wiley, Chichester (Ciba Found Symp 200) p 71–88

Davis EE 1984a Development of lactic acid-receptor sensitivity and host-seeking behaviour in newly emerged female *Aedes aegypti* mosquitoes. J Insect Physiol 30:211–215

Davis EE 1984b Regulation of sensitivity in the peripheral chemoreceptor systems for host-seeking behaviour by haemolymph-borne factor in *Aedes aegypti*. J Insect Physiol 30:179–183

Davis EE 1988 Structure–response relationship of the lactic acid-excited neurons in the antennal grooved-peg sensilla of the mosquito *Aedes aegypti*. J Insect Physiol 34:443–449

Ellin RI, Farrano RL, Oberst FW et al 1974 An apparatus for the detection and quantitation of volatile human effluents. J Chromatogr 100:137–152

Feinsod FM, Spielman A 1979 An olfactometer for measuring host-seeking behavior of female *Aedes aegypti* (Diptera: Culicidae). J Med Entomol 15:282–285

Foster WA 1995 Mosquito sugar feeding and reproductive energetics. Annu Rev Entomol 40:443–474

Geier M, Sass H, Boeckh J 1996 A search for components in human body odour that attract females of *Aedes aegypti*. In: Olfaction in mosquito–host interactions. Wiley, Chichester (Ciba Found Symp 200) p 132–148

Grant AJ, O'Connell RJ 1996 Electrophysiological responses from receptor neurons in mosquito maxillary palp sensilla. In: Olfaction in mosquito–host interactions. Wiley, Chichester (Ciba Found Symp 200) p 233–253

Kline DL 1994 Olfactory attractants for mosquito surveillance and control: 1-octen-3-ol. J Am Mosq Control Assoc 10:280–287

Klowden MJ, Lea AO 1979a Abdominal distention terminates subsequent host-seeking behaviour of *Aedes aegypti* following a blood meal. J Insect Physiol 25:583–585

Klowden MJ, Lea AO 1979b Humoral inhibition of host-seeking in *Aedes aegypti* during oocyte maturation. J Insect Physiol 25:231–235

Krotoszynski B, Gabriel G, O'Neill H 1977 Characterization of human expired air: a promising investigative and diagnostic technique. J Chromatogr Sci 15:239–244

McIver S 1978 Structure of sensilla trichodea of female *Aedes aegypti* with comments on innervation of antennal sensilla. J Insect Physiol 24:383–390

McIver SB 1974 Fine structure of antennal grooved-pegs of the mosquito, *Aedes aegypti*. Cell Tissue Res 153:327–337

Mitchell CJ, Briegel H 1989a Inability of diapausing *Culex pipiens* (Diptera: Culicidae) to use blood for producing lipid reserves for overwinter survival. J Med Entomol 26:318–326

Mitchell CJ, Briegel H 1989b Fate of the blood meal in force-fed diapausing *Culex pipiens* (Diptera: Culicidae). J Med Entomol 26:332–341

Muir LE, Kay BH, Thorne MJ 1992 *Aedes aegypti* (Diptera: Culicidae) vision: responses to stimuli from the optical environment. J Med Entomol 29:445–450

O'Meara GF 1985 Gonotrophic interactions in mosquitoes: kicking the blood-feeding habit. Fla Entomol 68:122–133

O'Meara GF, Larson VL, Mook DH 1993 Blood feeding and autogeny in the peridomestic mosquito *Aedes bahamensis* (Diptera: Culicidae). J Med Entomol 30:378–383

Pile MM, Simmonds MSJ, Blaney WM 1991 Odour-mediated upwind flight of *Culex quinquefasciatus* mosquitoes elicited by a synthetic attractant. Physiol Entomol 16:77–86

Simpson SJ, Raubenheimer D 1993 The central role of the hemolymph in the regulation of nutrient intake in insects. Physiol Entomol 18:395–403

Smith CN, Smith N, Gouck HK et al 1970 L-Lactic acid as a factor in the attraction of *Aedes aegypti* (Diptera: Culicidae) to human hosts. Ann Entomol Soc Am 63:760–770

Takken W 1991 The role of olfaction in host-seeking of mosquitoes: a review. Insect Sci Appl 12:287–295

Van Handel E 1976 The chemistry of egg maturation in the unfed mosquito *Aedes atropalpus*. J Insect Physiol 22:521–522

DISCUSSION

Boeckh: Certain factors in the haemolymph, such as serotonin, decrease the sensitivity of the L-lactic acid receptors. I would like your opinion of the nature of this inhibition. And have you done any haemolymph-transfer experiments?

Bowen: No, I haven't. This would be an interesting experiment to do. What inspired me to talk about nutritional preferences is the work on phytophagous insects by Simpson & Raubenheimer (1993). They found that changes in haemolymph quality affect nutritional preference, and they have suggested that this effect is mediated peripherally. However, not much has been done in this area, so I would be hard pressed to come up with a suitable hypothesis for the mechanisms that might be involved.

Boeckh: The regulation of the sensitivity of chemoreceptors by these factors is also important in general terms. I have not followed the literature very well, but I think there has been some speculation that some sort of neurotransmitter or neurohormone is involved.

Guerin: Modulation of chemoreceptor sensitivity could be nutrient derived or based on satiation. If you blow up the mosquito abdomen with water, do you get the same effect?

Bowen: That's a good point. We had difficulties with getting *Aedes bahamensis* females to host seek. They fed on sugar so frequently that distention could be a factor, but distention in the case of *Aedes aegypti* doesn't affect the peripheral receptors.

Galun: Those autogenous mosquitoes are atypical. In the case of *Culex pipiens molestus*, which are 100% autogenous, all of them lay eggs before they are ready to take a blood meal at the age of eight to 10 days.

Bowen: The mosquitoes that I worked with were prevented from laying eggs by placing a screen over their water source that was high enough to prevent oviposition but not drinking.

Galun: You showed that they seek the host at the age of two to three days. At this age they are normally not ready to lay eggs anyway.

Bowen: They started to lay eggs two to three days after emergence, and they took a blood meal even when gravid. By Day 6 they were fully gravid and they hadn't been allowed to lay eggs, but they took a blood meal and developed yet more eggs. I didn't see any evidence of egg absorption for at least 10 days after emergence.

Brady: I'm exercised by this insertion of the words 'host seeking', and I would like to ask Mary Bowen how she knows when a mosquito is host seeking?

Bowen: It's teleological, I agree.

Klowden: What would you recommend in its place?

Brady: That it is flying upwind, or that it is responding to CO_2 or L-lactic acid. Host seeking is a loaded term and does not help the behavioural analysis.

Boeckh: Perhaps one could define it if one knew the motivation of the animal.

Brady: But 'motivation' is teleology in itself.

Boeckh: But the term 'motivation' can be expressed in several terms— hormonal state or hunger, for example. All of these can be measured and they are not teleological. They provide an internal signal for the animal to do something, and this is what we might call the correlate of motivation.

Brady: But how do you know when a mosquito is 'hungry', which is an anthropomorphic description of human feelings?

Boeckh: You could find out if there is a correlation between the level of blood sugar, for example, and a certain sort of behaviour.

Klowden: You can't ask the question: is a mosquito hungry? However, you can ask: will it feed if it is offered food?

Gibson: There is a logical problem with using a single behavioural assay to assess the 'behavioural state' of an insect in that the stimulus used in the assay may be involved with more than one behaviour. For example, if you define 'host seeking' as a positive response to chemical X in the wind tunnel, and chemical X is actually present both at a host and at an oviposition site, then you cannot be certain whether, in nature, the mosquito would have ended up at the host (i.e. 'host seeking') or at the oviposition site (i.e. 'oviposition site seeking'). It can be difficult to find a single behavioural response which would unambiguously correlate with what is in essence a long sequence of stimuli and responses, i.e. the location of a host. With respect to electroantennograms, we already know that it would be foolish to conclude that an antennal response to a given chemical which is found on a host means that the insect will fly upwind toward the host if it encounters that chemical in the field. A positive electroantennogram result is taken to be just that and not as an identification of a host cue.

Bowen: The term 'host response' is not specific. The mosquito could be responsive to gustatory cues and be taking a blood meal. Alternatively, it could approach a host from a distance in response to olfactory cues. This is why Klowden uses 'host seeking' as the terminology for long-distance approach to a blood meal resource.

Gibson: This also implies that you know the insect would end up at a host rather than at an oviposition site, and you don't know that.

Brady: That is where the teleology comes in. How do you define host seeking? Because it feeds when it lands on a host? But what happens if it lands on something else on the way? What was it doing then? Resting site seeking? The point is that you cannot tell which, except by what happens at the end, and that is explaining causes by their ends, i.e. teleology.

Davis: There are some other considerations with respect to oviposition versus 'host seeking', if I could continue to use that term for just a moment. Consider a mosquito orienting towards an odour source. If the odour source is associated with an oviposition site, the female will not orient toward the odour if she is not gravid. Conversely, if she is gravid she may not respond to host odours. However, in the case of some odours, e.g. 1-octen-3-ol, a mosquito might use the same odour for both host orientation and oviposition site orientation. Gravid *Aedes triseriatus* are somewhat attracted to 1-octen-3-ol, whereas only non-gravid *Culex salinus* and *Culex quinquefasciatus* are attracted to it. Thus, a single odour species may mediate two different and mutually exclusive behaviours. We don't have enough information yet to determine if this difference is a species-specific use of a common chemical signal or whether a common signal may truly mediate different behaviours. If the latter is the case then clearly additional sensor information would be necessary for the female to distinguish between the two different sources of the odour (Kline et al 1990).

Gibson: But sometimes mosquitoes take two blood meals before they lay eggs. How do they know what to do the second time? How can they know where they've been and what they're trying to do when they are responding to that single chemical?

Davis: If you examine the ovaries after the first blood meal, you may find that it was not sufficient for oogenesis, so they can't respond to the oviposition site. In this case the animal remains host seeking. This is sometimes observed with females from nutritionally deprived larvae.

Lehane: You can only really get sensible answers to these sorts of questions if you do the relevant field work. Drawing inferences from behaviours observed in the laboratory may be misleading.

Davis: We could do the following series of experiments in the laboratory on the same insect: (1) define the exact behavioural patterns in terms of what odour stimuli the mosquito responds to and orients to; (2) look at the electrophysiology and assess how the receptors are responding to specific odours, some of which are specific to the behaviour in question; and (3) dissect the ovaries to see if the females have mature eggs and if they are inseminated. Because all these factors will affect what the insect will do with the information it receives.

Brady: In my opinion, conducting field experiments will not help this semantic argument at all, because people in the field use terms such as host seeking, mate seeking and resting site seeking all the time. These terms are a convenient shorthand. What you have just described is a legitimate use of the term 'host seeking', but you have to define it in the first place. In Mary Bowen's series of experiments the definition is simple: it is a mosquito that flies upwind, which is the objective description of what was happening. The danger arises when you start to impute to the insect cognitive states and say that it's trying to do particular things.

Klowden: Would describing a missile as a heat-seeking missile be teleological? In the same way that mosquitoes move in the direction of a host, how can a heat-seeking missile move in the direction of a source of heat without being teleological?

Brady: A heat-seeking missile is in a sense a teleological piece of technology anyway because it's designed precisely to respond only to heat.

Cardé: One issue is that behavioural assays vary widely. Practically every team has several of their own unique bioassay set-ups, some of which clearly rely on wind. Few of these include visual stimuli, which could be as important as the chemicals that evaporate from the skin. For example, the experiments of Harris & Rose (1990), show that in the onion maggot fly one can increase the effectiveness of many oviposition-stimulating compounds by making sure that the correct visual stimulus and the 'right' tactile stimulus to walk around on are present. We tend to just look at the chemistry, but to some extent trying to separate chemicals from the stimuli that the insect normally interacts with, such as heat and visual stimuli in the case of mosquitoes, can be misleading.

Davis: Regarding terminology, in the area of insect repellent research, the term repellent referred to an operational definition of a substance that reduced biting. It had an end product that one could measure some time after the event had occurred. However, no one looked at how the insect was behaving, so that many substances became known as insect 'repellents'. If one considers the strict definition of repellent, as a substance that causes orientation away from the source, one finds that many of these substances are not repellents. Therefore, we should be careful not to confuse classical behavioural descriptions with operational definitions.

References

Harris MO, Rose S 1990 Chemical, color and tactile cues influencing the ovipositional behaviour of the Hessian fly (Diptera: Ecidomyiidae). Environ Entomol 19:303–308

Kline DL, Takken W, Wood JF, Carlson DA 1990 Field studies on the potential of butanone, carbon dioxide, honey extract, 1-octen-3-ol, L-lactic acid and phenols as attractants for mosquitoes. Med Vet Entomol 4:383–391

Simpson SJ, Raubenheimer D 1993 The central role of the hemolymph in the regulation of nutrient intake in insects. Physiol Entomol 18:395–403

Endogenous factors regulating mosquito host-seeking behaviour

Marc J. Klowden

Department of Entomology, University of Idaho, Moscow, ID 83844–2339, USA

Abstract. The physiological state of the mosquito can modulate behaviours that are normally activated by external stimuli. Even though host stimuli may be present at certain times, the insect may not always express host-seeking behaviour, depending upon the physiological factors that predominate. Traditional views of the gonotrophic cycle characterize mosquitoes as engaging in blood feeding only once before depositing eggs. However, physiological state, including such factors as age, nutritional state, presence of eggs, mating condition, circadian rhythmicity and the number of gonotrophic cycles completed, can affect the expression of this behaviour.

1996 Olfaction in mosquito–host interactions. Wiley, Chichester (Ciba Foundation Symposium 200) p 212–225

Of the approximately one million species of insects that have been identified, only a small number are known to be vectors of disease agents. Indeed, the ability to transmit parasites is a specialized trait, requiring an insect to evolve morphological adaptations, such as mouthparts that are able to penetrate skin; physiological adaptations, including the production of enzymes to digest and utilize the blood; and behavioural adaptations, involving mechanisms to identify and locate a host amid the considerable background noise in the environment. Some vectors spend a considerable portion of their life cycles directly on or within the habitat of the host, but those that must find a new host for a blood meal in order to develop eggs require an efficient way to locate a source of blood in order to reproduce. Therefore, one of the most important behaviours that has evolved in many vectors is this ability to engage in host seeking.

Little is known about the actual process in the field, but host seeking appears to consist of a number of sequential preprogrammed behaviours that are activated in response to changing concentrations of host stimuli. Chemical cues are the most important long-range attractants, with visual cues, heat and tactile stimuli playing a role as the vector approaches the host (Takken 1991). However, host-seeking behaviour is not always activated when host stimuli are perceived. Along with mechanisms to find a host, mechanisms to prevent the expression of host-seeking behaviour when it is not in the best interests of the vector have also evolved. The environment offers an overabundance of stimuli but to survive and reproduce, the insect must act only on those stimuli that are biologically relevant at a particular time. Physiological state can provide information that allows the mosquito to make these decisions about which behavioural programmes are appropriate.

The gonotrophic cycle

Because of the periodic need for exogenous proteins for the synthesis of yolk, there is an intimate relationship between blood ingestion and egg maturation in most hematophagous dipterans. The gonotrophic cycle (defined as the period between one full blood meal and the next, and the consequent egg development and behavioural changes that result) has played a significant role in our assumptions about what mosquitoes are actually doing following blood ingestion. Age grading of populations, timing and placement of pesticide applications, and estimations of biting frequency and the resulting level to which populations must be suppressed to affect transmission rates, are all based on the implications of the cycle. However, these conclusions have been developed largely by modelling the behaviour of laboratory-reared mosquitoes fed with adequate diets. Field studies demonstrate that populations often feed several times during a single cycle, contrary to what the implications of the gonotrophic cycle would predict. In order to better understand why mosquitoes feed when they do, it is necessary to understand the physiological underpinnings of behaviour.

The different phases of the gonotrophic cycle are associated with profoundly different physiological states, and changes in physiological state can influence the ways that exogenous stimuli are able to initiate behaviour. In the model of insect behaviour shown in Fig. 1, environmental stimuli acting through sensory receptors create patterns in the CNS that can be associated with preprogrammed behaviours. Physiological state, by influencing the degree of endogenous stimuli present, appears to affect the way that the CNS integrates the environmental information and associates it with a biologically appropriate behaviour.

In addition to the teneral period immediately following emergence, I have divided the gonotrophic cycle into three periods, based on the physiological alterations that result from blood ingestion. These include the pre-blood meal phase, post-blood meal phase and gravid phase (Fig. 2). Environmental stimuli appear to affect the mosquito differently depending on its physiological state. I will focus on the changes in behaviour that occur during each phase.

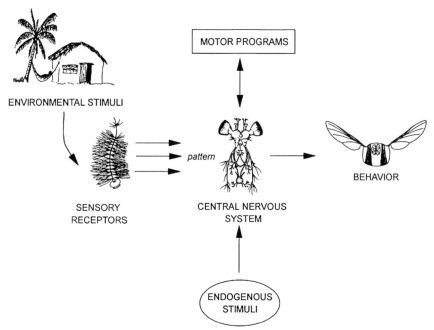

FIG. 1. Model for the interaction of exogenous and endogenous stimuli. Environmental stimuli impinging on the sensory receptors of the mosquito create a pattern in the signal that is sent to the CNS. The CNS evaluates the signal and associates the pattern with a particular motor programme that results in a behaviour. Endogenous physiological state affects the way these stimuli are associated with programmes.

Newly emerged mosquitoes

In many species the behaviours that together constitute eclosion are regulated by a circadian pacemaker, causing adult emergence at a particular time of the day. For the first 24–72 h after emergence, females generally refrain from seeking a host, apparently because the sensitivity of antennal receptors has yet to develop (Davis 1984). Similarly, newly emerged *Aedes aegypti* mosquitoes do not respond to mating attempts by males until juvenile hormone is released (Gwadz et al 1971). Juvenile hormone release is also necessary for the post-emergence development of biting behaviour in *Culex* mosquitoes (Meola & Petralia 1980), which contrasts with the *Aedes* sp. that have been examined (Meola & Readio 1988, Bowen & Davis 1988). Juvenile hormone that is present after emergence also mediates the maturation of ovarian follicles and several developmental events that prepare the fat body for vitellogenesis once blood is ingested (Raikhel 1992).

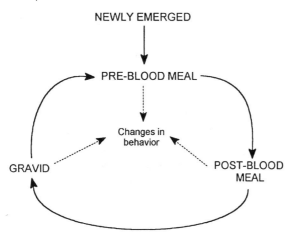

FIG. 2. The gonotrophic cycle of the mosquito can be divided into three phases centred around the ingestion of blood. Following a short teneral stage, the pre-blood meal, post-blood meal and gravid phases are all associated with changes in behaviour.

Male mosquitoes also require a period of post-emergence maturation before they can initiate mating attempts. In common with the males of most other dipterans, mosquitoes must rotate their terminalia 180° before they can mate (Roth 1948). *Ae. aegypti* males require about 24 h for this rotation, which results from the contraction of two pairs of opposing muscles within the rotating intersegmental region (Chevone & Richards 1976). The stimulus for muscle contraction is unknown; they are activated independently of any stimulus from the brain, but are slowed by the administration of methoprene to larvae and early pupae. Late pupae and adults are not affected by methoprene, suggesting that the commitment to rotation occurs during the early pupal stage (O'Donnell & Klowden 1996).

Male mosquitoes are attracted to the wing beat frequency of conspecific females for mating. Antennal hairs sense the vibrations from the female wing beat and activate the behavioural programmes associated with mating. In many species within the *Aedes* subgenus *Stegomyia*, these antennal hairs become erect within 12–24 h after emergence and remain so, allowing the male to mate at any time. In contrast, anopheline males undergo circadian periods of antennal erection mediated by the CNS, and they do not mate during the periods when these hairs are recumbent (Nijhout 1977).

Pre-blood meal phase

Once post-emergence maturation is complete, mosquitoes become more responsive to environmental stimuli, and preprogrammed behaviour patterns associated with those stimuli are expressed. Superimposed on this expression is a circadian rhythmicity that appears to result from one to several endogenous oscillators (Jones 1982). Activity rhythms, swarming and host-seeking behaviours are all based on an underlying circadian rhythmicity. As a preliminary to host seeking, the rhythmicity increases the female's activity and makes her more likely to encounter host stimuli during a species-specific window of opportunity, similar to the 'ranging' behaviour of the tsetse that occurs regularly in the absence of any host stimuli (Vale 1980). These rhythms are largely responsible for the tendency for some species to feed during the day, whereas others feed during crepuscular periods or at night. The rhythms can be altered by male accessory gland substances that are transferred from the male during mating (Jones & Gubbins 1978). Female mosquitoes express different behaviour patterns once they are mated.

Larval rearing conditions in the field are often less than optimal, and adults emerging from these conditions are smaller in size. A large proportion of mosquitoes reared in the laboratory under marginal conditions fail to engage in host-seeking behaviour (Klowden et al 1988), and those that do may have to ingest more than one blood meal to mature a batch of eggs (Feinsod & Spielman 1980). Smaller mosquitoes are also poorly represented in blood-fed field collections, suggesting that these smaller adults are not as successful as larger females in obtaining blood (Nasci 1986). Mosquitoes collected in the field contain reduced energy reserves compared to their counterparts reared in the lab (Day & Van Handel 1986).

Some mosquito species undergo an adult reproductive diapause in response to a changing photoperiod encountered during the larval stage. Diapausing females are characterized by both a hypertrophy of the fat body that is used as an energy store during the winter and a reduced tendency, induced by juvenile hormone deficiency, to engage in host-seeking behaviour (Case et al 1977). Carbohydrate ingested at this time is stored as lipid, but if blood does happen to be ingested, it is not used to synthesize more lipid (Mitchell & Briegel 1989). The reduced host-seeking behaviour of diapausing mosquitoes is correlated with a reduction in the sensitivity of sensory receptors that detect host stimuli (Bowen et al 1988).

In order to supplement metabolic reserves after emergence, both male and female mosquitoes ingest carbohydrate from flowers. Sugar-seeking behaviour appears to be initiated by floral scents (Healy & Jepson 1988) and is coincident with the circadian rhythmicity of general activity patterns (Yee & Foster 1992). *Ae. aegypti* that ingest large volumes of carbohydrate are subsequently less likely to engage in host-seeking behaviour (Foster & Eischen 1987).

Post-blood meal phase

The physiology and behaviour of mosquitoes change profoundly after a blood meal. The increased concentration of tryptic enzymes that digest the blood meal (Briegel & Lea 1975) involves a signal transduction pathway whereby the early trypsin activity activates transcription of the late trypsin gene (Barillas-Mury et al 1995). Other changes occur after a meal, including the increased ability to detoxify the pesticide dichlorodiphenyltrichloroethane (DDT), although the activity of several other pesticides is not diminished (Halliday & Feyereisen 1987).

The female is immediately affected in several ways by abdominal distension from the blood meal. As the abdomen increases in size, stretch receptors are triggered that mediate the termination of the ingestion of blood (Gwadz 1969) and mechanisms are initiated that excrete the large amount of water contained in the meal (Nijhout & Carrow 1978). Reproductive mechanisms are also enhanced when the blood meal is supplemented with additional distension (Klowden 1987). Upon reaching a certain volume threshold, abdominal distension curtails any further responsiveness to host stimuli. The threshold for volume varies with body size: smaller females resulting from marginal larval diets terminate their host-seeking behaviour with smaller ingested volumes (Klowden & Lea 1978). Females that are inflated with air or saline administered by enema also fail to engage in host-seeking behaviour, demonstrating that components specific to the blood meal, such as proteins, do not act as a stimulus (Klowden & Lea 1979a). A replete meal completely inhibits host seeking in *Ae. aegypti* and *Aedes albopictus* until eggs begin to develop, but *Anopheles gambiae* and *Anopheles freeborni,* which both have a limited gut capacity and can concentrate the blood meal, are not affected by this distension-induced inhibition to as great an extent (Klowden & Briegel 1994). This may explain their increased feeding and rates of pathogen transmission. It is not clear whether this inhibition of host-seeking behaviour by distension is mediated directly by the nervous system or by neurohormones released by distension.

The tendency to engage in host-seeking behaviour immediately following a blood meal is affected both by the age of the mosquito and its blood-feeding history. When 21 day-old mosquitoes ingested blood for the first time, it took less blood to terminate their host-seeking behaviour than for younger mosquitoes (Klowden & Lea 1980). However, if the older females had ingested prior blood meals there was no change in the distension threshold. Age is an important factor because the older segment of a mosquito population is the most likely to have acquired infectious agents.

Gravid phase

In *Ae. aegypti* and *Ae. albopictus*, as eggs develop and the volume of blood is reduced as digestion occurs, host-seeking behaviour remains unexpressed (Klowden & Briegel 1994). This oocyte-induced inhibition is the result of a second mechanism that involves a complex interplay between the ovaries, fat body and neurosecretory cells, as well as substances contributed by the male during mating. The defensive behaviour of mosquito hosts can be a powerful selective force (Edman & Kale 1971). This behaviour may have shaped the evolution of mechanisms that inhibit the approach to a host when it would not be of any reproductive benefit to the mosquito, and would certainly threaten her survival.

Blood ingestion triggers a cascade of endocrine events, including the release of neurosecretory hormones from the brain and 20-hydroxyecdysone from the ovaries. An endocrine signal from the ovaries is among the first of these events, which also initiates the second mechanism of host-seeking inhibition (Klowden 1981). Females ovariectomized before the blood meal or those that fail to develop eggs after blood ingestion never display oocyte-induced inhibition, demonstrating the requirement for an ovarian signal. When developing ovaries are removed within 8 h of blood ingestion, the inhibition of host seeking never develops, but if they remain for more than 12 h before being removed, the behavioural inhibition develops just as if they had always been present (Fig. 3A). This indicates the presence of a critical period for the release of an ovarian factor. By an as yet unidentified series of events, the ovarian signal activates the fat body to produce vitellogenin (the production of which is sustained by 20-hydroxyecdysone and sequestered by developing oocytes). The fat body also appears to trigger the release of a neuropeptide, *Ae. aegypti* head peptide I (Aea-HP-I) from neurosecretory cells in the CNS, which appears to directly inhibit host-seeking behaviour (Klowden et al 1987, Brown et al 1994). Transfusions of haemolymph from gravid females into non-gravid females inhibit the behaviour of recipients, as do injections of Aea-HP-I. Although one report attributed the behavioural inhibition directly to 20-hydroxyecdysone produced by the ovaries (Beach 1979), the large doses required and the timing of release suggest that the effect is pharmacological (Klowden 1982). The inhibition of behaviour is first evident by 30 h after blood ingestion, and 36–72 h after blood ingestion few gravid females respond to host stimuli (Fig. 3B). The hypothesis that defensive host behaviour was the selective pressure for the evolution of this mechanism is supported by the discovery that species of *Anopheles*, which generally feed during crepuscular periods or at night when hosts are less active and less defensive, continue to engage in host seeking and feed several times throughout the gonotrophic cycle (Klowden & Briegel 1994).

The humoral inhibition during oogenesis in *Ae. aegypti* is modulated by a number of physiological factors. Adult mosquitoes maintained on marginal diets are much more likely to engage in host seeking during the period that eggs are maturing compared to well-fed females (Klowden 1986). Inadequate adult nutrition can reduce the likelihood that a mosquito will develop eggs from a given volume of blood (Klowden 1986), and in the absence of egg development there will be no behavioural inhibition once abdominal distension from blood ingestion is reduced. The age of the mosquito also affects its expression of

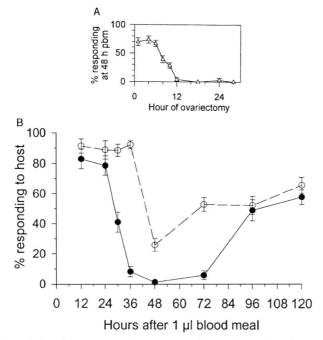

FIG. 3. (A) The effect of ovariectomy at times after the blood meal on the development of host-seeking inhibition at 48 h. When ovaries are removed up to 6 h after blood ingestion, no inhibition develops. When ovariectomy is delayed until after 8 h, host-seeking inhibition develops as if the ovaries were always present (data from Klowden 1981). (B) The percentage of *Aedes aegypti* mosquitoes responding to host stimuli after ingesting a 1 μl meal of blood. In mated females (●) the inhibition begins by 30 h and reaches a maximum between 36 and 72 h. Unmated females (○) show an initially delayed inhibition that never develops as strongly as in mated females (data from Klowden & Lea 1979b).

behavioural inhibition during oogenesis. Older mosquitoes, which have had more opportunity to acquire infectious agents, show less inhibition and are more likely to engage in host seeking while carrying eggs, a trend that is accentuated by prior gonotrophic cycles (Klowden & Lea 1984). The inactivation of the effect of male accessory gland substances over time may explain this change with age (Klowden & Fernandez 1996).

Mating modulates the inhibition of behaviour. Mated females are less likely to engage in host-seeking behaviour while gravid than are unmated females (Klowden & Lea 1979b; Fig. 3B). Males tend to aggregate in the vicinity of a host (Hartberg 1971), and the reduced inhibition of host seeking by unmated females may serve as a mechanism to allow them to return to the host for mating. Male accessory gland substances transferred during mating are responsible for this modulation, and the injection of accessory gland homogenates into unmated gravid females mimics the effect of mating (Fernandez & Klowden 1995). Male mosquitoes reared under marginal nutritional conditions have reduced amounts of total accessory gland proteins, and they also transfer reduced amounts of protein to females, resulting in increased host-seeking behaviour of the females to which they have mated (Fernandez & Klowden 1995). The male accessory gland substances appear to act specifically to alter metabolic priorities rather than by increasing the nutritional state of the female by providing needed precursors (Klowden 1993).

As the gravid phase proceeds, mosquitoes become more sensitive to oviposition site stimuli. The terrestrial female must change her behaviour to bring her to aquatic sites so that immatures may survive. This behavioural switch occurs as a result of another humoral factor that is released during oogenesis: transfusion of haemolymph from gravid to non-gravid females causes these non-gravid females to engage in pre-oviposition (Klowden & Blackmer 1987). Most mosquitoes are unable to lay mature eggs until they have mated, a behavioural block once again removed by male accessory gland substances (Leahy & Craig 1965). Likewise, pre-oviposition behaviour cannot occur unless male accessory gland substances signal that the female is mated and the eggs to be laid will be viable (Yeh & Klowden 1990). An important but unknown factor is the priority of these behaviours and how the gravid female responds when confronted with both host stimuli and oviposition site stimuli.

The inhibition of host-seeking behaviour is terminated following oviposition as a result of neural stimuli that can gauge the retention of eggs. Regardless of the number of eggs oviposited, host-seeking behaviour does not return if more than four eggs are retained in the ovary (Klowden 1981). Following oviposition, the female engages in host seeking and the ingestion of another meal of blood to begin a second gonotrophic cycle and the physiological changes that regulate her behaviour.

Endogenous control of behaviour

Following a brief maturation phase after emergence, host seeking becomes the default behaviour during a circadian window of mosquito activity. The expression of host-seeking behaviour is modulated by endogenous factors, such as nutritional and mating states, abdominal distension and several hormones that are released during ovarian maturation. Under ideal conditions, these mechanisms cause mosquitoes to remain refractory to host stimuli after a blood meal. However, given the changes in physiological state that occur after blood ingestion and the conditions experienced by mosquitoes under natural conditions, it is understandable how variations from the classical scheme can occur. The ability to exploit these factors may provide us with a novel way to control mosquito behaviour.

Acknowledgements

The research reported from the author's laboratory was sponsored in part by the National Institutes of Health (AI-24453) and the National Science Foundation (INT-8813376 and IBN-92243570).

References

Barillas-Mury C, Noriega FG, Wells MA 1995 Early trypsin activity is part of the signal transduction system that activates transcription of the late trypsin gene in the midgut of the mosquito, *Aedes aegypti*. Insect Biochem Mol Biol 25:241–246

Beach R 1979 Mosquitoes: biting behavior inhibited by ecdysone. Science 205:829–831

Bowen MF, Davis EE 1988 The effects of allatectomy and juvenile hormone replacement on the development of host-seeking behavior and lactic acid receptor sensitivity in the mosquito *Aedes aegypti*. Med Vet Entomol 3:53–60

Bowen MF, Davis EE, Haggart DA 1988 A behavioural and sensory analysis of host-seeking behaviour in the diapausing mosquito *Culex pipiens*. J Insect Physiol 34:805–813

Briegel H, Lea AO 1975 Relationship between protein and proteolytic activity in the midgut of mosquitoes. J Insect Physiol 21:1597–1604

Brown MR, Klowden MJ, Crim JW, Young L, Shrouder LA, Lea AO 1994 Endogenous regulation of mosquito host-seeking behavior by a neuropeptide. J Insect Physiol 40:399–406

Case TJ, Washino RK, Dunn RL 1977 Diapause termination in *Anopheles freeborni* with juvenile hormone mimics. Entomol Exp Appl 21:155–162

Chevone BI, Richards AG 1976 Ultrastructure of the atypic muscles associated with terminalial inversion in male *Aedes aegypti* (L.). Biol Bull 151:283–296

Davis EE 1984 Development of lactic acid-receptor sensitivity and host-seeking behaviour in newly emerged female *Aedes aegypti* mosquitoes. J Insect Physiol 30:211–215

Day JF, Van Handel E 1986 Differences between the nutritional reserves of laboratory-maintained and field-collected adult mosquitoes. Environ Entomol 2:154–157

Edman JD, Kale WH II 1971 Host behavior: its influence on the feeding success of mosquitoes. Ann Entomol Soc Am 64:513–516

Feinsod FM, Spielman A 1980 Nutrient-mediated juvenile hormone secretion in mosquitoes. J Insect Physiol 26:113–117

Fernandez NM, Klowden MJ 1995 Male accessory gland substances modify the host-seeking behavior of gravid *Aedes aegypti* mosquitoes. J Insect Physiol 41:965–970

Foster WA, Eischen FA 1987 Frequency of blood-feeding in relation to sugar availability in *Aedes aegypti* and *Anopheles quadrimaculatus* (Diptera: Culicidae). Ann Entomol Soc Am 80:103–108

Gwadz RW 1969 Regulation of blood meal size in the mosquito. J Insect Physiol 15:2039–2044

Gwadz RW, Lounibos LP, Craig GB Jr 1971 Precocious sexual receptivity induced by a juvenile hormone analogue in females of the yellow fever mosquito *Aedes aegypti*. Gen Comp Endocrinol 16:47–51

Halliday WR, Feyereisen R 1987 Why does DDT toxicity change after a blood meal in adult female *Culex pipiens*? Pest Biochem Physiol 28:172–181

Hartberg WK 1971 Observations on the mating behavior of *Aedes aegypti* in nature. Bull WHO 45:847–850

Healy TP, Jepson PC 1988 The location of floral nectar sources by mosquitoes: the long range responses of *Anopheles arabiensis* Patton (Diptera: Culicidae) to *Achillea millefolium* flowers and isolated floral odour. Bull Entomol Res 78:651–657

Jones MDR 1982 Coupled oscillators controlling circadian flight activity in the mosquito, *Culex pipiens quinquefasciatus*. Physiol Entomol 7:281–289

Jones MDR, Gubbins SJ 1978 Changes in the circadian flight activity of the mosquito *Anopheles gambiae* in relation to insemination, feeding and oviposition. Physiol Entomol 3:213–220

Klowden MJ 1981 Initiation and termination of host-seeking inhibition in *Aedes aegypti* during oocyte maturation. J Insect Physiol 27:799–803

Klowden MJ 1982 Nonspecific effects of large doses of 20-hydroxy-ecdysone on the behavior of *Aedes aegypti*. Mosq News 42:184–189

Klowden MJ 1986 Effects of sugar deprivation on the host-seeking behaviour of gravid *Aedes aegypti* mosquitoes. J Insect Physiol 32:479–483

Klowden MJ 1987 Distention-mediated egg maturation in the mosquito *Aedes aegypti*. J Insect Physiol 33:83–87

Klowden MJ 1993 Mating and nutritional state affect the reproduction of *Aedes albopictus* mosquitoes. J Am Mosq Control Assoc 9:169–173

Klowden MJ, Blackmer JL 1987 Humoral control of pre-oviposition behaviour in the mosquito, *Aedes aegypti*. J Insect Physiol 33:689–692

Klowden MJ, Briegel H 1994 Mosquito gonotrophic cycle and multiple feeding potential: contrasts between *Anopheles* and *Aedes* (Diptera: Culicidae). J Med Entomol 31:618–622

Klowden MJ, Fernandez NM 1996 Effects of age and mating on the host-seeking behavior of *Aedes aegypti* mosquitoes, submitted

Klowden MJ, Lea AO 1978 Blood meal size as a factor affecting continued host-seeking by *Aedes aegypti* (L.). Am J Trop Med Hyg 27:827–831

Klowden MJ, Lea AO 1979a Abdominal distension terminates subsequent host-seeking behaviour of *Aedes aegypti* following a blood meal. J Insect Physiol 25:583–585

Klowden MJ, Lea AO 1979b Humoral inhibition of host-seeking in *Aedes aegypti* during oocyte maturation. J Insect Physiol 25:231–235

Klowden MJ, Lea AO 1980 Physiologically old mosquitoes are not necessarily old physiologically. Am J Trop Med Hyg 29:1460–1464

Klowden MJ, Lea AO 1984 Blood feeding affects age-related changes in the host-seeking behavior of *Aedes aegypti* (Diptera: Culicidae) during oocyte maturation. J Med Entomol 21:274–277

Klowden MJ, Davis EE, Bowen MF 1987 Role of the fat body in the regulation of host-seeking behaviour in the mosquito, *Aedes aegypti*. J Insect Physiol 33:643–646

Klowden MJ, Blackmer JL, Chambers GM 1988 Effects of larval nutrition on the host-seeking behavior of adult *Aedes aegypti* mosquitoes. J Am Mosq Control Assoc 3:73–75

Leahy MG, Craig GB Jr 1965 Accessory gland substance as a stimulant for oviposition in *Aedes aegypti* and *Ae. albopictus*. Mosq News 25:448–452

Meola R, Readio J 1988 Juvenile hormone regulation of biting behavior and egg development in mosquitoes. Adv Dis Vect Res 5:1–24

Meola RW, Petralia RS 1980 Juvenile hormone induction of biting behavior in *Culex* mosquitoes. Science 209:1548–1550

Mitchell CJ, Briegel H 1989 Inability of diapausing *Culex pipiens* (Diptera: Culicidae) to use blood for producing lipid reserves for overwinter survival. J Med Entomol 26:318–326

Nasci RS 1986 The size of emerging and host-seeking *Aedes aegypti* and the relation of size to blood-feeding success in the field. J Am Mosq Control Assoc 2:61–62

Nijhout HF 1977 Control of antennal hair erection in male mosquitoes. Biol Bull 153:591–603
Nijhout HF, Carrow GM 1978 Diuresis after a blood meal in female *Anopheles freeborni*. J Insect Physiol 24:293–298
O'Donnell P, Klowden MJ 1996 Male terminalia rotation in *Aedes aegypti* is inhibited by methoprene, submitted
Raikhel AS 1992 Vitellogenesis in mosquitoes. Adv Dis Vect Res 9:1–39
Roth LM 1948 A study of mosquito behavior. An experimental laboratory study of the sexual behavior of *Aedes aegypti* (Linnaeus). Am Midl Natl 40:265–352
Takken W 1991 The role of olfaction in host-seeking of mosquitoes: a review. Insect Sci Appl 12:287–295
Vale GA 1980 Flight as a factor in the host-finding behaviour of tsetse flies (Diptera: Glossinidae). Bull Entomol Res 70:299–307
Yee WL, Foster WA 1992 Diel sugar-feeding and host-seeking rhythms in mosquitoes (Diptera: Culicidae) under laboratory conditions. J Med Entomol 29:784–791
Yeh CC, Klowden MJ 1990 Effects of male accessory gland substances on the pre-oviposition behaviour of *Aedes aegypti* mosquitoes. J Insect Physiol 36:799–803

DISCUSSION

Boeckh: What did you use as the host stimulus?

Klowden: Myself.

Geier: Which olfactometer did you use?

Klowden: I have a single-port olfactometer. It has a 1 m flight section along which the mosquitoes have to fly to reach me, and there is a trap section at the end into which the mosquitoes fly.

Geier: Did you get a good response from *Aedes aegypti?*

Klowden: Yes. And also from *Anopheles albimanus*, they're almost as good as *Ae. aegypti*.

Geier: You also said that you caught 80% of the released *Anopheles gambiae* in the trap. Over what time period was this performed?

Klowden: A 10 min exposure period.

Carlson: What would you predict would happen if you synthesized an oligonucleotide encoding *Ae. aegypti* head peptide I, put it under the control of a constitutive promoter and introduced it into *Ae. aegypti?*

Klowden: We're planning on doing this experiment. I hope that we will see some inhibition of host seeking, if we can get the levels of expression high enough. If we could get expression in all cells, then the accessory cells that surround the sensilla might express it to a high enough level to have an effect on the sensilla.

Carlson: Do you know if the peptide is initially synthesized as a pre-propeptide?

Klowden: No, there is no evidence for this, although Mark Brown (unpublished results 1994) has found some degradation products that are fragments of this peptide. It's localized in various parts of the CNS.

Curtis: In my opinion, transforming the cells with the oligomer so that they make the peptide will not be relevant to the control of malaria, because if the mosquitoes don't blood feed, they won't lay eggs. Therefore, the gene would not spread within the population. The human-feeding *Anopheles* are the dangerous ones. If we could produce a mutant that only has the sensory system for detecting animals and not humans, that mutant would be relatively fit because it would still be able to feed in most agricultural environments. Therefore, it is more likely that this mutant would spread within the population. This should be the aim for insect neurologists.

Galun: I completely agree with Chris Curtis. The current philosophy as presented to us by Boris Dobrokhotov states that it is impossible to control mosquitoes. Therefore, he suggests that mosquitoes should be undisturbed, and their natural population should be replaced with transgenic mosquitoes that are refractory to the transmission of malaria. However, if we produce a transgenic mosquito refractory to *Plasmodium falciparum*, will it also be refractory to *Plasmodium vivax* or *Plasmodium ovale*? It surely will not be refractory to the transmission of filariasis. *An. gambiae* is the major vector of filariasis in Kenya. There's no doubt that if the transgenic mosquito changes its host selection and become zoophilic rather than anthropophilic, the mosquito will be happy and the human disease it transmits will be curtailed.

Klowden: Perhaps more important than developing transgenic mosquitoes is to insert the gene for the inhibitor peptide into a virus that would infect a mosquito and exert control in that way.

Brady: You talked about your beautiful operations in a very modest way. What was the mortality rate of mosquitoes that had their ovaries ripped out and pushed back in?

Klowden: The mortality rate is virtually zero. It's a clean operation. On a bad day it might be as high as 10%.

Takken: Did you inject the peptide into *Culex*?

Klowden: No because *Culex* does not respond in the olfactometer. I injected the peptide into *Aedes albopictus* and I observed some inhibition, but unfortunately the *Ae. albopictus* controls that were injected with saline also did not respond.

Steinbrecht: How do you inject an amount as large as 1 μl into an animal as small as a mosquito without damaging the animal or losing substance that comes out again through the puncture?

Klowden: The 1 μl injections were administered as an enema into the gut. In fact, I can inject up to 6 μl in this way. The haemolymph injections involved only 0.2 μl.

Cardé: Why do you use sucrose? They don't encounter sucrose in their environment.

Klowden: Glucose, fructose and sucrose have all been detected in nectars.

Boeckh: What is the time between the injection of the peptide and the inhibition?

Klowden: This was achieved within 45 min.

Boeckh: How long does it take the receptors to respond?

Bowen: This is difficult to answer. I tried to begin recording as soon as possible after injection and testing for host-seeking behaviour, but the time taken to insert the electrodes and obtain recordings depended on whether it was a good day or a bad day.

Kaissling: Are all types of receptor cells inhibited?

Bowen: I didn't have time to look at all of them. I had to make the recordings quickly because the peptide has a short half-life. The way to avoid this may be to inject the peptide into mosquitoes that have had a blood meal because it's possible that the half-life of the peptide is longer in a blood-fed mosquito.

Kaissling: Is the inhibition of the receptor cells the only possible mechanism or are there other mechanisms; for example, behavioural inhibition?

Bowen: We don't know.

Curtis: On the question of whether *Anopheles* take extra blood meals, Gillies (1954) described the so-called 'pre-gravid' state, which referred to mosquitoes that were prepared to take a second feed in a first cycle. However, after they've laid, they become very regular. If you have never gone beyond the first laying, you may be looking at a biased situation.

Carlson: What is the target of the peptide?

Klowden: I have no idea. I would like to see it being bound by a certain receptor.

Hildebrand: Where in the mosquito is the peptide produced and released?

Klowden: It's been detected immunologically throughout the CNS, ventral nerve cord and the brain.

Boeckh: This does not necessarily mean that it is produced in those areas.

Galun: What made you study that particular peptide?

Klowden: Mark Brown had some peptides and asked me if I would like to use them in my bioassay, so I looked at the titre curves and thought that for this peptide the titres correlated well with the behavioural inhibition.

Carlson: How many peptides did you try before you found this one?

Klowden: Only four, so we were very lucky.

Reference

Gillies MT 1954 The recognition of age-groups within populations of *Anopheles gambiae* by the pre-gravid rate and the sporozoite rate. Ann Trop Med Parasitol 48:58–74

General discussion IV

Pickett: We've touched upon the possible role of molecular genetics in this area. What is the current situation in terms of the transformation of mosquitoes? Do we have suitable vectors for this process?

Dobrokhotov: Five years ago, the Programme for Research and Training in Tropical Diseases (TDR), together with the MacArthur Foundation and the Wellcome Trust, initiated new research projects on the prospects for malaria control by genetic manipulation of vectors (World Health Organization, unpublished document 1991). It has been recognized that an effective way to prevent the spread of vector-borne infections is to interfere with the vector-mediated parasite transmission. In the light of recent biotechnological advances in the area of host–parasite relationships and vector genetics, it has become possible to manipulate vector genomes so that they are incapable of parasite transmission, but can still survive in the wild (Crampton et al 1990, Collins & Besansky 1994).

Theoretically, the steps necessary for the engineering and release of a refractory (non-malaria competent) mosquito would involve the resolution of various laboratory and field problems, such as an understanding of: (1) the molecular basis of vectorial competence; (2) the development of genetic and molecular tools for transformation of refractory mosquitoes; and (3) the understanding of mosquito population dynamics to apply this method effectively in the field.

The results achieved during the few years of study showed the feasibility of such an approach. I can give you some examples of the projects that are in progress. High resolution genetic maps of the X chromosome of *Anopheles gambiae* have been created, and other chromosomes will be completed soon (Zheng et al 1993). This will enable the localization and cloning of genes controlling susceptibility or refractoriness of mosquitoes to malaria parasites. A number of phenotype marker genes have been cloned, and they may be incorporated into new DNA vectors for the efficient identification and selection of transformed mosquitoes (Besansky et al 1995). The cloning and definition of mosquito promoters, both constitutive and stage specific, is progressing well, so that newly introduced genes may be expressed at the desired level, and at the stage and in the tissue of choice (Muller et al 1993). Special efforts are currently being directed towards the development of mosquito germline transformation and expression systems, as well as drive mechanisms for population changes. Transposable elements, bacterial symbionts and retroviruses are being investigated for this purpose (Warren & Crampton 1994, Kidwell & Ribeiro 1992, Beard et al 1993, Robertson 1993).

An understanding of vector behaviour is also required in order to improve epidemiological studies, evaluate transmission indices and monitor the available anti-vector measures. Virtually nothing is known about the behaviour, and the neurobiological and chemical basis of the anthropophilic habits of the malaria vector. Molecular biology and genetic studies in this area could lead to a novel means of disrupting vector behaviour and the consequent reduction in vectorial capacity. Understanding mosquito olfaction mechanisms may contribute significantly to our knowledge of the patterns of *Anopheles* dispersal, the determinants of anthropophilic, endophagic and endophilic, non-blood feeding, and the mating performance of mosquitoes. I hope that this symposium will help to prioritize subjects for further studies in this area.

There are those who are sceptical and those who are enthusiastic about using transgenic mosquitoes. I am trying to promote a realistic, balanced point of view. This is a long-term study and we could achieve our goals within 10–15 years. However, we already have important spin-offs, in the fields of taxonomy, ecology and population studies of *Anopheles*, from the work I have described.

Curtis: It's all very well to develop a system for genetic transformation in the laboratory, but the problem of inserting and maintaining genes in huge wild populations is much more difficult to solve. There is an example in *Drosophila* where a transposable element seems to have spread throughout populations (Kidwell et al 1981), but we aren't even close to using them in *Anopheles*.

Pickett: If we are still some way off being able to transform mosquitoes, then we've got more time to think about which genes we might select as targets. This is where the studies on mosquito ecology and behaviour come in. In the plant world it looked as though it was going to take a long time to set up robust transformation systems, certainly for monocotyledonous plants, so during that time we began to select our targets carefully. Unfortunately, the transformation systems were developed much more quickly than we anticipated because of a large industrial involvement, and so now we're seeing transformed plants expressing inappropriate genes. The danger is that if you give genes to molecular genetics-oriented companies prematurely, if they've got the transformation techniques, they will insert them anywhere, and you may end up with herbicide-resistance cereal crops, including maize and other dicotyledonous crops, which may be environmentally damaging. Constitutive expression of *Baccillus thuringiensis* endotoxins, for example, could stimulate resistance. An analogous situation developing in the mosquito field would give insect molecular geneticists an equally bad name.

Dobrokhotov: This is one of the reasons why we are cautious when collaborating with industrial companies in this area.

Cardé: We tend to dislike variability in outcome when we do physiological and behavioural experiments, and we prefer setting up situations for which we obtain a 'yes' or 'no' answer. However, there are many variables concerning physiological state that cause outcomes to change, depending upon the history of the individual animal. If we do not investigate this aspect sufficiently, we will not unravel the complexity of behaviours as well as we might. We are also avoiding the question of the effect of still air on movement toward the source, which is obviously a critical issue in orientation to humans in houses. And we're also not considering the question of heritability and behaviour, i.e. whether a behaviour is genetically determined and how many genes might be involved. In my opinion it's likely to be polygenic, such as the situation in host plant resistance.

Boeckh: The more we study behaviour, the more variables we find. There are not enough quantitative behavioural studies. Mosquitoes probably use many cues, but we can't all study all of them: there has to be a division of labour. Take visual cues, for example. Several insects can see perfectly and even discriminate between colours at low levels of light. Therefore, one question is whether the so-called 'nocturnal' mosquitoes still fly when it is really dark. How do they orient in this situation? There might be other odours in addition to host odours that guide mosquitoes to certain areas or spots of interest.

Gibson: I have tested the light sensitivity of *An. gambiae* with a behavioural assay, called the optomotor response (Gibson 1995). The wavelength sensitivity of mosquitoes has been tested previously only by electroretinograms (Brammer & Clarin 1976), which record the electrical response in the retina to light at different wavelengths. The behavioural test is more sensitive, in the sense that I could show that mosquitoes could 'see' and respond to visual cues at wavelengths further into the red than expected from the electroretinograms and they could see at levels of white light as low as starlight. So far, we have assumed that mosquitoes orient in host-odour plumes by olfactory-mediated, visually-guided positive anemotaxis (Kennedy 1983), i.e. they turn and fly upwind when they encounter the appropriate odours. It is generally believed that wind direction is determined in flying insects by the effects of wind drift across the eye. In other words, the accepted wisdom is that flying insects depend on visual cues to determine wind direction. Having established the lighting conditions under which *An. gambiae* can and cannot see, I found, much to my surprise, that they continued to orient in an odour plume when it should have been too dark to see anything. Either my tests are still not sensitive enough, or mosquitoes have an alternative mechanism for navigating in the dark.

Mustaparta: Several sensory modalities may be important for insect behaviour. In heliothine moths we know that the pheromones are released at a certain period at dusk and that males are airborne during that period. The antennal lobe neurons in these species seem to receive input mainly from olfactory receptor neurons, but also from some mechanosensory neurons. At higher levels, however, visual inputs are also integrated with olfactory information. We have obtained recordings from neurons in the protocerebrum that do not respond to pheromones in the light, but that respond significantly to pheromones in the dark. One may speculate that the pheromone pathways in these insects are inhibited by visual input in strong light, and that weaker light at dusk does not inhibit the pheromone pathways.

Lehane: It has been mentioned that we may be able to change the host choice of the mosquitoes to drive them towards animals rather than people. I have some concerns about this strategy because our understanding of host choice is rather poor. For example, the disappearance of malaria from Europe was related to host choice. Prior to the first world war *Anopheles atroparvus* fed on humans regularly, not because they particularly liked feeding on them, but because they were the most abundant food source available. When the agricultural and husbandry practices were changed in the eighteenth and nineteenth centuries, more animals were kept near humans, so these mosquitoes gradually showed their true preferences and started feeding on animals, and malaria transmission declined and eventually ceased. Consequently, we have to be aware of the plasticity of host choice in the real world. Also, we have to be aware of the fact that humans are increasingly becoming the major 'blood crop', so I doubt whether we will be able to force *An. gambiae sensu stricto*, for example, into becoming an exclusive animal feeder.

Curtis: Apart from the possibility of backcrossing the genes that make *Anopheles quadriannulatus* zoophilic into the genetic background of *An. gambiae s.s.*, we might use our neurological expertise to pick a mutant in *An. gambiae* that lacks the mechanism for responding to human feet.

Lehane: But there is a huge niche containing hundreds of millions of people, and host selection is being driven by natural selection, so it is unlikely that one could successfully keep mosquitoes from feeding on people.

Curtis: If we could make *An. gambiae* only as dangerous as *An. atroparvus* was or indeed only as dangerous as *Anopheles culicifacies* is in India, where it's predominantly zoophilic, although it will sometimes bite humans, this would enormously alleviate the African problem.

Gibson: There may not be just one solution to the problem. For example, we need to reduce the vectorial capacity of the mosquitoes in various ways and protect people with bed nets. Nothing on its own will be sufficient.

Davis: We shouldn't dismiss potential control methods without trying them. If we don't try any of these methods, then they definitely won't work.

Galun: We have to combine information with epidemiological data. Maintaining filariasis endemicity requires high population densities and thousands of infective bites. In malaria the per cent of infected mosquitoes is low. Therefore, if we manage to manipulate host preference, in epidemiological terms, we may achieve our goal.

Curtis: The anthropophilic index, i.e. the percentage of bites on humans, appears as a squared term in the vectorial capacity, because the mosquito has to bite a human twice. Therefore, reducing the anthropophilic index of *An. gambiae* from 0.9 to 0.2 would have a huge effect on reducing the vectorial capacity, and would reduce Africa's problem to that of India, which is a manageable one.

Pickett: There is a way of doing that. Chris Curtis raised an important point when he said that there may be problems with reintroducing genetically manipulated mosquitoes into the wild-type population. However, one may be able to use an approach related to the insects' attack on humans because, if such an attack is employed to transmit a disease to the mosquito, then it would work in two ways to our advantage: (1) it would reduce the mosquito population by the disease effect; and (2) it would create a selection against biting humans. For example, the spraying of people with fungal pathogen spores that would themselves only affect mosquitoes would create such a situation. This sort of strategy would, at the same time, reduce the anthropophilic index of the target mosquito.

Dobrokhotov: For 20 years the TDR has supported research projects on the biological control of disease vectors. More than 50 potential agents have been evaluated but no viruses, fungi, microsporidia or protozoa have been found that are appropriate for this purpose. They all have a low level of infection and are not suitable for mass production necessary for the field applications. The success has been achieved only with the bacterial pathogens *B. thuringiensis* and *Bacillus sphaericus*, which are now used for vector and pest control.

Pickett: Are these all against larvae?

Dobrokhotov: Yes, some fungi have been tested against adult mosquitoes but, due to a low infection level, they could not control them. We have a number of tools for vector control; however, they are not sufficient to provide disease control—the incidences of malaria, filariasis and other tropical diseases are increasing each year. Therefore, we should pay more attention to the development of innovative approaches for vector control.

Carlson: A great deal is known about the mechanism of sex determination in *Drosophila*, and it is possible to change the sex ratio of a strain through genetic manipulation. Most of the damage in this field is caused by female mosquitoes, so I wonder whether altering the sex ratio would be an effective control strategy.

Dobrokhotov: The molecular basis of sex differentiation in some insects is currently under investigation. If we can regulate sex determination in *An. gambiae*, preference would be given to release transgenic males rather than blood-feeding females.

Curtis: A lot of work has already been done on the sex ratios of mosquitoes. Hickey & Craig (1966) obtained a meiotic drive factor on the Y chromosome (strictly speaking the chromosome carrying the male-determining gene), that does drive into populations to some degree. However, there are suppressor factors on what one could call the X chromosome (the chromosome with the female-determining gene) that are selected for in populations which contain a driving Y chromosome. We tested their competitiveness in the field in India (Grover et al 1976), and we found that they were able to compete for mates, even though they were genetically abnormal. There is some interest in meiotic drive in *Anopheles* now.

Lehane: I know we are a group of scientists, and as such, I know we like to stick with the science and avoid the politics. But we do need to remember that two things are necessary if we're going to control disease: we need an adequate technical package and we need an adequate political and economic package to go with it. For many of our diseases, we've got adequate technical packages, although that's not true for lowland malaria in Africa. However, we often don't have the political and economic packages to put them into place. We should remember that if we're going to develop any new techniques to control malaria in Africa, they have to suit the politicians and economists. Boris Dobrokhotov was correct in saying that we didn't have too much success with biological control agents, but we shouldn't get carried away with the idea that everything centres around malaria and Africa. There are many other vector-borne diseases out there for which we already have adequate technical packages.

References

Beard C, O'Neill S, Tesh R, Richards F, Aksoy S 1993 Modification of arthropod vector competence via symbiotic bacteria. Parasitol Today 9:179–183

Besansky N, Bedell J, Benedict M, Mukabayire O, Collins F 1996 Cloning and characterization of the *white* gene from *An. Gambiae*. Insect Mol Biol, in press

Brammer JD, Clarin B 1976 Changes in volume of the rhabdom in the compound eye of *Aedes aegypti*. J Exp Zool 195:33–40

Collins F, Besansky N 1994 Vector biology and the control of malaria in Africa. Science 264:1874–1875

Crampton J, Morris A, Lycett G, Warren A, Eggleston P 1990 Transgenic mosquitoes: a future control strategy. Parasitol Today 6:31–36

Gibson G 1995 A behavioural test of the sensitivity of a nocturnal mosquito, *Anopheles gambiae*, to dim white, red and infra-red light. Physiol Entomol 20:224–228

Grover KK, Suguna SG, Uppal DK et al 1976 Field experiments on the competitiveness of males carrying genetic control systems for *Aedes aegypti*. Entomol Exp Appl 20:8–18

Hickey WA, Craig GB 1966 Genetic distortion of sex ratio in the mosquito *Aedes aegypti*. Genetics 53:1177–1196

Kennedy JS 1983 Zigzagging and casting as a programmed response to wind-borne odour: a review. Physiol Entomol 8:109–120

Kidwell M, Ribeiro J 1992 Can transposable elements be used to drive disease refractorness genes into vector populations? Parasitol Today 8:325–329

Kidwell MG, Novy JB, Feeley SM 1981 Rapid unidirectional change of hybrid dysgenesis potential in Drosophila. J Hered 72:32–38

Muller H, Crampton J, della Torre A, Sinden R, Crisanti A 1993 Members of a trypsin gene family in *Anopheles gambiae* are induced in the gut by blood meal. EMBO J 12:2891–2900

Robertson H 1993 The *mariner* transposable element is widespread in insects. Nature 362:241–245

Warren A, Crampton J 1994 *Mariner.* Its prospects as a DNA vector for genetic manipulation of medically important insects. Parasitol Today 10:58–63

Zheng L, Collins F, Kumar V, Kafatos F 1993 A detailed genetic map for the X chromosome of the malaria vector *A. Gambiae.* Science 261:605–608

Electrophysiological responses from receptor neurons in mosquito maxillary palp sensilla

Alan J. Grant and Robert J. O'Connell

Worcester Foundation for Biomedical Research, 222 Maple Avenue, Shrewsbury, MA 01545, USA

Abstract. We recently completed an electrophysiological study of the receptor neurons found in the sensilla basiconica on the maxillary palps of mosquitoes. Our results describe a class of receptor neurons whose properties could provide the afferent input required for some aspects of CO_2-modulated host-locating behaviour. First, these neurons have apparent thresholds (150–300 ppm) which are at, or below, the concentration of CO_2 (300–330 ppm) normally reported for ambient air. Second, their concentration–response functions are steep, such that small (50 ppm) fluctuations in concentration elicit reliable changes in activity. Third, they behave like absolute CO_2 detectors in that their ability to respond to step increases in CO_2 concentration is little influenced by the background concentration of CO_2. And fourth, a linear extrapolation of the observed response function to the levels that might be expected near vertebrate hosts suggests that these neurons have sufficient dynamic range to cover those CO_2 concentrations that should be encountered during a large portion of the behaviour likely involved in host location. The mosquito CO_2 receptor neuron thus has an appropriately low threshold and a steep concentration–response function, it is not desensitized by ambient levels of stimulation, and it has a dynamic range appropriate for the distribution of CO_2 concentrations expected in the environment. In addition, this sensillum contains two other receptor neurons, neither of which respond to CO_2. One of these neurons responds to stimulation with very low doses of another behaviourally relevant compound, 1-octen-3-ol.

1996 Olfaction in mosquito–host interactions. Wiley, Chichester (Ciba Foundation Symposium 200) p 233–253

Since the early extirpation experiments of Roth and others (Jones & Madhukar 1976, Roth 1951, Omer & Gillies 1971), it has been known that chemoreceptor neurons in sensilla on the maxillary palps play an important role, together with those on the antenna, in the detection and processing of chemical stimuli implicated in initiating and modulating host location and feeding behaviours (Bowen 1991, Gillies 1980, Takken & Kline 1989, Kline et al 1991). However, physiological studies of the peripheral sensory system in mosquitoes have focused largely on the responses of olfactory receptor neurons in antennal sensilla. Although the antenna does contain a highly sensitive L-lactic acid receptor neuron, whose activity can be modulated by the behavioural repellent N,N-diethyl-m-toluamide (Davis 1977, 1985, Davis & Sokolove 1975), additional specialized, high sensitivity receptor neurons tuned to other behaviourally relevant compounds have not been found on this sensory appendage. Chemoreceptors for other synthetic and natural compounds do exist in antennal sensilla, but they generally require relatively large doses to elicit electrophysiological responses, suggesting that they may not play a role in providing the sensory inputs required for orientation behaviour. In the maxillary palps, a single early report (Kellogg 1970) described electrophysiological responses to CO_2. This neuron was found in sensilla basiconica of female mosquitoes. Other than this, little additional information has been available which bears directly on the receptor mechanisms involved with the detection and processing of CO_2 and other volatile stimuli by olfactory structures on the maxillary palps (Takken 1991).

Aedes aegypti sensory neurons in sensilla basiconica respond to small increments (50 ppm) in CO_2 concentration which could be appropriate for orientation behaviour (Grant et al 1995). This compound has long been implicated as an important chemical signal utilized by mosquitoes and other hematophagous insects in their host-locating and feeding behaviours (Gillies 1980, Bowen 1991, Galun 1977). CO_2 receptor neurons have been studied morphologically in mosquitoes (McIver 1972, 1982), and their physiological properties have been explored in several other arthropods, including Lepidoptera (Bogner et al 1986, Bogner 1990, Stange 1992, Stange & Wong 1993), Hymenoptera (Lacher 1964, Stange & Diesendorf 1973), other Diptera (Bogner 1992) and Arachnida (Steullet & Guerin 1992). In addition to the CO_2-sensitive neuron, the sensillum basiconicum is innervated by two other spontaneously active receptor cells. All three of these cells are developmentally related and therefore it is possible that they could all be involved in host-locating behaviour. Thus, in addition to characterizing the responses of the CO_2-sensitive neuron, we sought to determine effective stimuli for the other two companion cells in this sensillum.

Our initial objective in these studies was to explore the physiological properties of olfactory receptor neurons in the sensilla basiconica on the maxillary palps. Earlier studies (Kellogg 1970, Roth 1951, Omer & Gillies 1971) had suggested that this appendage is an important sensory structure (together with the antenna) for processing the chemical signals responsible for the olfactory-guided behaviours of mosquitoes. The results presented here deal with the responses of neurons in this sensillum to stimulation with metered amounts of CO_2 and other behaviourally active semiochemicals. We assume that a knowledge of the response properties of these olfactory receptor neurons will provide the framework within which to evaluate materials as potential control agents. This knowledge may in turn lead to the design of improved bio-rational methods of mosquito control. We assume also that methods of insect control which exploit the sensory capabilities of the target animal are likely to be cost effective, amenable to relatively simple trap designs and should increase the public's overall degree of protection against mosquito-borne diseases.

Methods

Details of the recording apparatus, computerized data processing and storage techniques have been described in detail in previous reports (O'Connell 1975, Grant & O'Connell 1986, Grant et al 1989) and are only summarized below.

Insects. We raised *Ae. aegypti, Anopheles stephensi, Culiseta melanura, Culex quinquefasciatus* and *Aedes taeniorhynchus* stocks in the insectary at American Biophysics Corp., Jamestown, RI, USA, under a 12:12h light:dark photoperiod at 21–27°C and 62–64% relative humidity. We fed *Aedes* larvae ground dry dog biscuits, *Anopheles* larvae ground tropical fish food, and *Culex* and *Culiseta* larvae lactose and yeast. We also obtained a few *Ae. aegypti* as adults from the University of Massachusetts at Amherst, MA, USA. We noted the age and general condition of each adult insect tested and provided them with *ad lib* 10% sucrose.

Insect preparation. To insure stable recordings, we immobilized the insects and positioned the maxillary palps in a manner that allows unobstructed access with the recording microelectrode. We immobilized adult insects on a 1cm^2 glass plate mounted in a plastic holder with a small amount of adhesive applied to the thorax, abdomen, legs and wings. We placed thin strips of double-sided sticky tape over the thorax to further reduce potential movements of the body produced by the strong thoracic muscles. We positioned the head and mouthparts on a transparent ledge with the sensilla on the maxillary palps facing the recording microelectrode and then secured the palps with a strip of sticky tape applied at their base. Once the insect was secured, transilluminated light allowed a clear view of the base of the sensillum through a compound microscope equipped with long working distance objectives.

Electrode manufacture and recording. The recording microelectrodes were made of tungsten wire, electrolytically sharpened to a tip diameter less than 0.5 mm and held in low drift, high gain (800 ×) micromanipulators (Ernst Leitz [Canada] Ltd., Ontario, Canada). We inserted the indifferent microelectrode into one eye under low power (40 ×) magnification, and the recording microelectrode under high power (500 ×) at the base of an individual sensillum on the palp. After we had positioned the electrodes, the electrical signals from the neurons within the sensillum were band-pass filtered (0.3–3.0 kHz), amplified (1×10^3) and sent in parallel to an audio monitor (AM7 [Grass Instrument Co., Warwick, RI, USA]), a storage oscilloscope (5113 [Tektronix, Inc., Wilsonville, OR, USA]) and a microcomputer for subsequent data acquisition, action potential discrimination and analysis, and storage (O'Connell et al 1973).

Stimulation with carbon dioxide. We directed two opposing gas streams toward the exposed palp: one carrying the background (225 ml/min) and the other the stimulus (150 ml/min). Two valves controlled access of the background and stimulus streams to the preparation. Initially, we selected the concentration of CO_2 that was to be delivered to the preparation from a bank of six gas cylinders (Matheson Gas Co., Gloucester, MA, USA), each containing a fixed amount of CO_2 (0, 150, 300, 350, 600 or 1000 ppm), purified oxygen (20%) and nitrogen (remainder). To establish the static concentration–response relationship in various background levels of CO_2, we exposed the sensillum to a randomly assigned background concentration and stimulated the preparation with successive 2 s pulses, controlled by the computer, of each of the CO_2 concentrations in the stimulus array. To insure that the transitions between the background and stimulus concentrations of CO_2 were as rectangular as possible, we activated the stimulus line 4 s before the opposed background stream was shut off to produce the 2 s stimulus pulse. During the 4 s pre-pulse interval the elevated flow rate in the background stream prevents the stimulus stream from reaching the preparation. This was easily verified, in each case, by monitoring the constancy of the receptor neuron activity during the pre-pulse period. We positioned a vacuum hood behind the preparation to exhaust potential contaminants from the area around the preparation.

Stimulation with odorants. We delivered volatiles from the stimulus cartridges to the antenna by passing a stream of filtered synthetic air (150 ml/min; 0 ppm CO_2) through the stimulus cartridge. Between stimulations, a purge stream of filtered dry synthetic air (225 ml/min; 0 ppm CO_2) bathed the antenna. We controlled the timing of the gas flows in both lines with the computer. A single 2 s pulse of the stimulus compound was presented to the preparation and the total number of impulses generated by each receptor neuron during this interval was determined. To insure that the stimulus profile had a relatively sharp onset and offset, we discarded the headspace of the odour cartridge by activating the stimulus line (150 ml/min), connected to the stimulus cartridge, 4 s before shutting off the opposed purge stream for the 2 s duration of the stimulus (Grant et al 1988). In this manner, the headspace of the stimulus cartridge was flushed, prior to the stimulation period, with a volume 16-fold larger than its internal volume.

Characteristics of maxillary palp sensilla

Both sexes of *Ae. aegypti* possess specialized olfactory sensilla on the maxillary palps. These sensilla basiconica, or palpal pegs, are approximately 10–20 mm in length and are enlarged distally to form a spoon-shaped structure, whose surface is covered with a series of pores similar to those seen in the cuticle of other chemosensory sensilla (McIver 1972). As first reported by Kellogg (1970), these sensilla are innervated by three receptor neurons, each of which produce typical biphasic action potentials in extracellular recordings. By our conventions, when multiple action potentials are observed in a recording, we label the neuron producing the largest amplitude action potential, the A neuron; the neuron producing the next largest amplitude action potential, the B neuron; and the neuron producing the third largest amplitude action potential, the C neuron. Our computer-based data acquisition and analysis techniques make it possible for us to discriminate reliably between the activity of these receptor neurons by utilizing information about both the amplitude and waveform of the action potentials they generate.

Concentration–response relationship for carbon dioxide in *Aedes aegypti*

The basiconica sensilla on the maxillary palps of female *Ae. aegypti* contain a receptor neuron (McIver 1972), which produces a large amplitude action potential (up to 300 mV peak to peak), that responds to stimulation with small amounts of CO_2 (Fig. 1A; inset 600 ppm). We established the average concentration–response relationship for this neuron type in a background containing 0 ppm CO_2. We stimulated each of 13 sensilla with six successive 2 s pulses, each pulse containing one of the six gas mixtures. We determined the order of presentation for each concentration in this six-pulse protocol randomly and separated each pulse from the next by 48 s of exposure to CO_2-free synthetic air. The neurons are silent in backgrounds without CO_2 and have response thresholds that range from 150 ppm (one neuron out of 13), to 300 ppm (11 neurons out of 13) to 600 ppm (one neuron out of 13). The average concentration–response function of the CO_2 receptor neuron is relatively steep, with stimulus increments as small as 50 ppm (e.g. 300–350 ppm) eliciting measurable increases in activity (Fig. 1A). This sensitivity suggests that mosquitoes might be able to discriminate between rather small changes in CO_2 concentration. Periodic stimulations with a fixed level of CO_2 do not reveal major shifts in sensitivity over time (Fig. 1B). In this experiment, we sampled the rate of action potential production to stimulation with either 300 ppm or 600 ppm CO_2 every 5 min for 60 min.

The A receptor neuron responds to rectangular pulses of CO_2 with a phasic-tonic pattern of action potential production (Fig. 2) similar, in some respects, to the temporal pattern seen in other chemoreceptor neurons (Grant &

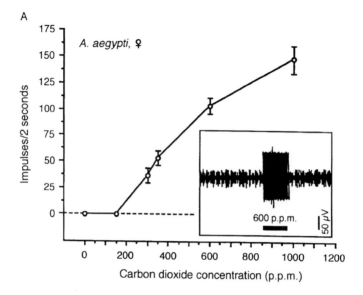

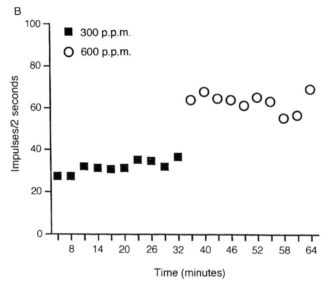

FIG. 1.   (A) Average ($\pm$ SEM, $n = 13$) number of action potentials during 2 s stimulus pulses of the indicated concentrations of CO_2 from 13 receptor neurons in the sensilla basiconica on the maxillary palps of 11 female *Aedes aegypti*. These responses were obtained in a synthetic air background containing 0 ppm CO_2. Inset shows a typical response to a 2 s stimulus of 600 ppm CO_2 from a background of 0 ppm CO_2. (B) The response to repeated 2 s stimuli separated by intervals of 3 min. The first 10 stimuli were steps to 300 ppm CO_2 from 0 ppm background levels. The second set of 10 stimuli represent steps from 0 ppm to 600 ppm CO_2.

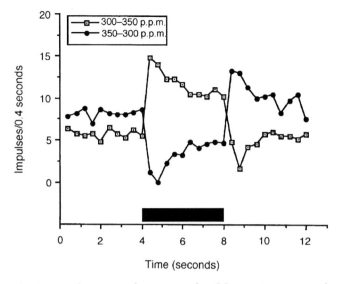

FIG. 2. The temporal pattern of response of a CO_2 receptor neuron in a sensillum basiconicum on the maxillary palp of a female *Aedes aegypti*. The open symbols represent the average response to a 4 s stimulus pulse up to 350 ppm CO_2 from a background concentration of 300 ppm. The solid symbols represent the average response to a 4 s stimulus pulse down to 300 ppm from a background level of 350 ppm CO_2. In both cases the number of action potentials were averaged in 400 ms time periods and represent the mean from a single sensillum stimulated 10 times with each pulse protocol. The data acquired during the 4 s purge of the stimulus line just before the onset of the stimulus is not displayed.

O'Connell 1986, Stange & Diesendorf 1973). It is apparent from the temporal pattern of discharge that these neurons are sensitive to both the magnitude and the direction of the step in CO_2 concentration. This pattern of response is similar to that observed in the CO_2 receptor neurons of other species (Stange & Diesendorf 1973), and serves to increase the range of action potential output for rapidly alternating shifts in concentration. This observation again suggests that these neurons may provide mosquitoes with the sensory input which would allow them to follow small, rapidly fluctuating CO_2 concentrations, such as might occur in nature (Elkinton & Cardé 1984, Gillies 1980).

Effect of background carbon dioxide concentration

Mosquitoes are likely to orient to sources of CO_2 in conditions where the background levels of CO_2 vary widely, e.g. from ambient (330 ppm) to the elevated concentrations likely encountered near a host. Thus it was important to determine if the specificity and sensitivity of the receptor neuron response to CO_2 (Fig. 1) would be altered by chronic exposure to different background levels of CO_2. We evaluated the relative sensitivity of the CO_2 receptor neuron in different background concentrations of CO_2 with the same six-pulse protocol used to establish the dose–response curve illustrated in Fig. 1A. However, in this case the preparation was exposed to a new randomly assigned background concentration of CO_2 for 5 min, after which a new six-pulse stimulus protocol was delivered. The entire CO_2 stimulus and background set was successfully presented to eight out of 13 sensilla.

The average number of impulses during a 2 s stimulus containing a given concentration of CO_2 is illustrated in Fig. 3. The magnitude of response appears to be independent of the background level, as long as the stimulus step is increasing (e.g. the stimulus concentration is greater than the background). For example, the magnitude of the response to a 1000 ppm pulse of CO_2 is similar in a background environment containing either 0, 150, 300, 350 or 600 ppm CO_2.

Comparative sensitivity to carbon dioxide and palpal morphology in other species

We have recorded from palpal sensilla in four other species of mosquitoes, in three different genera, to determine if taxonomic differences exist in the response characteristics of the CO_2 receptor neurons in the sensilla basiconica. In addition to *Ae. aegypti*, we recorded from *Ae. taeniorhynchus*, *An. stephensi*, *Cs. melanura* and *Cx. quinquefasciatus*. Females of these species possess sensilla containing receptor neurons that have similar physiological responses to stimulation with CO_2 (Fig. 4). The thresholds of the receptor neurons in all these species appear to be similar. The two most extensively studied species, *Ae. aegypti* and *An. stephensi*, also have similar slopes for their concentration–response curves, but it is not yet clear if the difference in the slopes of the response functions observed among the other species will continue as the sample of recordings is expanded in future studies. Although the CO_2 response curves are similar for these species, the morphology of the palp and the distribution of sensilla along the palp varies considerably, as was expected from studies in other species (Braverman & Hulley 1979). For example, there is interspecific variation in the external morphology of the maxillary palp and the distribution of sensilla along the palp. The palps of *An. stephensi* are significantly longer (about 1450 mm) than the palps of either *Ae. aegypti* (about 350 mm) or *Cx. quinquefasciatus* (about 300 mm). In addition, the sensilla basiconica in both *Ae. aegypti* and *Cx. quinquefasciatus* are restricted to the fourth subsegment from the head. *An. stephensi*, however, has sensilla basiconica distributed along the terminal three subsegments of the palp.

Response characteristics of the other chemoreceptive neurons

The presence of a receptor neuron with extreme sensitivity to CO_2 in the sensilla basiconica of the maxillary palps and the apparent lack of another

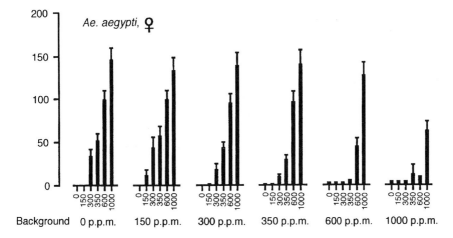

FIG. 3.   Mean response ($\pm$SEM, $n = 8$) of CO_2 receptor neurons in eight sensilla basiconica on the maxillary palps of seven female *Aedes aegypti*. The stimuli were 2 s pulses of six different concentrations of CO_2 delivered from each of the indicated background levels of CO_2.

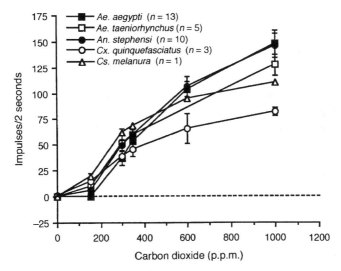

FIG. 4.   Mean ($\pm$SEM, $1 \leqslant n \leqslant 13$) number of action potentials during 2 s stimulus pulses of the indicated concentrations of CO_2 from sensilla basiconica on the maxillary palps of individuals from five different species of female mosquito (*Aedes aegypti*, *Aedes taeniorhynchus*, *Anopheles stephensi*, *Culex quinquefasciatus* and *Culiseta melanura*). The number of sensilla from which the averages were computed are indicated for each concentration–response function. All of these functions were established in a background environment containing 0 ppm CO_2.

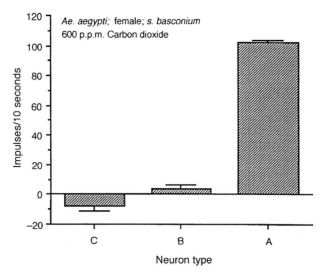

FIG. 5. Response from the three receptor neurons within a single sensillum basiconicum on a female *Aedes aegypti* to stimulation with a pulse of 600 ppm CO_2. Response values are expressed as the number of impulses that occurred during the two second stimulus period plus the 8 s post-stimulus minus the 10 s pre-stimulus period (mean $\pm$ SEM, n = 3 stimulations).

elsewhere on the mosquito (Omer & Gillies 1971, Gillies 1980, Takken 1991) suggests that this component of the mosquito olfactory system is intimately associated with the detection of at least this one stimulus involved with host-locating behaviour. The presence of two other receptor neurons in this sensory structure, which do not respond to CO_2 (Fig. 5), raises the possibility that other behaviourally active compounds may be effective stimuli for these neurons.

The literature contains many references to a range of compounds, in addition to CO_2, that have been shown to modulate various aspects of mosquito behaviour. For example, 1-octen-3-ol has been identified as a behavioural 'attractant' in several dipteran species, including mosquitoes (Takken & Kline 1989, Den Otter et al 1988, Kline et al 1990, 1991, Saini et al 1989). Figure 6 illustrates the electrophysiological responses elicited in the neurons of a single sensillum basiconicum by stimulation with increasing doses of 1-octen-3-ol diluted in mineral oil (0.0001 to 0.001 mg/ml) and delivered by a stream containing 0 ppm CO_2. It is clear from these preliminary recordings that neuron C, which produces the smallest amplitude action potential, is very responsive to this material. Responses are not observed in neuron B, which produces the medium amplitude potential or in neuron A, which is responsive to CO_2. None of the receptor neurons were responsive to stimulation with an odour cartridge containing the diluent alone (Fig. 6, top trace). As a point of

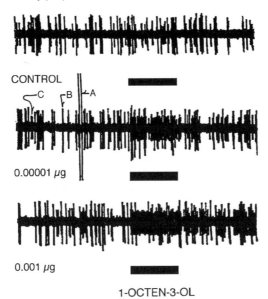

CONTROL

0.00001 µg

0.001 µg

1-OCTEN-3-OL

FIG. 6. Typical electrophysiological responses to stimulation with two doses (0.0001 µg and 0.001 µg) of 1-octen-3-ol and a control stimulus. The 2 s stimulus bar is indicated below each trace.

comparison, the neuron C response to 1-octen-3-ol appears to be as sensitive, if not more sensitive, than any of the responses to lepidopteran pheromones that we have observed in pheromone receptor neurons using similar protocols for stimulus preparation and delivery. Lepidopteran receptor neurons have generally been regarded as exemplars of the highly tuned and exquisitely sensitive neurons thought to be required for orientation to pheromones. Thus, based on the high degree of sensitivity, we speculate that neuron C of the mosquito sensilla basiconica may also be involved with host-locating behaviour.

Discussion

Mosquitoes live in an environment in which both the spatial and temporal distribution of olfactory stimuli are influenced by the scent release mechanisms of the host organism and by the vagaries of local winds and microhabitats. Chemical signals in the environment are therefore likely to appear to the orienting insect as complex mixtures with variable spatial and temporal distributions of individual components. To successfully navigate in such an environment, the mosquito must possess a sensory system capable of detecting and processing the type of complex signals found in nature. Thus, it is important to understand how sensory receptors respond to patterned stimuli designed to approximate those found in a natural environment. To this end, we have begun experiments to explore the response properties of the three receptor neurons found in maxillary palp sensilla, appendages that have previously been associated with orientation in mosquitoes.

The receptor neuron producing the largest amplitude action potential responds to CO_2. The response properties of this CO_2 receptor neuron suggests the presence of at least two different transduction mechanisms. On the one hand, there appears to be a tonic response mechanism that simply reads out the steady-state concentration of CO_2. The sensitivity of this receptor neuron falls within the range that might be encountered by a female mosquito orienting to a host organism. In addition, the slope of the concentration–response function for this neuron is steep, suggesting that small changes in concentration may be detectable. On the other hand, there is also a rate-sensitive transduction mechanism whose output rapidly follows both the direction and rate of change of CO_2 concentration. As is the case with cold receptor neurons in vertebrates (Darian-Smith et al 1973) and invertebrates (Loftus & Corbiére-Tichanué 1981) these later phasic properties may provide a sharpening mechanism which has the effect of greatly exaggerating the response output of the receptor neuron as stimulus concentration changes rapidly. Although we do not yet know the 'following' capabilities of the CO_2 receptor, the ability of the neuron to produce phasic bursts whose magnitude and sign are a function of the magnitude and sign of the change in CO_2 concentration surely must enhance the sensory signal transmitted to the CNS by small changes in concentration. This effect is similar to the 'Mach' effect seen in the output of visual receptor neurons (Hartline et al 1952), with the exception that the temporal properties of the chemoreceptors studied here are an intrinsic property of the receptor and do not appear to depend on lateral inhibitory processes as they do among neighbouring visual receptors. Depending on the nature of the central processing associated with these temporal inputs, the ability of the sensory system to follow small rapid changes in CO_2 concentration might be enhanced. This pattern of response would be particularly appropriate if mosquitoes oriented to hosts by following the rapidly changing increments in CO_2 concentration that are likely to be encountered in nature (Elkinton & Cardé 1984), as opposed to simply flying upwind.

Many chemicals have been discovered which appear to modify mosquito orientation. We have called attention to a sensillum on the maxillary palp, containing three primary receptor neurons which presumably share a common stem cell (Steinbrecht 1984) and are, therefore, developmentally related. These neurons also represent a set of sensory inputs which share a common extracellular environment, cuticular structure and location. Two of these neurons are clearly associated with detecting appropriately small amounts of two stimuli that are thought to modulate host-locating behaviour. As we discussed above, one neuron responds reliably with increased impulse activity to simulation with CO_2 and another responds strongly to stimulation with low doses of an identified behaviour attractant for mosquitoes, 1-octen-3-ol. The level of sensitivity seen in this latter neuron appears to be comparable to that seen in the exquisitely sensitive pheromone receptor neurons of many Lepidoptera. The high level of sensitivity to another behavioural attractant in maxillary palp sensilla further emphasizes the potential importance of this little-studied sensory appendage. In addition, the possibility exists that there may be a predictable relationship between the particular receptor neuron stimulated and the type of behavioural response elicited by that stimulus, as we have found in another insect species (Grant et al 1988). In other words, compounds which tend to act as 'attractants' may reliably elicit responses from only one of the three daughter neurons found in this sensillum and those which act as 'repellents' may elicit responses from another. We are aware that potentially large numbers of behavioural, endocrine and physiological factors may influence the behaviours actually produced by an organism in addition to the inputs provided by its sense organs. However, the sensory system responsible for processing particular stimuli places an initial filter on the range of environment energies that are detected by the organism and thus strongly influences those that ultimately serve to modulate behaviour.

There is ample behavioural evidence (Kline et al 1990, 1991, Mukwaya 1976, Takken 1991) to suggest that some of the chemical stimuli which drive host-locating and feeding behaviour in mosquitoes are species and gender specific. This also implies that the specificity may be derived from morphological and physiological differences in the maxillary palp sensilla and their primary receptor neurons (Braverman & Hulley 1979, McIver 1980). The mosquito, as a model system, provides an interesting opportunity to compare receptor systems among species that vary in both structure and life history. Often, comparative studies of this nature provide insight into those aspects of behaviour which might be subject to interference, and thus might be utilized in control strategies. In addition, the comparative approach will ultimately provide the information required to focus control strategies on particular pest species.

Collectively these experiments demonstrate that the maxillary palps of mosquitoes contain highly sensitive olfactory receptor neurons. Our observations reveal that behaviourally relevant stimuli for *Ae. aegypti* are potent stimuli for the receptor neurons in sensilla basiconica. The sensitivity and selectivity of these neurons rival that seen in antennal sensilla. Moreover, others (Braverman & Hulley 1979, McIver 1980) have noted that the number and distribution of particular classes of chemosensilla on the antennae and palps of various species of mosquito are related to differences in the animals' relative preferences for hosts. Armed with a knowledge of the peripheral response capabilities for more natural stimuli, control strategies might be designed that exploit these sensory capabilities and thus optimize the utility of bio-rational approaches.

Acknowledgements

We thank B. Wigton (American Biophysic Corp., Jamestown, RI, USA) and M. S. Mayer (USDA; ARS, Gainesville, FL, USA) for help and discussion. We also thank M. S. Mayer for graciously allowing us the use of his extensive collection of literature relating to mosquito biology. We thank J. Edman (University of Massachusetts at Amherst, MA, USA) for thoughtful discussion and adult *Ae. aegypti*. A portion of this work was presented at the 1994 Association for Chemoreception Sciences. Supported, in part, by the National Institute On Deafness and Communication Disorders; Grant DC00131 to R. J. O.

References

Bogner F 1990 Sensory physiological investigation of carbon dioxide receptors in Lepidoptera. J Insect Physiol 36:951–957

Bogner F 1992 Response properties of CO_2-sensitive receptors in tsetse flies (Diptera: *Glossina palpalis*). Physiol Entomol 17:19–24

Bogner F, Boppre M, Ernst K-D, Boeckh J 1986 CO_2 sensitive receptors on labial palps of *Rhodogastria* moths (Lepidoptera: Arctiidae): physiology, fine structure and central projection. J Comp Physiol A 158:741–749

Bowen MF 1991 The sensory physiology of host-seeking behavior of mosquitoes. Ann Rev Entomol 36:139–158

Braverman Y, Hulley PE 1979 The relationship between the numbers and distribution of some antennal and palpal sense organs and host preference in some *Culicoides* (Diptera: Ceratopogonidae) from southern Africa. J Med Entomol 15:419–424

Darian-Smith I, Johnson KO, Dykes R 1973 "Cold" fiber population innervating palmar and digital skin of the monkey: responses to cooling pulses. J Neurophysiol 36:325–346

Davis EE 1977 Response of the antennal receptors of the male *Aedes aegypti* mosquito. J Insect Physiol 23:613–617

Davis EE 1985 Insect repellents: concepts of their mode of action relative to potential sensory mechanisms in mosquitoes (Diptera: Culicidae). J Med Entomol 22:237–243

Davis EE, Sokolove PG 1975 Lactic acid-sensitive receptors on the antennae of the mosquito, *Aedes aegypti*. J Comp Physiol A 105:43–54

Den Otter CJ, Tchicaya T, van den Berg MJ 1988 Olfactory sensitivity of five species of tsetse (*Glossina* spp.) to 1-octen-3-ol, 4-heptanone, 3-nonanone and acetone. Insect Sci Appl 9:213–218

Elkinton JS, Cardé RT 1984 Odor dispersion. In: Bell WJ, Cardé RT (eds) Chemical ecology of insects. Chapman & Hall, New York, p 73–91

Galun R 1977 Responses of blood-sucking arthropods to vertebrate hosts. In: Shorey HH, McKelvey JJ Jr (eds) Chemical control of insect behavior. Wiley, New York, p 103–115

Gillies MT 1980 The role of carbon dioxide in host-finding by mosquitoes (Diptera: Culicidae): a review. Bull Entomol Res 70:525–532

Grant AJ, O'Connell RJ 1986 Neurophysiological and morphological investigations of pheromone-sensitive sensilla on the antenna of male *Trichoplusia ni*. J Insect Physiol 32:503–515

Grant AJ, O'Connell RJ, Hammond AM 1988 A comparative study of pheromone perception in two species of noctuid moths. J Insect Behav 1:75–96

Grant AJ, O'Connell RJ, Eisner T 1989 Pheromone-mediated sexual selection in the moth *Utetheisa ornatrix*: olfactory receptor neurons responsive to a male-produced pheromone. J Insect Behav 3:371–385

Grant AJ, Wigton B, Aghajanian J, O'Connell RJ 1995 Electrophysiological responses of receptor neurons in mosquito maxillary palp sensilla to carbon dioxide. J Comp Physiol A 177:389–396

Hartline HK, Wagner HG, MacNichol EFJ 1952 The peripheral origin of nervous activity in the visual system. Cold Spring Harbor Symp Quant Biol 17:125–141

Jones JC, Madhukar BV 1976 Effects of antennectomy on blood-feeding behavior of *Aedes aegypti* mosquitoes. Entomol Exp Appl 19:19–22

Kellogg FE 1970 Water vapour and carbon dioxide receptors in *Aedes aegypti*. J Insect Physiol 16:99–108

Kline DL, Takken W, Wood JF, Carlson DA 1990 Field studies on the potential of butanone, carbon dioxide, honey extract, 1-octen-3-ol, L-lactic acid and phenols as attractants for mosquitoes. Med Vet Entomol 4:383–391

Kline DL, Wood JR, Cornell JA 1991 Interactive effects of 1-octen-3-ol and carbon dioxide on mosquito (Diptera: Culicidae) surveillance and control. J Med Entomol 28:254–258

Lacher V 1964 Elektrophysiologische untersuchungen an einzelnen rezeptoren für geruch, kohlendioxyd, lufteuchtigkeit und temperatur auf den antennen der arbeitsbiene und der drohne (*Apis mellifica* L.). Z verlag Physiol 48:587–623

Loftus R, Corbiére-Tichanué G 1981 Antennal warm and cold receptors of the cave beetle, *Speophyes lucidulus* Delar., in sensilla with a lamellated dendrite. I. Response to sudden temperature change. J Comp Physiol A 143:443–452

McIver SB 1972 Fine structure of pegs on the palps of female culicine mosquitoes. Can J Zool 50:571–576

McIver SB 1980 Sensory aspects of mate-finding behavior in male mosquitoes (Diptera: Culicidae). J Med Entomol 17:54–57

McIver SB 1982 Sensilla of mosquitoes (Diptera: Culicidae). J Med Entomol 19:489–535

Mukwaya LG 1976 The role of olfaction in host preference by *Aedes* (*Stegomyia*) *simpsoni* and *Ae. aegypti*. Physiol Entomol 1:271–276

O'Connell RJ 1975 Olfactory receptor responses to sex pheromone components in the redbanded leafroller moth. J Gen Physiol 65:179–205

O'Connell RJ, Kocsis WA, Schoenfeld RL 1973 Minicomputer identification and timing of nerve impulses mixed in a single recording channel. Proc Inst Electrical Electronic Engineers 61:1615–1621

Omer SM, Gillies MT 1971 Loss of response to carbon dioxide in palpectomized female mosquitoes. Entomol Exp Appl 14:251–252

Roth LM 1951 Loci of sensory end-organs used by mosquitoes (*Aedes aegypti* L.) and *Anopheles quadrimaculatus* Say in receiving host stimuli. Ann Entomol Soc Am 44:59–74

Saini RK, Hassanali A, Dransfield RD 1989 Antennal responses of tsetse to analogues of the attractant 1-octen-3-ol. Physiol Entomol 14:85–90

Stange G 1992 High resolution measurement of atmospheric carbon dioxide concentration changes by the labial palp organ of the moth *Heliothis armigera* (Lepidoptera: Noctuidae). J Comp Physiol A 171:317–324

Stange G, Diesendorf M 1973 The response of the honeybee antennal CO_2 receptors to N_2O and Xe. J Comp Physiol A 86:139–158

Stange G, Wong C 1993 Moth response to climate. Nature 365:699
Steinbrecht AR 1984 Chemo-, hygro-, and thermoreceptors. In: Bereiter-Hahn J,
    Matoltsy AG, Richards KS (eds) Biology of the integument: invertebrates. Springer-
    Verlag, New York, p 523–553
Steullet P, Guerin PM 1992 Perception of breath components by the tropical bont tick,
    Amblyomma variegatum Fabricius (Ixodidae). 1. CO_2-excited and CO_2-inhibited
    receptors. J Comp Physiol A 170:665–676
Takken W 1991 The role of olfaction in host seeking of mosquitoes: a review. Insect Sci
    Appl 12:287–295
Takken W, Kline DL 1989 Carbon dioxide and 1-octen-3-ol as mosquito attractants. J
    Am Mosq Control Assoc 5:311–316

DISCUSSION

Boeckh: Is anything known about the role of 1-octen-3-ol from studying the behaviour of mosquitoes?

Takken: We showed in 1989 that a range of mosquito species is attracted to 1-octen-3-ol, but a strong response is nearly always only observed when CO_2 is also present (Takken & Kline 1989). This has subsequently been proven to be the case in several *Aedes* and *Culex* species in the USA, Australia and Southeast Asia (Kemme et al 1993, Vythilingam et al 1992). Therefore, it may be a common response for many *Aedes* species. We were not able to get a strong response in *Culex*.

Geier: We have tested 1-octen-3-ol in the Y-tube wind tunnel with and without CO_2 (M. Geier & B. Steib, unpublished work 1995). We did not observe any increase of flight activity or attraction in either case.

Costantini: One has to pay attention to the type of behaviour that is being measured. It is possible that if 1-octen-3-ol only has a negative kinetic effect, it would not be observed in a Y-tube wind tunnel.

Carlson: What levels of CO_2 do mosquitoes encounter in their environment?

Grant: The ambient level of CO_2 is about 350 ppm and the level in human breath is about 40 000 ppm.

I would like to point out that 1-octen-3-ol is the only compound tested. We have not yet identified how many other compounds may have similar effects, so we know little about the specificity of the receptor neurons. My results suggest only that one of the neurons producing the small amplitude action potentials are olfactory receptors which respond to 1-octen-3-ol.

Davis: So you can't say that these cells are only sensitive to 1-octen-3-ol. There may be other as yet unidentified compounds that give better stimuli.

Grant: My conclusions are that these cells can detect 1-octen-3-ol. I am not confident that we can conclude anything about what concentrations the doses represent because of the volatility of 1-octen-3-ol, i.e. even though the cells seem to respond to very low doses we don't actually know what concentrations are produced by the system. We're familiar with longer chain moth pheromones, which are very effective stimuli but they're also much less volatile.

Gibson: Can you estimate the concentrations?

Grant: Possibly, but the lepidopteran detectors are more sensitive than analytical detectors, so there's no way to measure the actual concentrations over the short time periods typically used for stimulation.

Davis: If the vapour pressure and the temperature are known, the amount of chemical in the vapour phase can be calculated. This is the 'flux' of odour molecules—i.e. moles per second—flowing over the antenna, and it may give a closer approximation of the concentration.

Guerin: The CO_2 receptors of mosquitoes, in common with those of moths, are on the mouth parts. In terms of activation and orientation, why are these sensilla on the maxillary palps and not on the antennae, which are bilaterally placed on the head to help the mosquito to fly in a particular direction? Also, why are they on the ventral side of the maxillary palps? I admit that the maxillae are also bilaterally placed, but there must be something more to this than meets the eye. In a hut there is probably a higher concentration of CO_2 than in the rest of the environment in which a mosquito has flown. After the mosquito has landed on the host, it is possible that the strategic location of these sensilla could be helpful for locating the feeding site because (1) irrespective of the background level of CO_2, the mosquito can still respond to any local flux of the stimulant; and (2) there may be a higher concentration of CO_2 at the biting site than that perceived on the approach to the host.

Grant: That's a good point. When the mosquito is probing, the palps are located near the skin surface. It is possible that information about the release of CO_2 very close to the skin surface is also important for the mosquito. I would like to make the point that there is a tremendous variation in the palpal morphology between species, yet the physiology of the receptors is similar wherever they are found on the palp.

Bowen: The amount of CO_2 released by the skin is well below 50 ppm, which is below the limit of detection of these receptors.

Hildebrand: Are these CO_2 receptors present anywhere else in the mosquito?

Grant: I don't know. Omer & Gillies (1971) removed the palps from the mosquitoes, and showed that these mosquitoes were not able to orient in a wind tunnel, suggesting that at least some of the CO_2 receptors are located on the palps.

Kaissling: It is interesting to compare these receptor studies with those of lepidopteran CO_2 receptors. They have a different morphology: the lepidopteran receptors are on the labial palps and not on the maxillary palps, and they are also located in a pit which opens at the tip of the palp. You mentioned the work of Stange (1992), who showed that lepidopteran CO_2 receptors are extremely sensitive. They respond to 1 ppm sinusoidal concentration changes at an atmospheric background of 350 ppm, and are able to follow a stimulus frequency of 10 changes of concentration per second. Stange produced sinusoidal concentration changes by putting the experimental chamber under sinusoidal pressure changes. The fact that the CO_2 receptors are located in the tips of the palps might be advantageous for monitoring the external CO_2 concentration. The palps point forward, so that the flying animal does not detect its own CO_2.

Grant: But in several species, including *Anopheles*, the distribution of these receptors extends down the palp, so that they are not only at the tip but also in the last couple of segments.

Steinbrecht: Honey-bees also have CO_2 receptors on the antennae, so the receptors on the palps are not necessarily involved alone.

Cork: I have several points to make. We had difficulties showing that 1-octen-3-ol was attractive to the tsetse fly, because the range of doses at which it was attractive was small. At high doses it reduced the catch and even at the optimal dose it only doubled the catch. It was not until we mixed 1-octen-3-ol with other host odour components that we were able to demonstrate its value as an attractant. It is therefore not too surprising that it has not yet been found to be attractive in mosquito bioassays, although it has been shown to be an attractant for *Aedes taeniorhynchus* in the field. Secondly, 1-octen-3-ol is produced and released by humans, so it is a mistake to think that it is only produced by other mammals such as oxen. And thirdly, I would like to point out that 1-octen-3-ol is optically active, so it is conceivable that a racemic mixture may not necessarily elicit a behavioural response.

Pickett: Obviously, if you observed differential activity to the enantiomers, then this would confirm that you were looking at an essential response.

Cork: Except that in the case of animal odour, 1-octen-3-ol occurs as a 40:60 ratio of enantiomers. Unfortunately, we do not know the enantiomeric composition of 1-octen-3-ol produced by humans.

Hildebrand: Have you found any stimulus that can drive the third cell?

Grant: No. But in preliminary experiments we have found other stimuli that drive the second cell. Without knowing much about the specificity of the system, it's difficult to know how important this might be. It seems clear that one of these neurons can detect 1-octen-3-ol.

Davis: Did you look at the effect of L-lactic acid on these cells?

Grant: No. We do know that citronellal stimulates these cells, but again we have the problem of not being able to measure the concentration. In my opinion, it's important to determine not only that the cell can detect it, but also at what levels it can detect it.

Davis: I was curious about the relationship between CO_2 and L-lactic acid because one of the questions I asked when I did the work with L-lactic acid-sensitive cells (Davis & Sokolove 1976) was: does the synergy occur at the level of the peripheral receptor neuron sensitive to L-lactic acid or does it occur in the CNS? Therefore, one could ask the same question for the CO_2 cell, i.e. does the synergy between L-lactic acid and CO_2 occur at the CO_2 receptor? This would be interesting to find out.

Grant: Also, we didn't control for age, mating status or blood meals. All of the insects tested were over three days old, and the maximum was 28 days. I don't know how any of these variables modulate the responses of these receptors.

Hildebrand: In at least some species, inputs from palps project to the antennal lobe. For example, the axons of the lepidopteran labial pit organ CO_2 receptor cells converge on a small glomerular complex in the antennal lobe (Kent et al 1986). Thus, those receptor cells project to a neuromere in the CNS where they don't 'belong'. This suggests that CO_2 information, and whatever else is detected by the labial pit organ receptors, is integrated with antennal olfactory information. I don't know if anyone is studying this integration, but it would be interesting to find out if there is interplay between antennal olfactory information and palp inputs in mosquitoes.

Grant: I would also like to mention that Kellogg (1970) found that these different cell types were excited by n-heptane and acetone vapours, and inhibited by n-amyl acetate. He didn't report the doses he tested or which of the cells responded to these compounds.

Anton: I have shown that in *Ae. aegypti* the maxillary palp axons project to different glomeruli than the antennal receptor neurons, which suggests that the interventions are occurring in the antennal lobe (Anton 1996, this volume).

Hildebrand: It would be interesting if you could stain the inputs from the palp sensilla specifically.

Carlson: The situation is the same in *Drosophila*, i.e. that the maxillary palp axons and antennal axons appear to project to different subsets of glomeruli in the antennal lobe (Stocker et al 1990).

Boeckh: This seems to be the general scheme for all holometabolous insects studied to date. It is different for hemimetabolous insects, such as cockroaches and locusts, because in these cases there is an ascending input from the maxillary palps to neuropils other than the antennal lobe glomeruli. An olfactory strand of this input runs into a microglomerular type of neuropil in the tritocerebrum adjacent to the antennal lobe. We suspect that this area of the brain in hemimetabolous insects might have been incorporated into the antennal lobe and formed a special type or series of glomeruli.

Guerin: Could this have anything to do with the motor output for biting?

Boeckh: There is a massive projection of the maxillary palps to the suboesophageal ganglion where many fibres branch profusely in 'their' neuromere. Only a certain strand follows the ascending pathway that has been mentioned above.

Guerin: That would constitute a very short neural path.

Kaissling: There is some degree of integration between CO_2 and odours at the peripheral level in termites. Ziesmann (1996) found sensilla in which the two cells respond to odours. The response of one of the two cells is blocked by CO_2, e.g. by its concentration inside the nest. When the termite goes outside, this cell is unblocked because of the lower level of CO_2 outside the nest.

Grant: Jürgen Boeckh can you comment on the internal morphology of the sensilla? Do sensilla-containing, CO_2-sensitive receptor neurons have ultrastructural similarities?

Boeckh: We have studied the ultrastructure of CO_2 receptors in the labial pit organ of an arctiid moth and found dendrites with fluted membranes. However, these sensilla possess walls with wall pores and thus resemble a common type of olfactory sensillum. In the tsetse fly the antennal CO_2 receptors are located within deep pits, where we have found a uniform population of strange-looking sensilla that do not have any wall perforations. CO_2 receptors in honey-bees are housed in sensilla that are different again. Their dendrites do not display any special features under the electron microscope.

Guerin: Is CO_2 perception affected by the greenhouse effect?

Kaissling: Stange & Wong (1993) detected an interesting compensation mechanism of CO_2 receptors in moths that relates to the atmospheric CO_2 level, presently 350 ppm. The response of CO_2 receptors does not depend on temperature, although CO_2 must dissolve in an aqueous medium before sensory transduction takes place, and the solubility of CO_2 strongly depends on temperature. The latter effect, however, is compensated by an opposite temperature dependence of the receptor cell physiology. The authors found that the compensation works only at the preindustrial level of CO_2, which is about 250 ppm.

Cardé: The 'wispiness' of the odour is important for orientation in male moths. Encountering filaments of pheromone at 5 Hz, for example, causes some species to switch from a zigzag path to one headed directly upwind. Your stimulus presentations were of 2 s duration. Have you looked at shorter fluctuations in signal strength to determine whether they are resolved and possibly therefore could be important in orientation?

Grant: No, I haven't looked at pulsed signals in mosquitoes. This needs to be looked at because some of the older papers suggested that continuous CO_2 stimulation is not effective in wind tunnel experiments. However, the intervals between pulses was rather long (of the order of seconds), which presumably mimics host respiratory rates. Higher frequency pulse trains that may mimic the filamentous nature of the plume have not been examined.

Cardé: A natural odour stimulus in wind should be discontinuous because of the effects of turbulence, even if the initial source size is relatively large.

Grant: The distances over which these plumes are effective may also be important. The maximum distances over which mosquitoes or moths will respond to odour sources are not clear. There have been reports in the kilometre range for Lepidoptera. Therefore, it's possible that one plume is acting over a much shorter distance than the other.

References

Anton S 1996 Central olfactory pathways in mosquitoes and other insects. In: Olfaction in mosquito–host interactions. Wiley, Chichester (Ciba Found Symp 200) p 184–196

Davis EE, Sokolove PG 1976 Lactic acid-sensitive receptors on the antennae of the mosquito, *Aedes aegypti*. J Comp Physiol 105:43–54

Kellogg FE 1970 Water vapour and carbon dioxide receptors in *Aedes aegypti*. J Insect Physiol 16:99–108

Kemme JA, van Essen PHA, Ritchie SA, Kay BH 1993 Response of mosquitoes to carbon dioxide and 1-octen-3-ol in southeast Queensland, Australia. J Am Mosq Control Assoc 9:431–435

Kent KS, Harrow ID, Quartararo P, Hildebrand JG 1986 An accessory olfactory pathway in Lepidoptera: the labial pit organ and its central projections in *Manduca sexta* and certain other sphinx moths and silk moths. Cell Tissue Res 245:237–245

Omer SM, Gillies MT 1971 Loss of response to carbon dioxide in palpectomized female mosquitoes. Entomol Exp Appl 14:251–252

Stange G 1992 High resolution measurement of atmospheric carbon dioxide concentration changes by the labial palp organ of the moth *Heliothis armigera* (Lepidoptera: Noctuidae). J Comp Physiol A 171:317–324

Stange G, Wong C 1993 Moth response to climate. Nature 365:699

Stocker RF, Lienhard MC, Borst A, Fischbach KF 1990 Neuronal architecture of the antennal lobe in *Drosophila melanogaster*. Cell Tissue Res 262:9–34

Takken W, Kline DL 1989 Carbon dioxide and 1-octen-3-ol as mosquito attractants. J Am Mosq Control Assoc 5:311–316

Vythilingam I, Lian CG, Thim CS 1992 Evaluation of carbon dioxide and 1-octen-3-ol as mosquito attractants. Southeast Asian J Trop Med Public Health 23:328–331

Ziesmann J 1996 The physiology of an olfactory sensillum of the termite *Schedorhinotermes lamanianus*: carbon dioxide as a modulator of olfactory sensitivity. J Comp Physiol A 178, in press

Responses of antennal olfactory receptors in the yellow fever mosquito *Aedes aegypti* to human body odours

Barbara Pappenberger, Martin Geier and Jürgen Boeckh[1]

Universität Regensburg, Institut für Zoologie, Lehrstuhl Boeckh, Universitässtrasse 31, D-93040 Regensburg, Germany

Abstract. Recent behavioural studies have demonstrated that human body odours which female *Aedes aegypti* find attractive exert their effects as complex mixtures of synergistically acting components. We have attempted to clarify the sensory mechanisms underlying the perception of these complex host odours by studying the responses of sensory cells underneath the A_3-type sensilla of the mosquito antenna to both a human skin wash extract and the extract's active chromatographic fractions. The reaction patterns show that the host stimuli elicit responses from several types of receptor cells in a typical across-fibre pattern mode. It seems as if this is another case where the essential message in a biologically significant odour consists of a complex pattern of compounds that is encoded in an according complex response pattern by a cooperating set of primary sensory neurons of different odour specificities.

1996 Olfaction in mosquito–host interactions. Wiley, Chichester (Ciba Foundation Symposium 200) p 254–266

Previous behavioural studies have demonstrated that CO_2 and L-lactic acid are potent compounds in human body odours which alert and attract female yellow fever mosquitoes (Acree et al 1968, Smith et al 1970). The sense organs for these compounds have been identified by means of electrophysiological recordings, and they have been classified according to the scheme of mosquito sensilla proposed by S. McIver on the basis of her fine structural studies (McIver 1974). CO_2-sensitive receptor cells have been found in the sensilla basiconica on the maxillary palps (Roth & Willis 1952, Kellogg 1970). In A_3-type sensilla, responses have been detected to both L-lactic acid and other human body odours, such as certain carbonylic acids (Davis & Sokolove 1975, Davis 1984, 1988). The CO_2- and L-lactic acid-sensitive cells have been

---

[1]This paper was presented by J. Boeckh.

investigated in detail in later studies by Grant et al (1995), Bowen (1991) and Davis & Bowen (1994). Recent behavioural studies on *Aedes* with odours taken directly from humans have revealed that there are other attractants in host odour. Human skin wash extracts are just as attractive as the natural host. Several chromatographic fractions of still unknown chemical nature have been isolated that together increase the potency of L-lactic acid to attract female mosquitoes (Geier et al 1996, this volume). L-Lactic acid itself is a constituent of one of these fractions. Consequently, an analysis of the responses of the olfactory sense organs to the different kinds of human body odours would be extremely valuable. Preliminary results have suggested that the A_3-type sensilla on the antennae are the most promising objects to study.

Materials and methods

We fastened adult female *Aedes aegypti* from stems raised by BAYER (Leverkusen, Germany) to holders, and held the antennae in place with sticky tape. We kept the preparation under a permanent current of fresh air and visualized it with an inverted microscope (Zeiss) via long distance optics at magnifications of up to $\times 430$. This enabled us to identify the different morphological types of antennal sensilla, including the A_3-type, by their length and shape according to features established by S. McIver (1972, 1973, 1974). Spot-checks with labelled sensilla in the scanning electron microscope reconfirmed the identification. A tungsten needle inserted into the joint between two distal antennal segments served as a reference electrode. We positioned the recording electrode (a saline-filled micropipette, $1-5\,\mu$m tip diameter) at the base of a given sensillum. Recorded impulses had amplitudes of $0.1-1.5$ mV. We fed the recordings into a computer for impulse selection so that amplitude histograms and stimulus-reaction relationships could be established.

Stimulation techniques

We delivered odours towards the recorded antenna either from a syringe olfactometer (Kafka 1970, Sass 1976) or from heatable glass tubes loaded with a small standard amount of odour solution on their inner walls. After evaporation of the solvent, we placed the tubes in front of an outlet and blew air through it towards the antenna. Syringes filled with fresh air or unloaded cartridges served as controls.

Host-related odour stimuli

(1) We pulled air from under the clothing of the experimenter or expired air into the 20 ml syringe of the olfactometer and applied it immediately. We

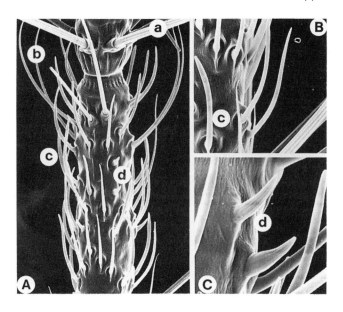

FIG. 1.   One of the 13 segments of the antenna of a female *Aedes aegypti* showing different types of sensilla. A, B and C represent different magnifications. a, sensillum chaeticum; b, A_1-type sensillum; c, A_2-type sensillum; d, A_3-type sensillum. Nomenclature according to McIver (1982). Magnifications: A = × 900, B = × 1500, C = × 3000.

    spotted samples of sweat, saliva and fresh or stored blood onto pieces of filter paper and stored them in glass vials in the syringes.

(2) We placed samples of methanol washes from human skin and certain behaviourally active chromatography fractions either in glass vials in the syringes or in the glass tubes.

(3) L-Lactic acid is presently the only active compound in the skin wash that has been identified. Therefore, we included L-lactic acid (at different concentrations in ethanol or acetone) in the test. We also tested a few other compounds: propanoic acid, butyric acid, valeric acid, caproic acid and amylamine (pentylamine) (Fluka Chemie AG, CH-9470 Buchs, Switzerland; Sigma-Aldrich Chemie GmbH, Deisenhofen, Germany); which were all dissolved and diluted in mineral oil or hexane and tested at various concentrations.

Results

Responses to different kinds of human body odours could be obtained only with cells from A_3-type sensilla (Fig. 1). Other sensillum types (A_1-type II and A_2-type I) did not elicit responses to either human body odours, L-lactic acid,

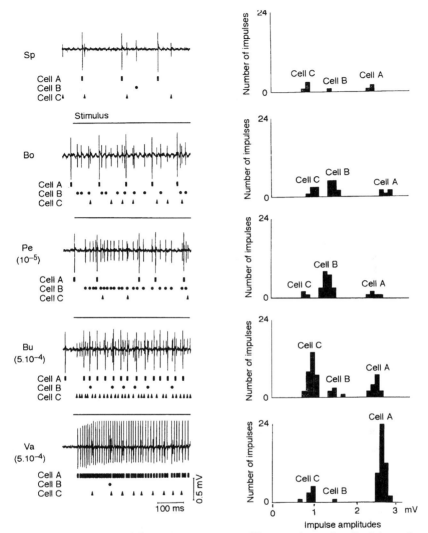

FIG. 2. Background activity and responses to different odour stimuli of three classes of impulse amplitudes in an A_3-type sensillum. The symbols below each recording indicate the different impulse types. Right hand column, amplitude histograms showing reactions of different cells. From top to bottom: Sp, spontaneous discharge; Bo, body odour; Pe, pentylamine (amylamine, primarily cell type B responds); Bu, butyric acid (cell types A and C respond); Va, valeric acid (cell type A responds).

short-chain carboxylic acids or amylamine, all of which elicited responses from cells underneath A_3-type sensilla (Fig. 1). In 30 recordings of A_3-type sensilla in 30 females, three clearly distinct classes of impulse amplitudes emerged from the histograms (Fig. 2).

TABLE 1 Response spectra of receptor cell types in A_3-type sensilla

Cell type	Stimulus[a]											
	Spontaneous discharge	room air	body odour	sweat	extract of saliva	amyl-amine[b]	heptyl-amine[b]	valeric acid[b]	caproic acid[b]	butyric acid[b]	propanoic acid[b]	L-lactic acid[c]
A (n = 30)	7±2	13±8	16±8	14+8	15±8	10±4	9±5	60±15	57±14	26±13	19±12	15±7
B (n = 10)	5±3	10±8	30±8	50±11	45±12	42±10	2±5	10±7	10±5	11±8	10±9	11±8
C (n = 10)	7±3	8±6	11±7	12±6	11±6	12±9	6±5	22±8	11±6	55±12	54±11	9±3

[a]Number of impulses measured over 500 ms following exposure to each stimulus ± SD.
[b]1 : 10 000 dilution.
[c]1 : 10 dilution.

spontaneous discharge

skin wash extract 10-1

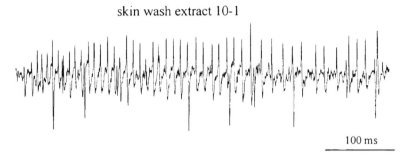

100 ms

FIG. 3. Spontaneous activity of cells underneath an A_3-type sensillum (top) and their reaction to human skin wash extract (bottom). Top: three receptor cells can be distinguished by their impulse amplitudes and time courses during spontaneous discharge. Bottom: discrimination of cells becomes more difficult due to modifications of impulse amplitudes and time courses.

Thus, three classes of receptor cells (A, B and C) could be established, each characterized by a certain odour spectrum (Fig. 2 and Table 1). Cell type B was the only one that responded clearly to human body odours and to amylamine. This cell type did not respond to L-lactic acid.

Different results were obtained in another study of 113 recordings from A_3-type sensilla that respond clearly to L-lactic acid or human skin wash extracts. In many of these cases four cells appeared in a given recording. Distinct differences were discerned in the amplitudes of the impulses from the different cells at low levels of activity, but this became difficult at higher impulse frequencies (Fig. 3). At higher impulse frequencies, impulses often altered their amplitudes and time courses, and they often coincided. Therefore, precisely defined cell types with consistent response characteristics from sensillum to sensillum have not been determined in this type of recording. The typical activity pattern that appeared in many recordings consists of excitatory responses of up to four receptor cells to the skin wash extract. Three of them responded in varying combinations and with different relative sensitivities to the behaviourally active fractions of the extract (Fig. 4). One of the receptor

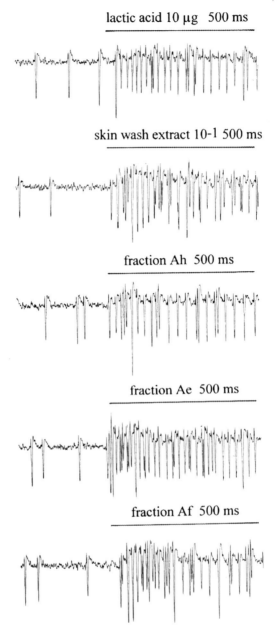

FIG. 4. Activity of cells from an A_3-type sensillum during stimulation with L-lactic acid, human skin wash extract and its behaviourally active fractions (Ah, Ae and Af). The L-lactic acid-sensitive cells respond to the Ah fraction, which contains L-lactic acid. These cells are also exited by fractions Ae and Af, which do not contain L-lactic acid.

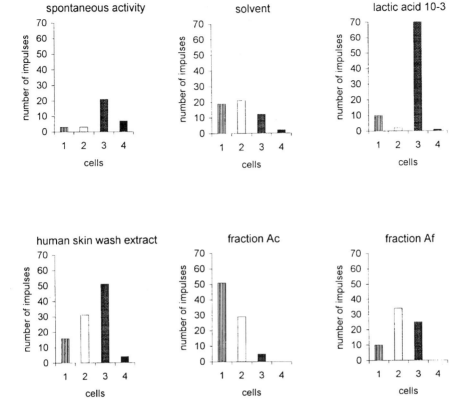

FIG. 5. The response of four receptor cells in an A_3-type sensillum to the human skin wash extract, its behaviourally active fractions (Ac and Af, which do not contain L-lactic acid) and L-lactic acid itself (1 : 1000 dilution).

cells (and sometimes two) responded to L-lactic acid, an L-lactic acid-containing chromatography fraction of the human skin wash extract (Ah) and at least two other fractions that do not contain L-lactic acid (Ae and Af) (Figs 4 and 5).

A third cell responded to a fraction called Ac, which does not contain L-lactic acid, and a fourth cell responded to one of the fractions that was not active in the behavioural test. The responses to the single fractions are not totally in accordance with the reaction picture of the different cells to the complete extract. For example, cell 1 in Fig. 5 has a greater response to fraction Ac than to the total extract. The extract contains many more substances than are in a single fraction, and these might interfere with the response of this cell or with the substances from the fraction. These apparent discrepancies will be resolved when the majority of the substances that have differential effects on the receptor cells have been characterized. The response magnitudes of a given cell were different for the different fractions.

There is also an overlap in the specificities of the different cells for the different fractions. The total response pattern of these recordings resembles the patterns obtained for other insects, such as the American cockroach *Periplaneta americana* (Sass 1978), where whole sets of classes of olfactory receptor cells participate in the responses to complex biologically significant odours. In these cases, each class responds according to its odour spectrum to certain components of the complete odour. The classes' spectra overlap and the across-fibre mode of coding prevails, so that a highly graded image of the actual blend of the components is preserved in the response patterns and a contour-rich image of the odour is transmitted to the brain (Sass 1976, Boeckh et al 1984). The complete odour spectra and actual structure–activity relationships of the responses of certain cells can be described only after the effective compounds have been isolated from the skin wash extract and the fractions. Only then will we be able to establish cell classes (if such classes exist) and study possible varieties in the receptor complement in the A_3-type sensilla and other olfactory sensilla. From this, we will be able to generate a final picture of the code.

In order to understand the synergism between different components of the host odour, we will need to test mixtures of fractions at the level of both the receptor cell and the CNS. The latter area is particularly *terra incognita*, but it is already being studied in detail by modern neuroanatomical and neurophysiological methods.

Acknowledgement

This study was performed in the course of a joint investigation of host odours for *Aedes aegypti* together with the research laboratories of Bayer AG, Leverkusen, Germany.

References

Acree F Jr, Turner RB, Gouck HK, Beroza M, Smith N 1968 L-Lactic acid: a mosquito attractant isolated from humans. Science 161:1346–1347
Boeckh J, Ernst KD, Sass H, Waldow U 1984 Anatomical and physiological characteristics of individual neurons in the central antennal pathway of insects. J Insect Physiol 30:15–26
Bowen MF 1991 The sensory physiology of host-seeking behaviour of mosquitoes. Ann Rev Entomol 36:139–158
Davis EE 1984 Development of lactic acid-receptor sensitivity and host-seeking behaviour in newly emerged female *Aedes agypti* mosquitoes. J Insect Physiol 30:211–215
Davis EE 1988 Structure–response relationship of the lactic acid-excited neurons in the antennal grooved-peg sensilla of the mosquito *Aedes aegypti*. J Insect Physiol 34:443–449

Davis EE, Bowen MF 1994 Sensory physiological basis for attraction in mosquitoes. J Am Mosq Control Assoc 10:316–325

Davis EE, Sokolove PG 1975 Lactic acid-sensitive receptors on the antennae of the mosquito, *Aedes aegypti*. J Comp Physiol A 105:43–54

Geier M, Sass H, Boeckh J 1996 A search for components in human body odour that attract females of *Aedes aegypti*. In: Olfaction in mosquito–host interactions. Wiley, Chichester (Ciba Found Symp 200) p 132–148

Grant AJ, Wigton BE, Aghajanian JG, O'Connell RJ 1995 Electrophysiological responses of receptor neurons in mosquito maxillary palp sensilla to carbon dioxide. J Comp Physiol A 177:389–396

Kafka WA 1970 Molekulare Wechselwirkungen bei der Erregung einzelner Riechzellen. Z vergl Physiol 70:105–143

Kellogg FE 1970 Water vapour and carbon dioxide receptors in *Aedes aegypti*. J Insect Physiol 16:99–108

McIver SB 1972 Fine structure of sensilla chaetica on the antennae of *Aedes aegypti* (Diptera: Culicidae). Ann Entomol Soc Am 65:1390–1397

McIver SB 1973 Fine structure of sensilla coeloconica of culicine mosquitoes. Tissue & Cell 5:105–112

McIver SB 1974 Fine structure of antennal grooved-pegs of the mosquito, *Aedes aegypti*. Cell Tissue Res 153:327–337

McIver SB 1982 Sensilla of mosquitoes (Diptera: Culicidae). J Med Entomol 19:489–535

Roth LH, Willis ER 1952 Possible hygroreceptors in *Aedes aegypti* and *Blatella germanica*. J Morphol 91:1–14

Sass H 1978 Olfactory receptors on the antenna of *Periplaneta*: response constellations that encode food odors. J Comp Physiol 128:227–233

Sass H 1976 Zur nervösen Codierung von Geruchsreizen bei *Periplaneta americana*. J Comp Physiol A 107:49–65

Smith CN, Smith N, Gouck HK et al 1970 L-Lactic acid as a factor in the attraction of *Aedes aegypti* (Diptera: Culicidae) to human hosts. Ann Entomol Soc Am 63:760–770

DISCUSSION

Mustaparta: The response to fraction Ac seemed to be stronger than the response to the whole extract. Can the concentrations of the stimuli be compared?

Geier: But the compounds in fraction Ac are probably more concentrated than in the head-space over the human body.

Mustaparta: Have you tested the whole extract and one fraction of that extract on the same receptor neuron?

Geier: Yes. We have tested fractions and the original extract on the same cell using glass cartridges. It is possible that the stronger response to a single fraction is caused by a higher volatility of the separated substances. This could be explained by the fact that compounds having a high molecular mass could reduce the volatility of other compounds in the extract.

Boeckh: The human skin extract contains these fractions. The recordings show that different cells respond in different ways to a given fraction and also to the total extract from the skin washes. There are no quantitative measurements as yet. The L-lactic acid-sensitive cell reacts to the fraction containing L-lactic acid but also, to a lesser degree, to other fractions.

Cork: I am surprised that cell A responded to the carboxylic acids tested but not to sweat, because sweat contains a wide range of carboxylic acids.

Boeckh: There are many observations that don't seem to fit into the picture as yet. We hope that the refinement of these gross fractions will clarify the situation.

Guerin: The interesting thing is that there are separate receptor cells for the different fractions. Your results are all relative at the moment because you do not have a firm identification of the products.

Takken: There are differences between the receptor cells for L-lactic acid as described by Davis (1988) and the cells in the mosquitoes you have studied. Have you considered that you may be working with completely different strains of *Aedes aegypti?*

Boeckh: Yes, we are working with different strains, so there might be a difference. We should compare our results with those from different strains.

Takken: In our work on *Anopheles gambiae*, we have all agreed to incorporate at least one mosquito from another strain in order to be able to compare results. I feel that this should also be done with *Ae. aegypti*.

Boeckh: This shouldn't be a problem. At the moment we are just developing the method. When we have refined the technique, we will look at other strains. We will also have to consider the health of the strains because of degenerative effects, for example.

Kaissling: Which of the cells respond to the repellent?

Boeckh: Ed Davis has shown that L-lactic acid cells are inhibited in the presence of L-lactic acid plus *N,N*-diethyl-*m*-toluamide (DEET), for example. We have not done this experiment, we have only used a selection of synthetic repellents, including DEET, that are highly excitatory substances for certain other sensilla which display strong reactions to DEET and to other repellents.

Galun: DEET is considered to be a synthetic material, but it has been found as an activator of the sex pheromone of the pink bollworm moth.

Boeckh: Was this observation made before DEET was used as a repellent?

Galun: It was used as a repellent as early as the 1940s, but was discovered as a pheromone activator only in 1968 (Jones & Jacobson 1968).

Boeckh: The other synthetic substances we used were synthesized according to molecular modelling studies. Many substances were made, some of which acted as repellents.

Galun: Do they also contain DEET?

Boeckh: No, their formulae are completely different.

Galun: More and more data are accumulating on the serious health problems caused by the continuous use of DEET. The quicker it is replaced the better.

Guerin: I have a comment and a question. My comment is that we are working on phlebotome vectors of leishmaniasis with a group at the University of Keele, and we have found that the antennae of these sandflies carry morphologically identical ascoid sensilla along their entire length but depending on the location on the antenna, these ascoid sensilla house different types of receptors. It's surprising that although there is only one pair of these large ascoid sensilla per antennal segment, their function differs from the tip to the base.

My question relates to the L-lactic acid receptors of mosquitoes. Are they mostly found in sensilla at a particular location on the antenna, i.e. are they more proximal or more distal?

Davis: McIver (1970) looked at the distribution of A_3-type sensilla on the *Ae. aegypti* antenna, and found that they tend to be distributed towards the more terminal segments—the last five or six segments contain 80% of the total number.

Guerin: Is there a dorsal/ventral pattern?

Davis: When you are preparing a mosquito with such a small antenna, the slightest rotation of the antenna may result in a markedly different orientation exposing a different set of sensilla that are in a position from which to obtain electrophysiological recordings. It is so difficult to control, and there are no apparent landmarks in the light microscope field of view (about $950 \times$) to be able to map the distribution of different types of grooved-peg sensilla.

Guerin: Jürgen Boeckh mentioned a subpopulation of sensilla that contain L-lactic acid receptors. Is there a preferential location of these sensilla?

Boeckh: I can't answer this question. First we need to complete our systematic recordings and then do labelling studies.

Bowen: My paper on the two types of sensilla basiconica is relevant to this and has recently been published (Bowen 1995).

Brady: We know that males have CO_2 receptors, do you know whether they have any L-lactic acid receptors?

Davis: Yes, they do.

I have a comment on the difficulty of obtaining and maintaining cells of one type or another. After I found the first L-lactic acid-excited cell, I worked on it for a number of weeks and then went on vacation for a couple of weeks. When I returned, all that I could find were L-lactic acid-inhibited cells. I don't understand how this happened. This probably involved an imperceptible change in position of the antenna.

Boeckh: It's important to remain aware of this sort of observation because we need to understand both the peripheral message to the brain and the receptor compliment. We must test the less-sensitive L-lactic acid receptors and the inhibited receptors to get the whole picture.

References

Bowen MF 1995 Sensilla basiconica (grooved pegs) on the antennae of female mosquitoes: electrophysiology and morphology. Entomol Exp Appl 77:233–238

Davis EE 1988 Structure–response relationship of the lactic acid-excited neurons in the antennal grooved-peg sensilla of the mosquito *Aedes aegypti*. J Insect Physiol 34:443–449

Dougherty MJ, Guerin PM, Ward RD 1995 Identification of oviposition attractants for the sandfly *Lutzomyia longipalpis* (Diptera: Psychodidae) in volatiles of faeces from vertebrates. Physiol Entomol 20:23–32

Jones WA, Jacobson M 1968 Isolation of N,N-diethyl-m-toluamide (Deet) from female pink ballworm moths. Science 159:99–100

McIver SB 1970 Comparative study of antennal sense organs of female culicine mosquitoes. Can Entomol 102:1258–1267

The multiple role of the pheromone-binding protein in olfactory transduction

Gunde Ziegelberger

Max-Planck-Institut für Verhaltensphysiologie, D-82319 Seewiesen, Germany

Abstract. Before airborne odorant molecules can stimulate the olfactory receptor cells of animals that live on land, they have to pass through an aqueous solution that contains high concentrations of soluble odorant-binding proteins (OBPs). In insect sensilla the role of these OBPs for signal transduction is becoming multifaceted. Sensillum lymph perfusion experiments in the moth *Antheraea polyphemus* implied a solubilizer and carrier function of the pheromone-binding protein (PBP) and led to the conclusion that it is the pheromone–PBP complex which activates the postulated receptors. Recent results have shown the presence of two redox states of the PBP and a shift in pheromone binding from the reduced to the oxidized form, depending on the presence of sensory hair material. Thus, PBP oxidation might occur simultaneously with receptor cell activation and might lead to deactivation of the pheromone–PBP complex terminating the pheromone stimulation.

1996 Olfaction in mosquito–host interactions. Wiley, Chichester (Ciba Foundation Symposium 200) p 267–280

A few years ago a family of odorant receptors was discovered in the rat by molecular cloning approaches (Buck & Axel 1991). Since then, odorant receptors of several vertebrate species have been sequenced, but the search for these G protein-coupled receptors is still being continued in invertebrates. Putative honey-bee odorant receptors have been described (Danty et al 1994), but because of the high homology to vertebrate odorant receptors, these results are under discussion and have not yet been confirmed by other groups. In contrast, the understanding of soluble odorant-binding proteins (OBPs) surrounding the receptive dendrites has increased (reviewed in Pelosi & Maida 1995). This paper will summarize the multiple role of these OBPs for olfactory transduction.

Pheromone-binding protein as a carrier

The best studied OBP is the pheromone-binding protein (PBP) of *Antheraea polyphemus* because of its high concentration in male antennae (Vogt &

Riddiford 1981, Klein 1987) and, thus, its availability for functional studies. In perfusion experiments of the aqueous sensillum lymph space, the PBP strikingly increased the electrophysiological response of the receptor cell when introduced together with the lipophilic pheromone (Van den Berg & Ziegelberger 1991). Thus, the PBP is able to act as solubilizer and carrier. The diffusion velocity of the pheromone is decreased when bound by the PBP, but it is compatible with the diffusion coefficient found for longitudinal pheromone transport in the olfactory hair of $3 \times 10^{-7} \text{cm}^2/\text{s}$ (Kanaujia & Kaissling 1985).

The pheromone–pheromone-binding protein complex as a receptor cell activator

Almost all incoming pheromone molecules will be bound by the PBP because of its millimolar concentration (10 mM; Vogt & Riddiford 1981) and its nanomolar dissociation constant (60 nM determined by Kaissling et al [1985], 640 nM measured by Du & Prestwich [1995]), so it seems likely that the pheromone–PBP complex activates the presumed receptor molecules. Therefore, the PBP might possess two binding sites: one for the odorant and the other for the receptor or another component of the receptor cell membrane. Thus, the moderate odour ligand selectivity of the PBP (De Kramer & Hemberger 1987, Du & Prestwich 1995) could contribute to the high specificity of the receptor cell response.

Pheromone-binding protein as a deactivator

A rapid pheromone deactivation has been postulated because the excitatory action of the pheromone occurs within a short, limited time period. The degrading enzyme found in the sensillum lymph could, in principle, be responsible for a rapid stimulus termination. Vogt et al (1985) calculated a half-life of about 15 ms for the isolated esterase and the free pheromone. *In vivo*, however, most of the pheromone is bound by the PBP (Kaissling 1986) and is probably protected from the enzyme (Vogt & Riddiford 1986). Kasang et al (1988) estimated the half-lives of ^{3}H-labelled pheromones in intact antennae of *A. polyphemus* as being about 3 min. This observation and the large variability of the esterase activity in individual males (including those with apparently no enzyme activity but with a normal electrophysiological response) suggest that the esterase is not responsible for the rapid deactivation of the pheromone. However, the esterase may be responsible for the final sequestration of the pheromone (Maida et al 1995). The above results, together with the finding that the pheromone is still present in the sensillum lymph after the electrophysiological response has terminated (Kanaujia & Kaissling 1985), led Kaissling (1986) to suggest the idea that the PBP deactivates the pheromone. Indeed, recent results on the presence of two redox states of the

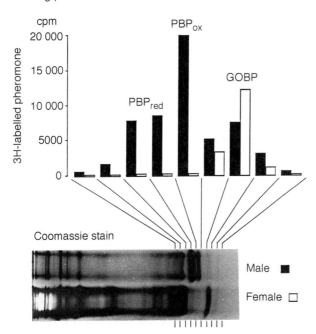

FIG. 1. Binding of ³H-labelled pheromones by male and female homogenates, analysed by native polyacrylamide gel electrophoresis after an incubation time of 45 min. Top: cpm profile of a 20% native gel (36–45 mm from start), cut into 1 mm slices. Bottom: Coomassie blue stain of the same samples (equivalents of about one male antenna without stem and three female antennae). GOBP, general odorant-binding protein; ox, oxidized; PBP, pheromone-binding protein; red, reduced.

PBP support the hypothesis that PBP, in addition to its carrier function, also plays a role in stimulus termination (Ziegelberger 1995).

Following native polyacrylamide gel electrophoresis of homogenates of male antennal branches under non-reducing conditions (Fig. 1) two PBP bands are observed. These bands are barely detected in females. Peptide mapping of the two PBP bands revealed that there were no differences in the amino acid sequences. However, chemical treatment of the two PBP populations with 2-nitro-5-thiocyanobenzoic acid (NTCB) led to cyanylation and cleavage at free sulfhydryl groups of cysteines and resulted in different disulfide formations. The upper, more slowly migrating PBP band gave rise to several peptides, indicating that at least two cysteines have free sulfhydryl groups (PBP_red band). In contrast, the lower PBP band was not cleaved by NTCB, implying that all six cysteine residues are involved in disulfide formation (PBP_ox band). All insect OBPs that have been sequenced so far have six conserved cysteine residues (reviewed in Breer et al 1992), suggesting that disulfide bonds are crucial for their physiological function.

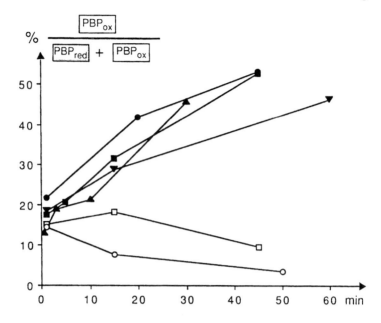

FIG. 2.   Percentage of pheromone binding to the oxidized pheromone-binding protein (PBP) band with respect to total binding. The proportion of oxidized PBP (PBP_{ox}) increases in the sensory hair preparations (filled symbols) with increasing incubation time, whereas no redox shift was observed with purified PBP (open symbols). red, reduced. Different symbols represent independent experiments.

Do the two redox states of PBP differ in pheromone binding? In kinetic studies with isolated sensory hairs, the ^{3}H-labelled pheromone was bound mainly by the PBP_{red} band at short incubation times. Increasing the incubation time resulted in a shift in the ratio of bound pheromone to the two PBP bands towards the PBP_{ox} band (Fig. 2). No redox shift was observed with purified PBP: it needed the presence of sensory hair material which contains the dendritic membrane of the receptor cells. Thus, the oxidation of the PBP_{red} might depend on the interaction of the pheromone–PBP_{red} complex with receptor proteins and could accompany a change in function—from a reduced PBP acting as carrier and stimulator to an oxidized PBP serving as deactivator for the pheromone (Fig. 3). This suggests a certain amount of PBP turnover (Vogt et al 1989), which seems to take place in the auxiliary cells of the sensilla. In immunocytochemical studies of *A. polyphemus*, vesicles containing PBP are frequently found in all three auxiliary cells (Steinbrecht et al 1992).

The half-life of the active pheromone *in vivo* (estimated from the decline of the receptor potential) was 0.5–2 s, depending on stimulus concentration

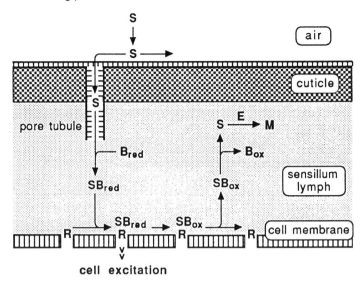

FIG. 3. Model of perireceptor events. Adsorbed pheromone molecules (S) enter the sensillum via the pore tubules and are transported by the reduced pheromone-binding protein (B_{red}) across the aqueous sensillum lymph. The pheromone–reduced pheromone-binding protein $(PBP)_{red}$ complex (SB_{red}) activates the receptor molecules (R) resulting in PBP oxidation and receptor cell excitation. The oxidized PBP (B_{ox}) is a rapid pheromone deactivator, and the final degradation to inactive metabolites (M) is achieved by the sensillum lymph specific esterase (E). Coomassie blue and silver staining of sensillum lymph proteins suggest that the ratio of PBP to E is at least 10 000 : 1.

(Kaissling 1995). Thus, PBP oxidation (which occurs in the minute range) in the hair homogenate is too slow for the postulated rapid pheromone deactivation. This discrepancy could, however, be due to the more than 200-fold dilution in the homogenate as compared to the living hair.

Odorant-binding proteins and odour discrimination

OBPs are divided into three groups according to their amino acid sequence homology and their presence in either males or both sexes: the PBPs, which are predominant in the antennae of male moths; and two groups of so-called general odorant-binding proteins (GOBPs), which occur in both sexes (Vogt et al 1991). Binding proteins from different subclasses show a complex and distinct distribution pattern among different sensillum types and are not co-localized within the same sensillum. Immunocytochemical studies on *A. polyphemus* show that PBP is found exclusively in the sensilla trichodea innervated by pheromone receptor cells, whereas GOBP is present in the sensilla basiconica tuned to general odours (Steinbrecht et al 1992, Laue et al 1994, Schneider et al 1964; see also Steinbrecht 1996, this volume). Therefore, it is possible that the soluble OBPs are already involved in the first step of stimulus recognition before the highly specific receptor cell is activated. This notion is supported by the observation that the number of OBPs found in a given species is increasing steadily. Five different OBPs have been sequenced in *Drosophila* (Pikielny et al 1994, McKenna et al 1994) and similar results are obtained in vertebrates, where up to eight OBPs have been identified in the Old World porcupine *Hystrix cristata* (Felicioli et al 1993).

However, at least in the case of *A. polyphemus*, OBPs are not associated with specific ligands. In this species, the GOBP of males and females binds, in a concentration-dependent manner, more labelled pheromone than does PBP. In Fig. 1, homogenates of male antennal branches and female antennae were incubated with the pheromone component ^{3}H-(E,Z)-6,11-hexadecadienyl acetate and analysed by native polyacrylamide gel electrophoresis. The labelled pheromone is bound in male homogenates by two PBPs and one GOBP, and in females by only a single GOBP. The binding sites of PBP and GOBP for the same pheromone are probably different because the two proteins share an amino acid sequence identity of only about 30%. Therefore, the classification of OBPs into PBPs and GOBPs is not substantiated by specific odorant binding, although the localization of a given OBP, as cited above, is correlated with the receptor cell specificity. It is worth mentioning that PBP is present in female homogenates because female silkmoths do not respond to their own pheromone in electrophysiological or behavioural studies. The natural stimuli of the PBP-containing sensilla of females have not been identified.

Effects of high concentrations of odorant-binding proteins

Odorant-binding proteins as buffers of toxic compounds

In addition to sex pheromones and general plant odours, toxic compounds also enter the sensilla through the pore tubules. If toxic compounds reach the dendrites, they might interfere with the pheromone response. It is conceivable that OBPs may also serve as scavengers, buffering unexpected compounds and maintaining a functioning sensillum.

Odorant-binding proteins as polyanions

The ionic composition of the sensillum lymph has been analysed by flame photometry and X-ray microanalysis, and has revealed a high potassium, low sodium and low calcium content (Kaissling & Thorson 1980, Steinbrecht & Zierold 1987). The total positive charges of the cation electrolytes are not balanced by the total negative charges of chlorine. Histochemical studies on olfactory sensilla of the blowfly, Calliphora (Gnatzy & Weber 1978) and estimations of the isoelectric points of OBPs (pI for PBP = 4.5, Klein 1987) suggest that the balance is most probably achieved by organic polyanions such as acid mucopolysaccharides and acidic OBPs.

Odorant-binding proteins keep pheromones in solution

The lipophilic nature of most odorants suggests that free odorant molecules are inserted in the receptor cell membrane. The binding of the incoming pheromone molecule by the OBP (at millimolar concentrations) might keep the pheromone in solution. It is not known whether a pheromone that is inserted in the lipid membrane of the receptor cell will activate the cell and how it could be deactivated.

Odorant-binding proteins protect pheromones from degrading enzymes

The binding of the pheromone by the PBP might prevent the pheromone from enzymatic degradation. Pheromone perception is very sensitive: one pheromone molecule is able to elicit a nerve impulse. Therefore, molecules should not be degraded before they activate the receptor cell. At present I am testing whether the addition of PBP slows down the degradation rate (as suggested by Vogt & Riddiford [1986]).

Conclusions

The presence of two redox states of the PBP in olfactory hairs of *A. polyphemus* and a receptor cell-dependent shift in pheromone binding to the oxidized form suggest a multiple role of the PBP in olfactory transduction. The novel feature of the proposed functional model is the stepwise handling of the adsorbed pheromone molecules: (1) the reduced PBP carries the pheromone across the aqueous sensillum lymph and stimulates the presumed receptor molecules, leading to both receptor cell excitation and PBP oxidation; (2) pheromone molecules are not degraded prior to cell activation, assuming that binding to PBP protects them from degradation; and (3) the slow pheromone degradation to an inactive metabolite is then achieved by sensillum lymph-specific esterases.

References

Breer H, Boekhoff I, Krieger J, Raming K, Strotmann J, Tareilus E 1992 Molecular mechanisms of olfactory signal transduction. In: Corey DP, Roper SD (eds) Sensory transduction. Rockefeller University Press, New York, p 94–108
Buck L, Axel R 1991 A novel multigene family may encode olfactory receptors: a molecular basis for odor recognition. Cell 65:175–187

Danty E, Cornuet J-M, Masson C 1994 Honeybees have putative olfactory receptor proteins similar to those of vertebrates. C R Acad Sci Ser III Sci Vie 317:1073–1079

De Kramer JJ, Hemberger J 1987 Neurobiology of pheromone reception. In: Prestwich GD, Blomquist GJ (eds) Pheromone biochemistry. Academic Press, Orlando, FL, p 433–472

Du G, Prestwich GD 1995 Protein structure encodes the ligand binding specificity in pheromone-binding proteins. Biochemistry 34:8726–8732

Felicioli A, Ganni M, Garibotti M, Pelosi P 1993 Multiple types and forms of odorant-binding proteins in the Old World porcupine *Hystrix cristata*. Comp Biochem Physiol B 105:775–784

Gnatzy W, Weber K M 1978 Tormogen cell and receptor-lymph space in insect olfactory sensilla: fine structure and histochemical properties in Calliphora. Cell Tissue Res 189:549–554

Kaissling K-E 1986 Chemo-electrical transduction in insect olfactory receptors. Ann Rev Neurosci 9:121–145

Kaissling K-E 1995 Pheromone deactivation on the antenna of the moth *Antheraea polyphemus*: its minimum velocity and its possible mechanism. In: Elsner N, Menzel R (eds) Neurobiology report, vol II: Proceedings of the 23rd Göttingen Neurobiology Conference, 1995. Thieme Verlag, Germany, p 362

Kaissling K-E, Thorson J 1980 Insect olfactory sensilla: structural, chemical and electrical aspects of the functional organization. In: Sattelle DB, Hall LM, Hildebrand JG (eds) Receptors for neurotransmitters, hormones and pheromones in insects. Elsevier, Amsterdam, p 261–282

Kaissling K-E, Klein U, De Kramer JJ, Keil TA, Kanaujia S, Hemberger J 1985 Insect olfactory cells: electrophysiological and biochemical studies. In: Changeux JP, Hucho F, Maelicke A, Neumann E (eds) Molecular basis of nerve activity: proceedings of the international symposium in memory of David Nachmansohn (1899–1983), Berlin, West Germany, October 11–13, 1984. de Gruyter, New York, p 173–183

Kanaujia S, Kaissling K-E 1985 Interactions of pheromone with moth antennae: adsorption, desorption and transport. J Insect Physiol 31:71–81

Kasang G, von Proff L, Nicholls M 1988 Enzymatic conversion and degradation of sex pheromones in antennae of the male silkworm moth *Antheraea polyphemus*. Z Naturforsch 43:275C–284C

Klein U 1987 Sensillum-lymph proteins from antennal olfactory hairs of the moth *Antheraea polyphemus* (Saturniidae). Insect Biochem 17:1193–1204

Laue M, Steinbrecht RA, Ziegelberger G 1994 Immunocytochemical localization of general odorant-binding protein in olfactory sensilla of the silkmoth *Antheraea polyphemus*. Naturwiss 81:178–180

Maida R, Ziegelberger G, Kaissling K-E 1995 Esterase activity in the olfactory sensilla of the silkmoth *Antheraea polyphemus*. NeuroReport 6:822–824

McKenna MP, Hekmat-Scafe DS, Gaines P, Carlson JR 1994 Putative *Drosophila* pheromone-binding proteins expressed in a subregion of the olfactory system. J Biol Chem 269:16340–16347

Pelosi P, Maida R 1995 Odorant-binding proteins in insects. Comp Biochem Physiol B 111:503–514

Pikielny CW, Hasan G, Rouyer F, Rosbash M 1994 Members of a family of Drosophila putative odorant-binding proteins are expressed in different subsets of olfactory hairs. Neuron 12:35–49

Schneider D, Lacher V, Kaissling K-E 1964 Die Reaktionsweise und das Reaktionsspektrum von Riechzellen bei *Antheraea pernyi* (Lepidoptera, Saturniidae). Z vergl Physiol 48:632–662

Steinbrecht RA 1996 Structure and function of insect olfactory sensilla. In: Olfaction in mosquito–host interactions. Wiley, Chichester (Ciba Found Symp 200) p 158–177

Steinbrecht RA, Zierold K 1987 The electrolyte distribution in insect olfactory sensilla as revealed by X-ray microanalysis. Ann N Y Acad Sci 510:638–641

Steinbrecht RA, Ozaki M, Ziegelberger G 1992 Immunocytochemical localization of pheromone-binding protein in moth antennae. Cell Tissue Res 270:287–302

Van den Berg MJ, Ziegelberger G 1991 On the function of the pheromone binding protein in the olfactory hairs of *Antheraea polyphemus*. J Insect Physiol 37:79–85

Vogt RG, Riddiford LM 1981 Pheromone binding and inactivation by moth antennae. Nature 293:161–163

Vogt RG, Riddiford LM 1986 Pheromone reception: a kinetic equilibrium. In: Payne TL, Birch M, Kennedy CEJ (eds) Mechanism in insect olfaction. Clarendon, Oxford, p 201–208

Vogt RG, Riddiford LM, Prestwich GD 1985 Kinetic properties of a sex pheromone-degrading enzyme: the sensillar esterase of *Antheraea polyphemus*. Proc Natl Acad Sci USA 82:8827–8831

Vogt RG, Kohne AC, Dubnau JT, Prestwich GD 1989 Expression of pheromone binding proteins during antennal development in the gypsy moth *Lymantria dispar*. J Neurosci 9:3332–3346

Vogt RG, Prestwich GD, Lerner MR 1991 Odorant-binding protein subfamilies associate with distinct classes of olfactory receptor neurons in insects. J Neurobiol 22:74–84

Ziegelberger G 1995 Redox-shift of the pheromone-binding protein in the silkmoth *Antheraea polyphemus*. Eur J Biochem 232:706–711

DISCUSSION

Cardé: I had a question relating to the specificity of the proteins that bind to pheromones. Did you test whether aldehyde binds to the binding proteins?

Ziegelberger: At the moment we only have radiolabelled acetate, but we are also trying to obtain radiolabelled aldehyde, the second main pheromone component of *Antheraea polyphemus*. In different experimental set-ups the odour-binding specificity was moderate. We have shown in our experiments on antennal homogenates that the so called 'general odorant-binding protein' (GOBP) binds more labelled pheromone than does the pheromone-binding protein (PBP). We have also used hexanol as a general plant odour and benzoic acid, which stimulates receptor cells in the sensilla trichodea of female *Bombyx mori*, but we did not succeed in getting any binding.

Mustaparta: Are multiple PBPs found in the same species or does one PBP bind to many different pheromone components?

Ziegelberger: Two different PBPs have been identified in *Antheraea pernyi* and *Lymantria dispar*. In all other species studied, only one PBP has been found so far. Other PBPs might exist at lower concentrations. There are only two approaches that can be used to identify these binding proteins: ligand-binding assays and finding genes with sequence homology to known PBPs.

Mustaparta: Have Breer's group identified more than one PBP in the silkmoth you have used in this study?

Ziegelberger: No, Raming et al (1989) published the sequence of only one PBP gene in *A. polyphemus*. I asked them if they had found a second one, as in *A. pernyi*, but they said they hadn't. However, this doesn't exclude that more than one PBP is present. I showed you some Coomassie-stained gels. If these gels are silver stained, you can see that more proteins with low molecular weight are actually present. These are good candidates for unidentified binding proteins.

Steinbrecht: There are only one or two well-characterized and sequenced PBPs in each moth species. However, with more sophisticated separation methods (i.e. isoelectric focusing), Rosario Maida in our group has shown that more PBPs are present. For example, in *B. mori* he found as many as four PBPs that cross-react with the antiserum raised against a PBP of *A. polyphemus*. These PBPs differ mainly in their acidity. The number of identified PBPs per species will probably increase with improved methodology. In *Drosophila*, five PBP-like proteins have been found by molecular cloning methods (Pikielny et al 1994, McKenna et al 1994).

Guerin: What is responsible for transforming the binding protein from its reduced state to its oxidized state?

Ziegelberger: PBP oxidation does not occur with purified PBPs and pheromone in Ringer's solution. It only occurs in the isolated sensory hair homogenates, which contain sensillum lymph and receptive dendrites. Thus, oxidation of the reduced PBP might be receptor cell mediated.

The interaction of the pheromone–PBP complex with the receptor may cause structural changes that lead to close proximity of free sulfhydryl groups of cysteines and, thus, to disulfide formation.

Boeckh: What is the speed of these events and how do they relate to each other?

Ziegelberger: Oxidation of PBP in the homogenate occurs more slowly than the postulated rapid pheromone deactivation. If you take the dilution factor of 200 in the homogenate into account, then the redox shift *in vitro* comes quite close to the half-life of the active pheromone *in vivo*, as estimated from the decline of the receptor potential (Kaissling 1995).

Hildebrand: What is the consensus about the specificity of reception? The number of putative binding proteins described in the literature is growing rapidly. At the same time there has been a Herculean effort to try to find odour receptor molecules in insects, and to my knowledge no lab has yet succeeded. Therefore, on the one hand we have a hypothetical but elusive set of receptors, and on the other a growing family of binding proteins. Is the take-home message that the specificity of reception may reside with the binding proteins?

Steinbrecht: At least some of the specificity lies within the binding proteins. This is probably more true for the PBPs than for the GOBPs because, although the GOPBs have been characterized into two subgroups, they are not particularly diverse within each subgroup. The PBPs are more diverse, even in closely related species. Immunocytochemical results (Steinbrecht 1996, this volume) from my lab suggest that pheromone-sensitive sensilla trichodea in some moth species cross-react with our antibody against PBP of *A. polyphemus*, but not in other, closely related species. Gunde Ziegelberger (unpublished results 1995) has found that in immunoblots of *Spodoptera* and *Agrotis* there is cross-reaction with a minor band, but not with the major, male-specific, low molecular weight protein. By and large, cross-reactivity appears to be high when similar pheromones are used, and is low or absent, when the pheromone is very different from that of *A. polyphemus*.

Kaissling: J. Hemberger in my lab measured the degree of competition between radiolabelled pheromone and cold pheromone, and other compounds for binding to PBP (De Kramer & Hemberger 1987). He found that the specificity of the PBP was less than that of the receptor cell. For example, the saturated pheromone derivative, which doesn't contain double bonds, binds with only a 10-fold lower affinity to the binding protein, but it is a million times less effective on the receptor cells. The alcohol derivative of the pheromone binds about 300-fold less strongly to the PBP—as confirmed by Du & Prestwich (1995)—but is about 10^5-fold less effective at exciting the receptor cell than the pheromone.

Ziegelberger: Usually, when we talk about the binding specificity of the PBP, we only think about the binding site of the protein to the pheromone ligand. However, the pheromone–PBP complex may also have a binding affinity for the receptor. This may represent a second type of binding specificity.

Boeckh: There are reports that vertebrate olfactory receptor cells *in vitro* react to many substances, but *in vivo* they are more specific. When the olfactory binding protein is removed, something is absent that keeps the cell from being bombarded with all kinds of molecules. In such a system, where millions of cells are exposed to the odorants, the binding proteins might have a crucial role. Then it would not make sense to have one binding protein per odorant.

Carlson: In terms of the diversity of odour and binding proteins, one point that has become clear from work in both moths and *Drosophila* is that different members of this family are expressed in different subsets of sensilla (Laue et al 1994, Pikielny et al 1994, McKenna et al 1994).

Secondly, I would like to point out that GOBPs have not yet been directly demonstrated to bind to general odorants.

Boeckh: But this has been shown in vertebrates.

Guerin: Does this suggest that in the mosquito, for example, sensilla which house L-lactic acid-excited and L-lactic acid-inhibited receptors could have a similar ultrastructure but contain different types of binding proteins?

Carlson: It's possible that different odorant-binding proteins are expressed in different sensilla. We don't yet know if all of them bind to the same odorants.

Different members of the family might bind to different types of odorants, and if that's the case, their differential distribution in different sensilla of the olfactory system could play an important role in olfactory coding.

Boeckh: We should keep in mind that in each sensillum, the sensory cells and the accessory cells are derived from the same stem cell. It is possible that a specific binding protein, which is produced by an accessory cell, has a relationship with receptor molecules that are produced by the receptor cell.

Hildebrand: For the benefit of those not familiar with the development of a sensillum, it can be summarized as follows. The cells of the sensillum all arise from one epidermal precursor cell, which gives rise to the cells that become the sensory neurons as well as the cells that form the socket and the hair. Therefore, your point is that the precursor cell may have some degree of determinacy, leading to the co-ordinated expression of receptor molecules and binding proteins.

Ziegelberger: I originally thought that sensillum lymph would be identical in all sensilla, like that of mucus in the vertebrate nose. But we have learned from the immunocytochemical studies that different subclasses of OBPs occur in different sensilla. Thus, the specificity of the cell response could partly be due to the presence of a certain OBP.

Boeckh: But in vertebrates there are many receptor cells that have different specificities. Therefore, there must be something else which confines the cells of a given specificity to certain areas in the epithelium. It is possible that some mucous-producing cells produce different binding proteins in different parts of the olfactory area. We need to do more localization studies.

Hildebrand: It may be worth emphasizing here, again for those who are more removed from this immediate field, our reasons for considering these binding proteins in depth. It is clear that in the primary events of olfactory reception, there are many levels of molecular intervention, such as: (1) the pathway of diffusion of the odorants into the sensilla and the components that line the sensillum wall pores; (2) the binding proteins; and (3) the receptor cell dendrites, which have many molecular components involved with signal transduction. We emphasize the OBPs because out of this list, they are the molecules about which we have the most information.

We may not all have the same level of enthusiasm or confidence in the notion that genetic transformation of mosquitoes is the most promising way to go. But in thinking about how to pursue that avenue, we ought to try to identify genes that we can access and modify. The OBPs constitute a family of proteins that is likely to be represented in mosquitoes and be important for olfaction reception. Therefore, we should urge investigators to study these proteins in mosquitoes.

Guerin: I have difficulty with the term 'general odorant-binding protein'. It's only as 'general' as the compounds we have so far identified as stimuli.

Carlson: And there's no direct evidence that they actually bind odorants.

Pickett: This term is most unfortunate because we also do not have any clear evidence that general receptors exist, only the assertion by earlier workers.

Hildebrand: Hanna Mustaparta has shown that in moth antenna there are receptor cells as specific for plant volatiles as pheromone receptor cells are for pheromone components (see Mustaparta 1996, this volume). These are not broadly tuned cells.

Guerin: I agree, and if they bind specific stimuli, could we then use them to identify those stimuli?

Ziegelberger: Binding assays turned out to be problematic because of the hydrophobic nature of most odorants and possibly also because of the millimolar PBP concentration *in vivo*, which one cannot obtain with *in vitro* assays. Pheromone binding to PBP and GOBP survives the conditions of native gel electrophoresis, suggesting that the binding is relatively stable. In sensillum lymph perfusion experiments (Van den Berg & Ziegelberger 1991) we found that bovine serum albumin (BSA) also 'bound' pheromone. My guess is that the pheromone prefers to stick to any protein rather than staying in water. However, this binding to BSA is not stable because the pheromone does not stay bound to BSA during native gel electrophoresis.

Guerin: How did you bind the pheromone to the binding protein in the first place?

Ziegelberger: Usually our pheromones are dissolved in hexane. For the biochemical studies we put the pheromone in a glass tube, let the hexane evaporate and then add the homogenate. In this artificial system the PBP and GOBP seem to pick up the pheromone from the glass wall.

Pickett: In my presentation I showed that in *Culex quinquefasciatus* there is a highly sensitive cell for 3-methylindole (Pickett & Woodcock 1996, this volume). If you transfer the analogy of your model system to the mosquito, and I'm sure we could find similar cells specific for particular compounds in *Anopheles gambiae*, then we could provide either photoaffinity-labelled or radiolabelled 3-methylindole. If anyone here is interested in developing the biochemistry, then we could provide the chemistry within weeks.

Boeckh: That's a very good offer.

References

De Kramer JJ, Hemberger J 1987 Neurobiology of pheromone reception. In: Prestwich GD, Blomquist GJ (eds) Pheromone biochemistry. Academic Press, Orlando, FL, p 433–472

Du G, Prestwich G 1995 Protein structure encodes the ligand binding specificity in pheromone-binding proteins. Biochemistry 34:8726–8732

Kaissling K-E 1995 Pheromone deactivation on the antenna of the moth *Antheraea polyphemus*: its minimum velocity and its possible mechanism. In: Elsner N, Menzel R (eds) Neurobiology report, vol II: Proceedings of the 23rd Göttingen Neurobiology Conference, 1995. Thieme Verlag, Germany, p 362

Laue M, Steinbrecht RA, Ziegelberger G 1994 Immunocytochemical localization of general odorant-binding protein in olfactory sensilla of the silkmoth *Antheraea polyphemus*. Naturwiss 81:178–180

McKenna MP, Hekmat-Scafe DS, Gaines P, Carlson JR 1994 Putative *Drosophila* pheromone-binding proteins expressed in a subregion of the olfactory system. J Biol Chem 269:16340–16347

Mustaparta H 1996 Introduction IV: coding mechanisms in insect olfaction. In: Olfaction in mosquito–host interactions. Wiley, Chichester (Ciba Found Symp 200) p 149–157

Pickett JA, Woodcock CM 1996 The role of mosquito olfaction in oviposition site location and in the avoidance of unsuitable hosts. In: Olfaction in mosquito–host interactions. Wiley, Chichester (Ciba Found Symp 200) p 109–123

Pikielny CW, Hasan G, Rouyer F, Rosbash M 1994 Members of a family of Drosophila putative odorant-binding proteins are expressed in different subsets of olfactory hairs. Neuron 12:35–49

Raming K, Krieger J, Breer H 1989 Molecular cloning of an insect pheromone-binding protein. FEBS Lett 356:215–218

Steinbrecht RA 1996 Structure and function of insect olfactory sensilla. In: Olfaction in mosquito–host interactions. Wiley, Chichester (Ciba Found Symp 200) p 158–177

Van den Berg MJ, Ziegelberger G 1991 On the function of the pheromone binding protein in the olfactory hairs of *Antheraea polyphemus*. J Insect Physiol 37:79–85

General discussion V

Guerin: Could you speculate on what might be the difference between the transduction pathways of the L-lactic acid-inhibited receptor and the L-lactic acid-excited receptor?

Boeckh: Each effect could be based upon a different cascade of signals between the receptor and the ionic channels. Such a situation is easy to imagine, but it might not be true in reality.

Guerin: But how would it achieve opposite effects?

Boeckh: It would depend on which pathway the receptor was connected to. One intracellular chain of events may couple a receptor to an anionic channel, which becomes gated and the cell becomes hyperpolarized and inhibited.

Mustaparta: In the lobster an individual receptor neuron can be excited and inhibited by different odours. These responses are linked to different cascade reactions involving inositol-1,4,5-trisphosphate and cAMP. However, we do not know if these reactions are due to the activation of different receptor proteins linked to a specific G protein.

Hildebrand: Most of the mechanisms that have been studied involve relatively non-selective cation channels. These channels are activated via second messengers, resulting in the depolarization and excitation of the cells. An intracellular signalling cascade that activates an anion-selective channel, such as a Cl^- channel, is likely to have an opposite, inhibitory effect.

Steinbrecht: The discrimination between excitation and inhibition by different concentrations of L-lactic acid could be sharpened by having two cells that react in an opposite manner. Another example of such antagonistic pairs of receptor cells occurs in humidity-sensitive sensilla: one cell is excited with increasing humidity, whereas the other one is inhibited.

Geier: It's possible that the cell which is inhibited by L-lactic acid is activated by another, unknown component of the host odour. The cells that are inhibited by L-lactic acid have not been thoroughly investigated to date.

Bowen: There are fewer L-lactic acid-inhibited cells, so we have fewer recordings from these cells. They could also function in the following way: when the sensitivity of the L-lactic acid-activated cells shifts from high to low, the net afferent input could be depressed even further by the L-lactic acid-inhibited cell. In this way, the inhibited cells would act as a noise reduction switch.

Boeckh: We found that some of our dose–response curves were optimum curves. It is possible that the receptors are inhibited by high concentrations of

L-lactic acid and excited by lower concentrations. This might reflect an overload situation. Does this make sense in terms of the absolute shift in the sensitivity? I am sceptical of whether optimum concentration dose–response curves exist because after a closer look, many of them turned out to be artefactual.

Davis: But this is not the case for the L-lactic acid-inhibited cells. This is not an overload response. Trying to get enough L-lactic acid in the vapour phase is difficult, so to get an overload condition would be unnatural. Any response to an artificially elevated L-lactic acid intensity could be a pharmacological rather than a physiological reaction. The rate of L-lactic acid emanation from a human does not produce an inhibition of the response to L-lactic acid in the L-lactic acid-excited neuron (Smith et al 1970, Davis & Bowen 1994).

Boeckh: I was not referring to L-lactic acid-inhibited cells. The cells that I mentioned were basically excited by L-lactic acid and then possibly 'over-stretched'. They start responding at low doses and reach 50% of their maximum activity at a very low concentration. They are overloaded and not necessarily inhibited *sensu stricto*!

Mustaparta: The excitation and inhibition of these cells is also interesting in terms of detecting information from pheromones and plant odours. Although it is difficult, in general, to detect the inhibitory responses of pheromone receptor neurons, due to a low spontaneous activity, it is observed in several species as an inhibitory cross-effect of plant odours on pheromone receptor neurons and vice versa. A similar distinction may exist in the antennal lobe.

Hildebrand: We should be looking for mixture effects at the level of primary reception. We are used to finding little or no evidence for that in the case of sex pheromone receptor cells in moths. However, Ache et al (1988) have shown that olfactory receptor cells of decapod crustaceans exhibit mixture suppression.

Also, White et al (1990) have found that in gustatory pathways of insects, there are interactions between receptor cells in the periphery. We should bear in mind that this could also occur in the olfactory system.

Boeckh: We often observe in olfactory sensilla that if one cell is excited vigorously, then the other cells decrease their activity.

Mustaparta: We have also observed this in the heliothine moth. Stimulation with increased concentrations of the major pheromone component reduces the responses of the neighbouring neuron to the interspecific signal. This supports the idea that there is an interaction between the receptor neurons.

Boeckh: Sylvia Anton has also shown this in the locust (Anton & Hansson 1996).

Pickett: I would like to make a point relating to the role of chemistry in this area. One or two things have been said by various speakers which I would like to moderate, as a chemist. We often think, quite correctly, that sucrose and amino acids are involatile, but it's worth remembering that they are usually associated with volatile compounds. Therefore, for example, it's possible for us to detect sweetened tea without tasting it because of the small amounts of isomaltol that is associated with the sucrose. Also, we often use high levels of CO_2, which has two origins: fermentation and combustion. In both situations the CO_2 contains a different range of minor impurities. For example, CO_2 produced from fermentation contains dimethylsulfide, which was one of the compounds that was referred to as active in Alan Cork's presentation (Cork 1996, this volume). L-Lactic acid itself is also not very pure, in fact it's usually even sold together with water. I've noticed some young biologists working in the area of hematophagous insects who are not aware of this water content in their commercially obtained L-lactic acid. These points need to be made to newcomers to the area who, for example, may be surprised that lysine elicits a response from olfactory receptors. This is because of the minor hydrocarbon amine contaminants.

Galun: Lysine was found by A. W. A. Brown to attract mosquitoes. He later found that this response was actually due to the CO_2 released from his lysine preparation (Brown & Carmichael 1961).

Hildebrand: Now you have made me nervous. Do we know with certainty that the L-lactic acid cells are responding to L-lactic acid?

Pickett: I'm not disputing that. We've done some more work on the L-lactic acid response of biting midges and it's certainly there. I was just making the point that when you use something from a bottle you should be critical of what's in the bottle relative to what it says on the bottle.

Boeckh: This is also important for the analyses of the skin wash extract, for example. The extract stays active as an attractant for a long time, but it is vital to analyse the head-space and to find out what really reaches the sense organs. There may be a matrix that releases small amounts of particular secondary products. If you only analyse the chemistry of the extract, you may miss the important compounds that are released into the head-space.

Davis: There are several publications on the chemistry of L-lactic acid. It's not a simple compound. Depending upon conditions, it can exist as lactones, dimers and other forms. I have always been cautious of what's really going on when this is being blown over a mosquito.

Boeckh: Yes, it's difficult to identify the component that is the 'real' attractant.

Cardé: But the same situation applies to pheromones. For a time we thought we had characterized the complete blend of the oriental fruitmoth pheromone, but it turned out that the acetate mixture was contaminated with 0.3% of (Z)-8-dodecen-1-ol, which was crucial to behavioural activity; and, as it turns out, was the third pheromone component. On a somewhat less serious note I remember Vince Dethier's story from Africa during World War II. He passed time by trying to requisition anhydrous water from headquarters back in the USA. One of the replies was returned with the query, 'what purity'?

Hildebrand: My favourite part of that story is that after several documents were exchanged about the purity, quantity, etc., a request of 'what container?' came back to Dethier. Of course, he never received his order!

References

Ache BW, Gleeson RA, Thompson HA 1988 Mechanisms for mixture suppression in olfactory receptors of the spiny lobster. Chem Senses 13:425–434
Anton S, Hansson BS 1996 Antennal lobe interneurons in the desert locust *Schistocerca gregaria* (Forskål): processing of aggregation pheromones in adult males and females. J Comp Neurol, in press
Brown AWA, Carmichael AG 1961 Lysine and alanine as mosquito attractants. J Econ Entomol 54:317–324
Cork A 1996 Olfactory basis of host location by mosquitoes and other haematophagous Diptera. In: Olfaction in mosquito–host interactions. Wiley, Chichester (Ciba Found Symp 200) p 71–88
Davis EE, Bowen MF 1994 Sensory physiological basis for attraction in mosquitoes. J Am Mosq Control Assoc 10:316–325
Smith CN, Smith N, Gouck HK et al 1970 L-Lactic acid as a factor in the attraction of *Aedes aegypti* (Diptera: Culicidae) to human hosts. Ann Entomol Soc Am 63:760–770
White PR, Chapman RF, Ascoli-Christensen A 1990 Interactions between two neurons in contact chemosensilla of the grasshopper, *Schistocerca americana*. J Comp Physiol A 167:431–436

Genetic and molecular studies of olfaction in *Drosophila*

Daria S. Hekmat-Scafe and John R. Carlson[1]

Department of Biology, Yale University, Kline Biology Tower, PO Box 208103, New Haven, CT 06520–8103, USA

Abstract. Drosophila melanogaster, an insect amenable to convenient molecular and genetic manipulation, has a highly sensitive olfactory system. A number of *Drosophila* olfactory mutants have been isolated and characterized. The *smellblind* mutant has a defect affecting a voltage-gated Na$^+$ channel. The *norpA* mutant, defective in a phospholipase C, has a reduced response to odorants in one type of olfactory organ, providing genetic evidence for use of the inositol-1,4,5-trisphosphate signal transduction pathway in olfaction. Since the *norpA* gene is also required for phototransduction, this work demonstrates overlap in the molecular genetic basis of vision and olfaction. Interestingly, genetic analysis indicates that some olfactory information flows through a pathway which does not depend on *norpA*. Some mutants, such as *ptg*, *acj6* and *Sco*, show odorant specificity, in the sense that some odorant responses are greatly reduced, whereas others are little affected, if at all. Some, but not all, mutations affect both larval and adult olfactory responses. Two tightly-linked *Drosophila* genes encode homologues of moth pheromone-binding proteins (PBPs). Genetic analysis may help determine whether PBPs facilitate transit of pheromones to or from olfactory receptor neurons. Information from *Drosophila* could be useful in designing means of controlling mosquitoes. It may also be possible to study olfactory genes, such as those encoding PBPs, from other insects by mutating them, introducing them into *Drosophila* and analysing their function *in vivo*.

1996 Olfaction in mosquito–host interactions. Wiley, Chichester (Ciba Foundation Symposium 200) p 285–301

Chemical communication is widespread among insects, reflecting its fundamental importance for survival and reproduction (Hildebrand 1995). Pheromones, species-specific chemical messengers, are used to co-ordinate the complex behaviours of attraction, courtship and mating. Volatile chemicals can also play a role in such phenomena as insect oviposition and feeding.

This chapter describes some studies of olfactory perception in *Drosophila melanogaster*, an organism with well-defined genetics and molecular biology. It

---

[1]This paper was presented by John R. Carlson.

is our hope that a better understanding of how *Drosophila* perceives odorants may permit the development of more refined methods to control agricultural pests and vectors of human disease, such as mosquitoes.

Olfactory system organization

Drosophila adults respond to a wide variety of volatile chemicals, which they perceive primarily through the third segment of the antenna as well as through a second olfactory organ, the maxillary palp (Fig. 1; Carlson 1991, Ayer & Carlson 1992). The third antennal segment is covered with about 500 sensory hairs, of three morphological classes (sensilla basiconica, sensilla trichodea and sensilla coeloconica), each of which is distributed in a characteristic pattern on the antennal surface (Venkatesh & Singh 1984). Sensilla basiconica are also present on the maxillary palp (Singh & Nayak 1985). At least two of these sensillar types, the sensilla basiconica and sensilla trichodea, contain multiple pores that permit entry of volatile odorants (Stocker 1994). Sensilla basiconica respond to general (food-derived) odorants (Siddiqi 1987), as has been observed in other insects. It is not known whether *Drosophila* sensilla trichodea are pheromone responsive, as is the case in moths (Stengl et al 1992). Olfactory receptor neurons of *Drosophila*, like those of other insects, extend dendrites into the aqueous sensillar lymph

FIG. 1.   Scanning electron micrograph of a *Drosophila melanogaster* head. The third antennal segment (arrow) and maxillary palp (arrowhead) are indicated. Bar = 100 μm. Reproduced from Riesgo-Escovar et al 1995 with permission.

within the sensory hair and project axons to condensed synaptic glomeruli in the brain (Stocker 1994). The three different classes of sensory hairs project to different glomeruli (Stocker et al 1990).

Drosophila larvae also display a strong olfactory response (Monte et al 1989). Airborne odorants are believed to be perceived primarily by the antennal organ, which is located at the anterior tip of the larva. As in the adult olfactory system, the olfactory neurons extend dendrites into a sensillar fluid-filled space surrounded by a porous cuticle and project axons to the brain. The antennal organ is histolyzed during metamorphosis, and the adult antenna and maxillary palp develop subsequently from a common imaginal disc.

General olfactory mutants

smellblind mutations in a Na⁺ channel gene

smellblind mutations in a Na^+ channel gene

The original *smellblind* (*sbl*) mutation was identified in a screen for mutants defective in an olfactory-based learning paradigm (Aceves-Pina & Quinn 1979). Subsequently, *sbl* mutants were shown to display a defective olfactory response at both the larval and adult stages. *sbl* larvae show reductions in chemotaxis towards several attractive odorants, and in contact chemosensory avoidance of NaCl (Lilly & Carlson 1990). *sbl* adults display an anosmia (loss of smell) to several classes of structurally unrelated compounds (Ayyub et al 1990). *sbl* mutants are defective in a number of courtship behaviours: *sbl* females do not slow down in response to male courtship, resulting in a protracted courtship latency (Gailey et al 1986); and *sbl* males, unlike their wild-type counterparts, continue to court mated females (Tompkins et al 1983). It is possible that the *sbl* product is required in adults for the perception of pheromones as well as general odorants.

We have shown that *sbl* mutations are alleles of the Na^+ channel gene, *paralytic* (*para*) (Lilly et al 1994). Both *sbl* and *para* map to the same region of the X chromosome, and some strong *sbl* mutations fail to complement those in *para*. Furthermore, two *sbl* alleles have molecular lesions in the *para* transcription unit. These results indicate that Na^+ channel function is required in some way for the response of *Drosophila* to chemical stimuli. We note that voltage-sensitive Na^+ channels, such as that encoded by the *para* gene, are the major targets of dichlorodiphenyltrichloroethane (DDT) and the pyrethroid insecticides (Knipple et al 1994).

Mutants displaying both visual and olfactory deficits

Mutants displaying both visual and olfactory deficits

A number of *Drosophila* mutants exhibit both olfactory and visual deficits. Some of these mutants contain lesions in components associated with the inositol-1,4,5-trisphosphate (InsP₃) pathway, providing genetic evidence that in *Drosophila*, olfaction, like vision, most likely occurs at least in part through the InsP₃ signalling pathway.

A screen for *Drosophila* mutants that failed to enter a trap containing an attractive olfactory stimulus yielded a number of *olfactory trap abnormal* (*ota*) mutants, some of which also displayed defects in visual system physiology (Woodard et al 1989). One of these, *ota1*, was subsequently shown to be allelic to the mutant *retinal degeneration B* (*rdgB*), which undergoes light-induced retinal degeneration (Woodard et al 1992). The *rdgB* gene has been cloned (Vihtelic et al 1991) and its product shown to encode a membrane-bound phosphatidylinositol transfer protein possessing Ca^{2+}-binding sites (Vihtelic et al 1991, 1993). Phosphatidylinositol transfer proteins provide the receptor membrane with the phosphatidylinositol precursor necessary to maintain the phosphatidylinositol (4,5)-bisphosphate (PIP$_2$) that is hydrolyzed to InsP$_3$ (and diacylglycerol) by phospholipase C (Martin 1995).

The *rdgB* protein is expressed in photoreceptor cells, chemosensory neurons of the antennae and maxillary palps, and sensory processing regions of the central brain (Vihtelic et al 1991, 1993, Riesgo-Escovar et al 1994). One possibility is that the *rdgB* product responds to an increase in intracellular Ca^{2+} (which occurs during phototransduction and possibly olfaction) by providing phosphatidylinositol to phosphatidylinositol kinases at the receptor cell membrane. Electrophysiological recordings from antennae and maxillary palps (electroantennograms and electropalpograms, respectively) reveal that

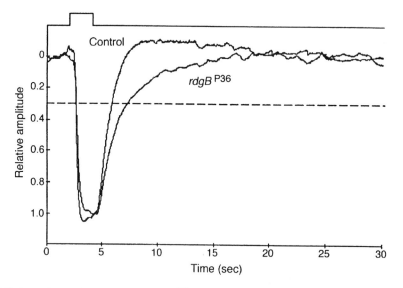

FIG. 2.   The maxillary palps of *rdgB*[P36] mutant flies exhibit delayed recovery following odour stimulation. Each trace represents the average of 90 recordings from nine flies. The dotted line indicates the amplitude used to determine 2/3 recovery time. The upper trace indicates stimulus delivery; its position is approximate. Reproduced from Riesgo-Escovar et al 1994 with permission.

rdgB flies display a delayed recovery (Fig. 2) to olfactory stimuli (Woodard et al 1992, Riesgo-Escovar et al 1994). This could reflect abnormal membrane restoration, suggesting that in *Drosophila*, olfaction occurs, at least in part, through the InsP$_3$ pathway. This conclusion is consistent with work in other invertebrates, as well as vertebrates, that has also implicated the InsP$_3$ signal transduction cascade in olfactory response (Breer et al 1990, Ache 1994).

The *Drosophila* visual mutant *no receptor potential A* (*norpA*) encodes a phospholipase C-β, which is expressed in the retina and maxillary palp, but has not been detected in the antenna (Bloomquist et al 1988, Riesgo-Escovar et al 1995). *norpA* flies display a reduced electropalpogram amplitude to all odorants tested (Riesgo-Escovar et al 1995). This provides further evidence for an InsP$_3$ requirement in olfactory signal transduction in *Drosophila*, and for overlap in the molecular and genetic underpinnings of vision and olfaction. Interestingly, electropalpogram recordings from a *norpA* mutant which produces no detectable product reveals that, for all odorants, the electropalpogram amplitude is reduced but not entirely eliminated (Fig. 3). Hence, some olfactory information likely flows through a *norpA*-independent pathway. This might involve either a different phospholipase C or a phospholipase C-independent pathway, such as a cyclic nucleotide pathway, which has been implicated as a second means of olfactory signal transduction in vertebrates (Dionne 1994).

Olfactory mutants with chemically specific anosmias

Analysis of *Drosophila* mutants with chemically specific olfactory deficits has revealed that odorants are perceived through at least two distinct pathways.

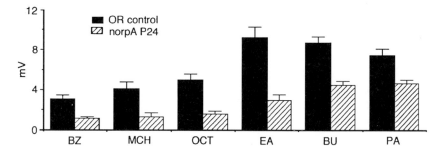

FIG. 3.   Olfactory response is reduced in the maxillary palp of *norpA*[P24] flies. Shown is the response amplitude $\pm$ SEM ($n = 17$). Odorant stimuli are generated by passing air over the following solutions: benzaldehyde (BZ, diluted 1:10), 4-methylcyclohexanol (MCH, diluted 1:10), 3-octanol (OCT, undiluted), ethyl acetate (EA, diluted 1:100), 1-butanol (BU, diluted 1:10) and propionic acid (PA, diluted 1:10). Differences between *norpA*[P24] and control flies are significant for all odorants ($P < 0.05$). Reproduced from Riesgo-Escovar et al 1995 with permission.

Lesions at the *pentagon* (*ptg*) locus produce an anosmia to benzaldehyde (scent of almond) but not to other odorants tested (Helfand & Carlson 1989). The basis of the benzaldehyde specificity is unknown. It is known that benzaldehyde differs from many other chemicals in a number of respects. It is a repellent at all concentrations tested, whereas most other odorants are attractants at lower concentrations but repellents at higher concentrations (Ayyub 1990). Furthermore, although most odorants produce only ipsilateral excitation, benzaldehyde produces both contralateral and ipsilateral excitation (Rodrigues 1988).

Interestingly, another *Drosophila* mutant displays specificity which is, in a sense, reciprocal: it responds strongly to benzaldehyde but displays a reduced response to other odorants tested (McKenna et al 1989, Ayer & Carlson 1992). A mutation in the olfactory gene *abnormal chemosensory jump 6* (*acj6*) displays reduced electroantennogram and electropalpogram amplitudes to most odorants tested, but demonstrates a strong response to benzaldehyde and another repellent, methylcyclohexanol, in the maxillary palp (Ayer & Carlson 1992, P. Clyne & J. Carlson, unpublished results 1992).

It will be of interest to determine from molecular analysis whether the reciprocal anosmias produced by these mutations are reflected in non-overlapping patterns of expression of the two gene products. It has been shown previously by 2-deoxyglucose labelling that distinct glomeruli in the brain are responsive to different odorants (Rodrigues 1988). It would be particularly interesting if *ptg* and *acj6* are expressed differentially in olfactory hairs housing receptor neurons with different odorant sensitivities or in the glomeruli to which such neurons project.

A different profile of olfactory deficits is produced by the dominant *Scutoid* (*Sco*) mutation (Dubin et al 1995). *Sco* flies display a reduced electroantenno-gram response to ethyl acetate and two short-chain ketones (acetone and 2-butanone) but respond normally to aldehydes (including benzaldehyde), alcohols and organic acids. *Sco* is believed to be an antimorphic allele of *no-ocelli* (*noc*), whose most visible phenotype is a loss of characteristic macrochaetae mechanosensory bristles. *Sco* reduces the maximal electro-antennogram response to short-chain acetate esters and ketones at all concentrations tested. The molecular mechanism underlying this olfactory alteration is currently under investigation.

Mutants defective in larval and adult olfactory response

The olfactory organs of *Drosophila* larvae and adults have a common function but different developmental origins. The larval antennal organ is histolyzed during metamorphosis, and the adult antenna develops from an imaginal disc.

Although some of the olfactory mutations described (e.g. *ptg* and *Sco*) specifically affect adults (Helfand & Carlson 1989, Dubin et al 1995), others affect larvae as well. As indicated previously, viable *sbl* mutations in the Na$^+$ channel gene *para* disrupt both larval chemotaxis towards a number of attractants and larval chemosensory avoidance of NaCl (Lilly & Carlson 1989). In addition, both *acj6* and *rdgB* are required for the larval response to ethyl acetate (Ayer & Carlson 1991, Woodard et al 1992).

Drosophila pheromone-binding protein homologues

Pheromone-binding proteins

In moths pheromone-binding proteins (PBPs) are present at high concentrations in the aqueous lymph surrounding the dendrites of pheromone-sensitive olfactory receptor neurons, where they are believed to solubilize hydrophobic pheromones by forming pheromone–PBP complexes (Hildebrand 1995, Stengl et al 1992). Their function is unknown: they have been proposed to facilitate transit of hydrophobic molecules to or from the olfactory receptor, and possibly to play a role in chemosensory coding (Kaissling 1986, Prestwich et al 1995). Multiple, related PBPs have been found in various moth species, and in one example, two PBPs expressed in the moth *Antheraea pernyi* have been shown to have different binding preferences for two biologically relevant pheromones (Prestwich et al 1995). Because a pheromone–PBP complex might be the actual ligand for olfactory receptors, PBPs could influence how the receptor 'sees' a pheromone. The availability of PBP homologues in an organism such as *Drosophila*, which offers a rich array of genetic and molecular techniques, should allow an investigation of PBP function not available in other systems.

Characterization of Drosophila pheromone-binding protein homologues

A family of genes whose predicted products share sequence similarity with moth PBPs has been isolated in subtractive screens for cDNAs expressed specifically in the *Drosophila* adult olfactory system (McKenna et al 1994, Pikielny et al 1994). We have concentrated on two such olfactory-specific genes, *OS-E* and *OS-F*, which are tightly linked and structurally similar. Sequence analysis revealed that, like their moth counterparts, the predicted *OS-E* and *OS-F* primary translation products are small (16.1 and 17.3 kDa), have a signal sequence and contain six cysteine residues located in stereotypic positions (Fig. 4). The predicted mature *OS-E* and *OS-F* products are 72% identical at the amino acid level. *Drosophila lebanonensis*, which diverged from *D. melanogaster* about 62 million years ago (Ashburner 1989) has a gene that appears to encode a protein closely related to both OS-E and OS-F (D. Hekmat-Scafe, M. A. Moise & J. Carlson, unpublished results 1995).

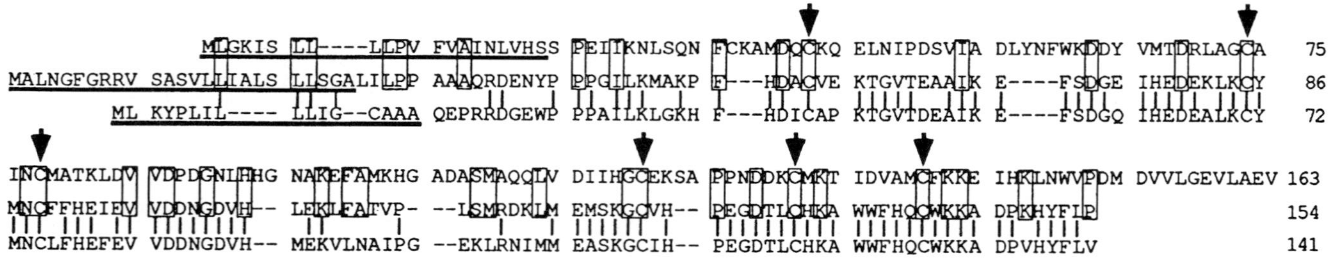

```
              MDGKIS ----LLPV FVAINLVHSS PEIIKNLSQN FCKAMDQCKQ ELNIPDSVIA DLYNFWKDDY VMTDRLAGCA   75
MALNGFGRRV SASVLIHALS LLSGALILPP AAAQRDENYP PPGILKMAKP E---HDACVE KTGVTEAAIK E----FSDGE IHEDEKLKQY   86
ML KYPLIL---- LLIG--CAAA QEPRRDGEWP PPAILKLGKH F---HDICAP KTGVTDEAIK E----FSDGQ IHEDEALKCY   72

INCMATKLIV VDPDGNLLHG NAKEFAMKHG ADASMAQQIV DIIHGEKSA PPNDDKCMKT IDVAMCFKKE DVVLGEVLAEV  163
MNCFFHEIEV VDDNGDVH-- LEKLEATVP- --LSMRDKLM EMSKGQVH-- PEGDTLCHKA WWFHQCWKKA DEKHYFIP     154
MNCLFHEFEV VDDNGDVH-- MEKVLNAIPG --EKLRNIMM EASKGCIH-- PEGDTLCHKA WWFHQCWKKA DPVHYFLV     141
```

FIG. 4. Sequence alignment between the pheromone-binding protein (PBP) Aper-1 from the moth *Antheraea pernyi* (top) and the *Drosophila* olfactory-specific proteins OS-F (middle) and OS-E (bottom). Amino acid identities between OS-F and the moth PBP (29%) are indicated with boxes; identities between OS-E and OS-F (68%) are indicated with lines. Conserved cysteines are indicated with arrowheads. Predicted signal peptides are underlined. Reproduced from McKenna et al 1994 with permission.

Interestingly, OS-E and OS-F show sequence similarity not only to moth PBPs and the related moth general odorant-binding proteins (GOBPs), but also to two proteins secreted by the male accessory sex gland of the beetle *Tenebrio molitor* (McKenna et al 1994, Paesen & Happ 1995). Hence, OS-E and OS-F could represent *Drosophila* members of a class of proteins whose various members function in a number of different tissues, such as the vertebrate lipocalin family of hydrophobic molecule transporters.

Both *OS-E* and *OS-F* map to the cytological position 83CD, with the two transcription units oriented in the same direction about 1 kb apart. Both *OS-E* and *OS-F* have two introns inserted at identical codons. The adjacent location, sequence similarity and similar intron positions of *OS-E* and *OS-F* suggest that the two genes arose by gene duplication.

Localization of *OS-E* and *OS-F* transcripts and proteins revealed that both are expressed in most, but not all, sensilla trichodea (McKenna et al 1994, D. Hekmat-Scafe, R. A. Steinbrecht & J. Carlson, unpublished results 1995), a spatially restricted class of olfactory sensory hairs. PBPs are known to be expressed in the long sensilla trichodea of the moth *Antherea polyphemus* (Steinbrecht et al 1992, Laue et al 1994). OS-E and OS-F are also expressed in some of the small sensilla basiconica. Sensilla basiconica respond to general odorants in *Drosophila* (Siddiqi 1987) and, in *A. polyphemus*, express GOBPs in most cases (Laue et al 1994). High resolution immunocytochemical electron microscopy showed that, like moth PBPs (Steinbrecht et al 1992), OS-E and OS-F are present in the secretory apparatus of sensillar auxiliary cells, from which they are presumably secreted into the sensillum lymph surrounding the olfactory receptor neurons (R. A. Steinbrecht, D. Hekmat-Scafe & J. Carlson, unpublished results 1995).

Based on their proximity, *OS-E* and *OS-F* could potentially share one or more enhancer elements. Immunocytochemical electron microscopy employing subtracted, affinity-purified antisera on alternating sections, revealed that all sensilla which express OS-E coexpress OS-F. Because of this coincident expression, and as some vertebrate OBPs are known to form dimers (Pelosi & Maida 1990), it is conceivable that OS-E and OS-F are present as homodimers and/or heterodimers, each of which could potentially have different binding properties. Formation of heterodimers of spatially restricted PBPs might provide one means of combinatorial coding of olfactory information.

The similar patterns of expression of OS-E and OS-F with that of the moth PBPs, together with their sequence similarities, suggests that these abundant, spatially restricted proteins may play similar roles in the two organisms. It has been difficult to determine the function of moth PBPs *in vivo* because of the absence of moth mutants that overexpress or underexpress particular PBPs. *Drosophila* strains with altered numbers of OS-E and OS-F genes can be created by molecular and genetic manipulations (D. Hekmat-Scafe & J. Carlson, unpublished results 1995). Electrophysiological testing of the odorant responses of such strains may provide clues about the nature and specificity of OS-E and OS-F function. In addition, it may be possible to assay *in vivo* the function of PBPs from various species by introducing the corresponding genes into *Drosophila* and testing the odorant response of transformant flies electrophysiologically. Through these experiments we hope to gain insight into the roles of PBPs in pheromone perception and chemosensory coding in *Drosophila* and other insects.

Acknowledgements

This work was supported by research grant R01 DC 02174–10 from the National Institute on Deafness and Other Communication Disorders, National Institutes of Health, by the National Science Foundation and by the Human Frontiers Science Program. D. H. was supported in part by a National Institutes of Health Research Service Award (DC00139). We thank R. A. Steinbrecht for a fruitful collaboration and much enlightening discussion.

References

Aceves-Pina E, Quinn WG 1979 Learning in normal and mutant *Drosophila*. Science 206:93–95
Ache BW 1994 Towards a common strategy for transducing olfactory information. Semin Cell Biol 5:55–63
Ashburner M 1989 *Drosophila*: a laboratory handbook and a laboratory manual. Cold Spring Harbor Laboratory Press, Cold Spring Harbor, New York
Ayer RK, Carlson J 1991 *acj6*: a gene affecting olfactory physiology and behavior in *Drosophila*. Proc Natl Acad Sci USA 88:5467–5471
Ayer RK, Carlson J 1992 Olfactory physiology in the *Drosophila* antenna and maxillary palp: *acj6* distinguishes two classes of odorant pathways. J Neurobiol 23:965–982
Ayyub C, Paranjape J, Rodrigues V, Siddiqi O 1990 Genetics of olfactory behavior in *Drosophila melanogaster*. J Neurogenet 6:243–262
Bloomquist BT, Shortridge RD, Schneuwly S et al 1988 Isolation of a putative phospholipase C gene of Drosophila, norpA, and its role in phototransduction. Cell 54:723–733
Breer H, Boeckhoff I, Tareilus E 1990 Rapid kinetics of second messenger formation in olfactory transduction. Nature 345:65–68
Carlson J 1991 Olfaction in *Drosophila*: genetic and molecular analysis. Trends Neurosci 14:520–524
Dionne VE 1994 Emerging complexity of odor transduction. Proc Natl Acad Sci USA 91:6253–6254
Dubin AE, Heald NL, Cleveland B, Carlson JR 1995 *Scutoid* mutation of *Drosophila melanogaster* specifically decreases olfactory responses to short-chain acetate esters and ketones. J Neurobiol 28:214–233
Gailey DA, Lacaillade RC, Hall JC 1986 Chemosensory elements of courtship in normal and mutant, olfaction-deficient *Drosophila melanogaster*. Behav Genet 16:375–405

Helfand S, Carlson J 1989 Isolation and characterization of an olfactory mutant in *Drosophila* with a chemically specific defect. Proc Natl Acad Sci USA 86:2908–2912

Hildebrand JG 1995 Analysis of chemical signals by nervous systems. Proc Natl Acad Sci USA 92:67–74

Kaissling K-E 1986 Chemo-electrical transduction in insect olfactory receptors. Ann Rev Neurosci 9:121–145

Knipple DC, Doyle KE, Marsella-Herrick PA, Soderlund DM 1994 Tight genetic linkage between the kdr insecticide resistance trait and a voltage-sensitive sodium channel gene in the housefly. Proc Natl Acad Sci USA 91:2483–2487

Laue M, Steinbrecht RA, Ziegelberger G 1994 Immunocytochemical localization of general odorant-binding protein in olfactory sensilla of the silkmoth *Antheraea polyphemus*. Naturwiss 81:178–180

Lilly M, Carlson J 1990 *smellblind*: a gene required for *Drosophila* olfaction. Genetics 124:293–302

Lilly M, Kreber R, Ganetzky B, Carlson JR 1994 Evidence that the *Drosophila* olfactory mutant *smellblind* defines a novel class of sodium channel mutations. Genetics 136:1087–1096

Martin T 1995 New directions for phosphatidylinositol transfer. Current Biol 5:990–992

McKenna M, Monte P, Helfand S, Woodard C, Carlson J 1989 A simple chemosensory response in *Drosophila* and the isolation of *acj* mutants in which it is affected. Proc Natl Acad Sci USA 86:8118–8122

McKenna MP, Hekmat-Scafe DS, Gaines P, Carlson JR 1994 Putative *Drosophila* pheromone-binding proteins expressed in a subregion of the olfactory system. J Biol Chem 269:16340–16347

Monte P, Woodard C, Ayer R, Lilly M, Sun H, Carlson J 1989 Characterization of the larval olfactory response in *Drosophila* and its genetic basis. Behav Genet 19:267–283

Paesen GC, Happ GM 1995 The B proteins secreted by the tubular accessory sex glands of the male mealworm beetle, *Tenebrio molitor*, have sequence similarity to moth pheromone-binding proteins. Insect Biochem Mol Biol 25:401–408

Pelosi P, Maida R 1990 Odorant-binding proteins in vertebrates and insects: similarities and possible common function. Chem Senses 15:205–215

Pikielny CW, Hasan G, Rouyer F, Rosbash M 1994 Members of a family of Drosophila putative odorant-binding proteins are expressed in different subsets of olfactory hairs. Neuron 12:35–49

Prestwich GD, Du G, LaForest S 1995 How is pheromone specificity encoded in proteins? Chem Senses 20:461–469

Riesgo-Escovar JR, Woodard C, Carlson JR 1994 Olfactory physiology in the *Drosophila* maxillary palp requires the visual system gene *rdgB*. J Comp Physiol A 175:687–693

Riesgo-Escovar J, Raha D, Carlson JR 1995 Requirement for a phospholipase C in odor response: overlap between olfaction and vision in *Drosophila*. Proc Natl Acad Sci USA 92:2864–2868

Rodrigues V 1988 Spatial coding of olfactory information in the antennal lobe of *Drosophila melanogaster*. Brain Res 453:299–307

Schneider D 1992 100 years of pheromone research. Naturwiss 79:241–250

Siddiqi O 1987 Neurogenetics of olfaction in *Drosophila melanogaster*. Trends Genet 3:137–142

Singh RN, Singh K 1984 Fine structure of the sensory organs of *Drosophila melanogaster* meigen larva (Diptera: Drosophilidae). Int J Insect Morphol Embryol 13:255–273

Singh RN, Nayak SV 1985 Fine structure and primary sensory projections of sensilla on the maxillary palp of *Drosophila melanogaster* meigen (Diptera: Drosophilidae). Int J Insect Morphol Embryol 14:291–306

Steinbrecht RA, Ozaki M, Ziegelberger G 1992 Immunocytochemical localization of pheromone-binding protein in moth antennae. Cell Tissue Res 270:287–302

Stengl M, Hatt H, Breer H 1992 Peripheral processes in insect olfaction. Ann Rev Physiol 54:665–681

Stocker RF 1994 The organization of the chemosensory system in *Drosophila melanogaster*: a review. Cell Tissue Res 275:3–26

Stocker RF, Lienhard MC, Borst A, Fischbach KF 1990 Neuronal architecture of the antennal lobe in *Drosophila melanogaster*. Cell Tissue Res 262:9–34

Tompkins L, Siegel RW, Gailey DA, Hall JC 1983 Conditioned courtship in *Drosophila* and its mediation by association of chemical cues. Behav Genet 13:565–578

Vihtelic TS, Hyde DR, O'Tousa JE 1991 Isolation and characterization of the *Drosophila* retinal degeneration B (*rdgB*) gene. Genetics 127:761–768

Vihtelic TS, Goebl M, Milligan S, O'Tousa JE, Hyde DR 1993 Localization of *Drosophila* retinal degeneration B, a membrane-associated phosphatidylinositol transfer protein. J Cell Biol 122:1013–1022

Venkatesh S, Singh R 1984 Sensilla on the third antennal segment of *Drosophila melanogaster* meigen. Int J Insect Morphol Embryol 13:51–63

Woodard C, Huang T, Sun H, Helfand S, Carlson J 1989 Genetic analysis of olfactory behavior in *Drosophila*: a new screen yields the *ota* mutants. Genetics 123:315–326

Woodard C, Alcorta E, Carlson J 1992 The *rdgB* gene of *Drosophila*: a link between vision and olfaction. J Neurogenet 8:17–32

DISCUSSION

Pickett: You mentioned a voltage-gated Na^+ channel, which I found very interesting. What would you like us to do?

Carlson: Have you looked at the olfactory response of any of the mutants that are resistant to pyrethroid insecticides?

Pickett: Yes, but not in *Musca domestica*. We've looked at aphids, and we will be doing some work on lepidopterans with Alan McCaffery at Reading University. We've have cloned the voltage-gated Na^+ channel from super-resistant and susceptible flies, and we are expressing them in an electro-physiologically compatible expression system. Is this useful for you?

Carlson: I would be interested in knowing where those mutations map in the gene that encodes the Na^+ channel, and also if the flies that contain those mutations have abnormal olfactory responses.

Pickett: We will do the electrophysiology experiments and let you know the results. We know the location of the mutations in the functional parts of the Na^+ channel. Because it's a voltage-gated channel there's no natural substrate, so we don't know at the moment where the pyrethroids interact. We expect the key mutations to occur within those regions, but this may not necessarily be the case: the mutation may exert its effect at another stereochemical site.

Gibson: How far are we from doing similar mutation analyses in mosquitoes?

Carlson: I'm not the best person to answer this because I'm still trying to get to grips with mosquito genetics. However, some types of genetic analysis are easier to perform than others. For example, screening for X-linked mutations is relatively easy because males only have one copy of this chromosome, so that you can identify recessive mutations more easily. This may be a good place to start.

Curtis: But the problem is that it's the females that bite and are the vectors. One may be able to identify some autosomal mutants by taking individual daughters and inbreeding their progeny. Is that a reasonable suggestion?

Carlson: Yes. And you already have some markers. For example, one of your markers is a white-eyed X-linked mutant. There have been efforts to clone the *white* gene in *Anopheles*, and mating disruption as a possible means of control may also be possible. A remarkable observation in *Drosophila* is that a kind of mating disruption occurs when the *white* gene is misexpressed (Zhang & Odenwald 1995). When the *white* gene is coupled to a heat shock promoter— so that under heat shock conditions there is a high level of *white* expression in many cell types—the males exhibit a kind of homosexual behaviour. They start following each other head-to-tail in long lines and sometimes the lines circularize. They show little interest in females, in fact the females form a cluster in the corner while the males are doing this. This may be an interesting experiment to do in mosquitoes.

Cardé: Do you understand the basis of that behaviour?

Carlson: The *white* gene is believed to encode a subunit of transporter proteins that sit in the membrane and let through guanine and tryptophan. Guanine and tryptophan are the precursors of eye pigments, so that's why the mutant has white eyes. The reason for this bizarre behaviour is less clear. It may have something to so with tryptophan also being a precursor of the neurotransmitter serotonin.

Hildebrand: Has anyone looked at serotonin levels in the white mutant?

Carlson: I'm not aware of any published data.

Brady: Do these *Drosophila* males sing like the wild-type males?

Carlson: They vibrate their wings, but I don't know if they produce a normal song. We have been interested in the sensory basis of this behaviour. Audrey Hing in my lab surgically ablated the wings, maxillary palps and antennae, and found no effects on the level of this homosexual behaviour, but when she turned off the lights the level of homosexual behaviour decreased, so it may be mediated in part by visual cues.

Hildebrand: Have you made any double mutants?

Carlson: No, We haven't done that yet, but we intend to. We're still in the process of mapping the mutation that affects nitric oxide synthase (NOS). I can tell you that it's X-linked but we don't know its exact location yet.

Hildebrand: Some olfactory receptors use a cyclic nucleotide pathway, whereas others use the inositol-1,4,5-trisphosphate (InsP₃) pathway.

Carlson: This is true for vertebrates (Boeckhoff et al 1990, Ronnett et al 1993). Ziegelberger et al (1990) and Boeckhoff et al (1993) have found that if moth antennae are stimulated with large amounts of pheromone, cGMP is produced.

Ziegelberger: From the relatively slow time course of this increased production and its localization, we concluded that cGMP is not involved in the transduction process generating the receptor potential, but it might play a role in adaptation processes.

Carlson: NO is known to stimulate guanylyl cyclase, which makes cGMP, and cGMP is produced in the presence of high levels of odorant. Persaud et al (1988) have shown that if 8-bromo-cGMP is administered to frog olfactory cells, the odorant-evoked currents are suppressed. Likewise, Boeckhoff et al (1993) have shown that cGMP suppresses the response of moth antennae to pheromone. This all fits together, i.e. NO increases cGMP, which suppresses olfactory response, which could give rise to adaptation.

Baumann et al (1994) have found expression of a cGMP-sensitive channel in *Drosophila* antennae, and Zufall & Hatt (1991) have evidence for an effect of cGMP on a pheromone-dependent ion channel in moths. Ruth et al (1993) have shown that a cGMP-dependent protein kinase can inhibit InsP₃ production in another system. Another point that comes to mind is the evidence which suggests that olfactory neurons may talk to each other at some level. Therefore, it is possible that a signal causes the production of NO which then travels to the next neuron and tells it that it's being activated.

Mustaparta: This may be the mechanism for the reduced responses we see in neighbouring neurons when we elicit strong responses in a receptor neuron.

Carlson: Different events might occur over different time-scales and at different levels in different cells. It's also possible that cGMP affects primary responses in some cells.

Steinbrecht: This cross-talk could be just an electrical contact between the receptor cells. For example, there is peripheral interaction between taste sensilla. Interactions are also observed in thermo-/hygrosensitive sensilla but not in a single-walled olfactory sensilla (see Steinbrecht 1989 and references therein).

Curtis: Are all the mutants that you have studied recessive, so that you have to make them homozygous before you can detect them?

Carlson: One of the mutants is not recessive.

Hildebrand: Was that a plant?

Carlson: No, an animal!

Curtis: How do you make the recessive mutants homozygous?

Carlson: One of the advantages of working with *Drosophila* is that there are chromosomes, called balancer chromosomes, which are marked with dominant markers so you can easily keep track of them. By crossing a fly homozygous for a particular mutation to a fly with a balancer chromosome, you can generate heterozygotes, and when these are crossed the F2 progeny that don't have the markers will be homozygous for the mutation.

There are mutations in *Drosophila* that specifically alter a response to some odorants and not others. For example, we have a mutant strain that has a defect in the behavioural response to benzaldehyde but it has a normal response to ethyl acetate and propionic acid. We also have a strain with defects that are to some extent reciprocal. Another example is the dominant *Scutoid* mutant, which has a decreased response to short-chain acetate esters and some ketones, and a normal response to aldehydes, organic acids and alcohols.

Gibson: How do you perform mutagenesis?

Carlson: We use a chemical mutagen that increases the mutation rate to about 1:1000. There's a standard simple procedure for mutagenizing *Drosophila*, which I'm sure could be adapted for mosquitoes.

Curtis: Sorting through 1000 mosquito progenies would be no small task.

Carlson: But it would help if you started with an enrichment of some sort.

Curtis: Presumably, these mutants would have to be made homozygous, so you would have to inbreed 1000 progenies.

Carlson: Or you could pick out the dominant mutations.

Curtis: But those are a lot rarer than recessives.

Carlson: Yes, but whether a mutation is fully recessive or not depends on the sensitivity of your assays. Semi-dominant mutations are not that rare if you've got a very sensitive assay. However, I agree that it's easier to find recessive mutations.

Davis: George Craig at the University of Notre Dame (Indiana, USA) did a lot of genetic manipulation of mosquitoes. He found a mutant with a third antenna on the end of its proboscis, for example. Would there be any benefit of using his old data to find some olfactory mutants?

Carlson: Yes. Does he have mutations that affect the structure of the antenna?

Davis: Yes, I think so.

Carlson: There is one interesting morphological mutation in *Drosophila* which eliminates the sensilla basiconica, leaving the numbers of sensilla trichodea and sensilla coeloconica substantially intact.

Guerin: Could we link up your talk with the pheromone-binding protein (PBP) story? What's the likelihood of finding a probe that we can use to manipulate olfaction?

Carlson: We've taken one of the cloned odorant-binding proteins (OBPs) and put it under the control of a heat shock promoter. We introduced this into *Drosophila*, and we're in the process of seeing if the OBPs are produced in response to heat shock. We're also thinking about expressing a moth PBP in *Drosophila* and finding out whether that *Drosophila* now responds to moth pheromones. It is also possible to do structure–function studies, where you can change critical residues, for example the cysteine residues, in the gene encoding the OBP. You may then be able to introduce the modified protein into *Drosophila* and determine whether the structural alterations have functional consequences.

Kaissling: You described a mutant that has only one copy of the PBP genes. Does having only one copy result in a reduced amount of PBPs?

Carlson: Preliminary results with Western blots suggest that there is a reduced amount of PBPs in those flies that have only one copy of the genes.

Kaissling: If the amount of PBPs is reduced by one half, would you really expect there to be such a large effect on the electroantennogram? Is it possible that the mutation affects other mechanisms which produce this effect?

Carlson: We still have to determine whether the olfactory defect is a consequence of decreasing the number of these PBP genes. However, we have shown that adding a third copy of the PBP genes does not have an effect. This suggests to me that wild-type flies may contain an optimal amount of these PBPs.

Kaissling: What happens if there are no copies?

Carlson: We've been trying to determine that but we haven't yet succeeded in constructing flies with no copies. We have deleted a large region, which resulted in a homozygous lethal, so we're now trying to make smaller deletions.

References

Baumann A, Frings S, Godde M, Seiferd R, Kaupp U 1994 Primary structure and functional expression of a *Drosophila* cyclic nucleotide-gated channel present in eyes and antennae. EMBO J 13:5040–5050

Boeckhoff I, Taveilus E, Strotmann J, Breer H 1990 Rapid activation of alternative 2nd messenger pathways in olfactory cilia from rats by different odorants. EMBO J 9:1953–1458

Boeckhoff I, Seifert E, Göggerle S, Lindemann M, Krüger BW, Breer H 1993 Pheromone-induced second messenger signaling in insect antennae. Insect Biochem Mol Biol 23:757–762

Persaud KC, Heck GL, DeSimone SK, Getchell TV, DeSimone JA 1988 Ion transport across the frog olfactory mucosa: the action of cyclic-nucleotides on the basal and odorant-stimulated state. Biochim Biophys Acta 944:42–62

Ronnett G, Cho H, Hester LD, Wood SF, Snyder SH 1993 Odorants differentially enhance phosphoinositide turnover and adenyl cyclase in olfactory receptor neuronal cultures. J Neurosci 13:1751–1758

Ruth P, Wang GX, Boekhoff I et al 1993 Transfected cGMP-dependent protein kinase suppresses calcium transients by inhibition of inositol 1,4,5-trisphosphate production. Proc Natl Acad Sci USA 90:2623–2627

Steinbrecht RA 1989 The fine structure of thermo-/hygrosensitive sensilla in the silkmoth *Bombyx mori*: receptor membrane substructure and sensory cell contacts. Cell Tissue Res 255:49–57

Zhang SD, Odenwald WF 1995 Misexpression of the white (ω) gene triggers male-made courtship in *Drosophila*. Proc Natl Acad Sci USA 92:5525–5529

Ziegelberger G, Van den Berg MJ, Kaissling K-E, Klumpp S, Schultz JE 1990 Cyclic GMP levels and guanylate cyclase activity in pheromone-sensitive antennae of the silkmoths *Antheraea polyphemus* and *Bombyx mori*. J Neurosci 10:1217–1225

Zufall F, Hatt H 1991 Dual activation of a sex pheromone-dependent ion channel from insect olfactory dendrites by protein kinase C activators and cyclic GMP. Proc Natl Acad Sci USA 88:8520–8524

Synthesis and future challenges: the response of mosquitoes to host odours

Willem Takken

Department of Entomology, Wageningen Agricultural University, PO Box 8031, 6700 EH Wageningen, The Netherlands

Abstract. There is ample evidence that host seeking in mosquitoes is mediated by semiochemicals emanating from the host. Olfactory cues (kairomones) are detected through an intricate pathway, beginning with sensilla located on the antennae (odour) and palpi (CO_2). Age and physiological state of the mosquito determine whether detection of kairomones results in a behavioural response. Only a few kairomones have been described so far. CO_2 is a kairomone for most mosquito species and signifies the presence of a potential host because of its occurrence in the volatile emissions of all vertebrates. Other chemicals are likely to play a role in mosquito–host interaction as well, notably L-lactic acid, fatty acids and 1-octen-3-ol. Species-specific host preference is thought to be olfactory based and related to the presence of specialized sensilla or perception at the central olfactory pathway. Host specificity is genetically determined, as demonstrated by inherited differences within members of the *Anopheles gambiae* complex. In view of the available evidence, challenging research areas are: (1) the function of olfactory receptors and the level of specialization; (2) the identification of general and host-specific kairomones; (3) demonstration of behavioural responses to laboratory-identified olfactory cues in the field; (4) studying whether mosquito host-locating behaviour can be manipulated by kairomones; (5) the genetics of the regulation of olfactory behaviour; and (6) determining whether there is an olfactory basis for the evolution of mosquito–host interactions.

1996 Olfaction in mosquito–host interactions. Wiley, Chichester (Ciba Foundation Symposium 200) p 302–320

The hypothesis that mosquito–host interactions are mediated by olfactory cues has been convincingly proven by a wide range of studies on different mosquito species (for reviews see Bowen 1991 and Takken 1991). Contributions in this volume provide new evidence for this phenomenon and challenge the reader to embark on studies that might elucidate the poorly understood mechanism of olfaction-mediated host-seeking behaviour in this group of important insects. Although fundamental questions on mosquito–host interactions are addressed, the underlying reasons for these questions are the need for the development of new strategies for mosquito control. Being vectors of debilitating and often lethal agents of parasitic disease in humans and animals, mosquitoes are the prime target of health interventions designed to reduce the risk of disease transmission and improve the quality of life throughout the world. It has frequently been shown that interruption of human–vector contact is the single most effective measure for the prevention or control of a vector-borne disease (Macdonald 1957, Anderson & May 1992). Traditional methods of vector control—such as spraying of insecticides or removal of breeding sites—have, after initial successes, failed to cause significant changes in risk for various reasons (Bruce-Chwatt 1985). New methods of vector control are required, which should incorporate sustainability and target specificity, and they should rely as little as possible on recurrent costs. This strategy requires an approach based on manipulation of the interaction between the vector and its host. Although it is known that several cues may affect mosquito–host interactions—e.g. physical, visual and acoustical cues—olfaction is the dominant process affecting host-seeking behaviour in mosquitoes. This volume addresses fundamental aspects of olfaction-mediated host-seeking behaviour of mosquitoes from molecular genetics through behavioural responses to host odours. Where relevant, this is discussed in relation to studies in other groups of insects.

Mosquito biology and behaviour

Mosquitoes are obligatory blood feeders, blood being a requirement for egg production. Only females feed on blood: the males are sustained on plant juices. Several species can produce a first egg batch without a blood meal (autogeny), but successive eggs can only be produced after ingestion of a blood meal, which is obtained from vertebrates. Between blood meals, mosquitoes may imbibe nectar from plants (Foster 1995), the carbohydrate contents of which is used as an energy reserve for metabolic processes. The behaviour of mosquitoes is tuned to obtain maximal success in foraging for blood. This increases fitness, and the number of eggs produced is positively correlated with meal size (Briegel 1990). Foraging behaviour is governed by endogenous and exogenous factors, such as physiological age, nutritional state, mating condition, circadian rhythm, ambient conditions and cues emitted by the host (Fig. 1). Host location only occurs when the female mosquito has reached a physiological stage in which she is receptive to cues which will direct her to the host. The mechanism of host location is optomotor anemotaxis, which is governed by physical, visual and olfactory stimuli. Temperature and humidity cause attraction and alighting at short range of the host, especially as convection currents, which are used by the mosquito to arrive at a landing site (Laarman 1958). These stimuli do not play a role at greater distances from the host. Visual stimuli are used for in-flight orientation and may also be

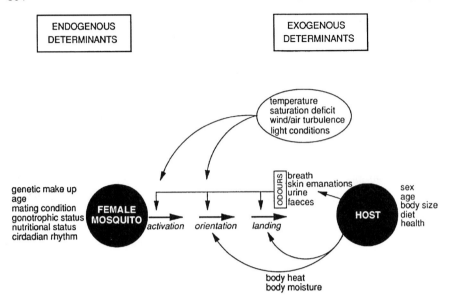

FIG. 1.   Diagram of endogenous and exogenous factors affecting host location.

attractants. They are recognized by luminous reflectance and contrast, but few conclusions can be made concerning spectral sensitivity (Muir et al 1992, Allen 1994). On moonlit nights, nocturnal mosquitoes are more active than on dark nights (Rubio-Palis 1992). Moonlight is thought to affect in-flight orientation. Several species respond to infrared light, and *Anopheles gambiae* orients to moving stripes on a white background under infrared light conditions up to 850 nm (Gibson 1995). Olfactory stimuli are the most important group of behavioural stimuli acting over short (Laarman 1958, de Jong & Knols 1995) and long distances (Gillies & Wilkes 1969). Details of this behaviour are discussed below.

Malaria transmission and mosquito behaviour

Human malaria is transmitted exclusively by mosquito species of the genus *Anopheles*. Because of their association with this important infectious disease, the genetics, biology and ecology of anopheline mosquitoes have been studied extensively. In recent years it has been shown that many *Anopheles* species consist of species complexes, which are morphologically (nearly) identical but may differ greatly in behaviour (Service 1988). These behavioural differences, in particular the degree of anthropophily and endophily within one group of malaria vectors that often occur sympatrically, affect the transmission of malaria because the vectorial capacity is determined by the association between humans and vectors. The *Anopheles gambiae* complex, which includes Africa's most important malaria vectors, consists of six sibling species that differ in host preference as well as in other behavioural characteristics (White 1974, Coluzzi 1992). *Anopheles quadriannulatus* is zoophilic (Gillies & Coetzee 1987), rarely attacking man, whereas *Anopheles gambiae sensu stricto* is highly anthropophilic (White 1974). Other members of the complex vary in degree of anthropophily and may be considered opportunistic in their feeding preference. Within *Anopheles arabiensis* there are also large geographical variations in host preference, which are genetically determined (Coluzzi et al 1985). From the studies on *An. gambiae* it is assumed that other sibling complexes also exhibit marked differences in host preference, as has been suggested for the *Anopheles funestus* group (J. D. Charlwood, personal communication 1995). Exploitation of these behavioural differences between anopheline mosquitoes will provide new avenues for the control of vector-borne diseases.

Olfactory control of behaviour

Optomotor anemotaxis in insects is often odour mediated. This behaviour can be undertaken as a strategy for locating a mate, food or oviposition site, and the olfactory stimuli used are produced by a potential mate, food source (plant or animal) or oviposition substrate, respectively. Odour-mediated insect flight has been best studied in moths responding to pheromones produced by the opposite sex. In moths the strategy of locating an odour source is determined by the aerial distribution of odour as set by turbulent diffusion and initial size of the odour source, wind speed, variability in direction and possibility of using visual cues. The actual flight path is usually a zigzagged (counterturning) upwind flight, where the amplitude of the zigzag is influenced by odour concentration and the frequency of encountering small-scale filaments of pheromone (Mafra-Neto & Cardé 1994, Cardé 1996, this volume). Although host-seeking mosquitoes may adopt a strategy similar to that used by flying moths, it is realized that there is a vast difference in source size, which for mosquitoes is many times that of the point source of moth pheromone. This difference in size may necessitate a different, and as yet unknown, relationship between the patterns of odour encounter and resulting manoeuvres.

In recent years much progress has been made on the identification of chemicals involved in the host location of insects that feed or oviposit on vertebrates. A combined use of gas chromatography and electrophysiology to identify compounds that are active at the sensory level, mass spectrometry for characterization and wind tunnel studies on behavioural responses have revealed kairomones for tsetse flies, *Glossina*, the New World screwworm, *Cochliomyia hominivorax*, and the highly anthropophilic *An. gambiae s.s.* This strategy has led to the recent discovery of the attraction of *An. gambiae s.s.* to fatty acids, which are common on and specific for human skin (Knols et al 1996). Wind tunnel bioassays have also been used to elucidate behavioural responses of *Aedes aegypti* to L-lactic acid and other skin emanations, and they have recently shown that distant fractions of skin extracts are required in addition to L-lactic acid to elicit the same degree of attraction as a complete skin extract (Geier et al 1996, this volume). Although several kairomones have been identified, few studies have so far demonstrated their attractiveness to mosquitoes in the field (Kline 1994), with the notable exceptions of CO_2, which is a kairomone for many mosquito species (Gillies 1980) and 1-octen-3-ol, which, in the presence of CO_2, attracts a wide range of zoophilic mosquitoes (Takken & Kline 1989).

Mosquito species exhibit marked differences in the selection of biting sites on their hosts and this behaviour is highly conservative for several species. It was recently found that selection of biting sites by *An. gambiae s.s.* and *Anopheles atroparvus* is odour mediated (De Jong & Knols 1995). From these findings it has been suggested that mosquito species which respond primarily to exhaled air are generalists, with CO_2 being the most important olfactory attractant; whereas those responding more to skin emanations might be specialists, having developed a preference for species-specific chemicals produced by the skin. This is supported by the reliability–detectability theory developed by Vet & Dicke (1992). Studies on biting sites may reveal clues for the source of production of mosquito kairomones.

Much attention is devoted to kairomones, although the research leading to the identification of chemicals that attract mosquitoes may also result in compounds which are repellent (allomones). For instance, the attraction of cattle urine for tsetse flies was attributable to phenolic compounds present in the urine. However, a minor compound, 2-methoxyphenol, was found to be repellent (Bursell et al 1988). Recent work has suggested that compounds which mask the attractiveness of specific skin products for livestock insects might be isolated from animal emanations (Pickett & Woodcock 1996, this volume).

Oviposition behaviour in several mosquito species is strongly mediated by olfactory cues (Pickett & Woodcock 1996, this volume). In *Culex quinque-fasciatus* and *Culex tarsalis* pheromones are present in the apical droplets of newly deposited eggs. The evolutionary function of these pheromones is not well understood: it is argued that they may be advantageous as well as disadvantageous for the mosquito as indicators of favourable breeding sites and, possibly, attractants for parasites and predators. Several mosquito species are also attracted to semiochemicals produced by microbial activity in the oviposition site. It has been observed that some species use both pheromones and attractants of microbial origin (Bentley & Day 1989), although the interactions of these stimuli at the sensory and behavioural levels are not yet understood.

Olfactory systems and mechanisms

Receptors for olfactory stimuli in mosquitoes are located on the antennae and maxillary palpi. The latter contain the sensilla that are receptive to CO_2. These were found to have a large activity range, covering those CO_2 concentrations that may be encountered during host-seeking behaviour near vertebrate hosts and also from a distance (Grant & O'Connell 1996, this volume). Although the morphology of the sensory system on the antennae of mosquitoes has been studied in detail (McIver 1982), the functions of most sensilla need to be determined. Olfactory receptors on the antennae include sensilla basiconica, or grooved pegs, and sensilla trichodea. There is almost no information on the sensitivity or specificity of these receptors to olfactory compounds. In *Ae. aegypti* the grooved pegs contain neurons that are sensitive to L-lactic acid and short-chain carboxylic acids (Pappenberger et al 1996, this volume). In *An. gambiae s.s.*, however, these sensilla did not respond to these stimuli, although another type of sensillum was responsive to several carboxylic acids (J. Meijerink, unpublished data 1995).

Studies on pheromone perception in moths show that olfactory molecules are bound to the cell wall of the olfactory neuron by odorant-binding proteins (OBPs), which are present in the sensillum lymph surrounding the sensory dendrites. The existence of different OBPs suggests that these proteins play a role in stimulus recognition. The physiological role of OBPs is becoming multifaceted and has been developed into a functional model. According to the model, a pheromone-binding protein (PBP) increases the electrophysiological response of the receptor cell when introduced simultaneously with a pheromone. A redox shift from the oxidized to the reduced form of PBP might enable the PBP to carry the pheromone to the dendritic membrane, and receptor cell-mediated oxidation might lead to deactivation of the pheromone–PBP complex (Ziegelberger 1996, this volume). This model might become universal for all insects. It would enable studies on physiological responses to olfactory stimuli in mosquitoes, and explain why some sensilla are not responsive to some stimuli whereas others are. Recent work on central olfactory pathways in insects has enabled the study of the integration of inputs from receptors on the maxillary palpi and antennae at the central level. The CNS of *Ae. aegypti* contains 20–25 glomeruli in males and females. Afferent projections from the antennae and maxillary palpi have been identified (Anton 1996, this volume). These basic studies on the electrophysiology of mosquito olfaction provide the background for future studies on the translation of perception of olfactory stimuli into behavioural responses.

Behavioural responses to external stimuli are modulated by endogenous factors, and they determine whether host-seeking behaviour is expressed. To some extent this behaviour can be influenced by manipulation of hormonal factors that control the responsiveness to external stimuli. For instance, prolonging of the effect of a haemolymph-borne factor, which is released in response to a humoral signal from a vitellogenic ovary, inhibits host-seeking behaviour in *Ae. aegypti* (Klowden et al 1987). This inhibition is accompanied by a decrease in the sensitivity of the peripheral L-lactic acid receptors. Molecular studies on the genetics of mosquitoes may reveal the genes that express production of hormones at the CNS level and other regulatory sites. The recent discovery of genes that encode for olfactory responses in *Drosophila* opens new avenues for future manipulation of olfaction-mediated host-seeking behaviour in mosquitoes.

Future challenges

The principle of odour-mediated, host-locating behaviour in mosquitoes has been widely accepted. Although many individual aspects of this behaviour are not yet understood, there are good prospects for manipulation of some of these steps which would open the way for new methods of vector control. For example, molecular manipulation of the genes expressing the production of OBPs might change the sensitivity for specific host odours rendering the insect non-receptive to that host. Alternatively, the development of mosquito attractants provides opportunities for introducing novel methods for the monitoring of vector populations or even vector control through mosquito removal. These and other possibilities are challenging research questions which can be solved with currently available techniques.

Recent advances in the study of mosquito genetics, sensory physiology and behaviour have provided clear evidence for the existence of species-specific behaviour, mediated by olfactory stimuli produced by the host. It is, however, not known how this specific behaviour is regulated and which genes are responsible for the expression of this behaviour. By studying the differences in behaviour of closely related species, new insights in the olfactory-mediated, host-locating behaviour of mosquitoes may be revealed.

Although the morphology of receptors on antennae and maxillary palpi has been well described, their function and physiology, as well as the neurological pathways of olfactory transmission, are poorly understood. Recent studies on the CO_2 receptors on the maxillary palpi and L-lactic acid receptors on the antennae have shown that progress can be made, particularly on the way olfactory cues are bound to the neuron cell membrane, as well as on the sensitivity and specificity of these associations. With the discovery of new kairomones, such studies should be expanded to include work on the integration of sensory information at the CNS level. It may be expected that interruption of the molecular bindings of kairomones is a first step to the development of novel methods of modification of mosquito host-seeking behaviour.

New work on olfactory genetics in *Drosophila* suggests that the identification of genes that code for the recognition of specific olfactory molecules is feasible. Therefore, recombinant DNA techniques can be used to make genetic constructs that lack specific behavioural characters. These studies may be expanded to mosquitoes and used to elucidate the processes underlying the response to host odours. Furthermore, studies on the genetics and physiology of endogenous control of host-seeking behaviour may result in new prospects for the manipulation of the humoral–neurological processes, for instance by the prolonging of hormonal levels that suppress host seeking (Klowden 1990).

The known range of mosquito kairomones in relation to host seeking is small and there is no synthetic equivalent of a vertebrate odour that attracts mosquitoes in equal numbers to those attracted to a live host. The recent confirmation that the highly anthropophilic *An. gambiae s.s.* is indeed attracted to human odours, that CO_2 plays a minor role in this process (Costantini et al 1996, Mboera et al 1996) and that fatty acids cause positive behavioural responses at the receptor and behavioural levels (Knols et al 1996) provide evidence that specific classes of attractive organic compounds can be identified. These studies have also shown that the mosquitoes respond to synthetic mixtures of compounds and that specific physiological and environmental conditions must be met before responses to host odours are turned into host-seeking behaviour (W. Takken, M. Braks, H. Otten & B. G. J. Knols, unpublished work 1995). Future work should be targeted towards the identification of the complex package of olfactory cues that mediate the mosquito's host-seeking behaviour. A combined approach of chemical analyses, sensory physiology and behavioural assays is currently the best option for making advances in this area. Studies on mosquito siblings of known differences in host preference might elucidate the differential responses to candidate odours and reveal species-specific kairomones.

Strong evidence for long-distance olfactory attraction of mosquitoes to host odours is available from West Africa. However, these studies were all done with live hosts or CO_2, a kairomone which is not suitable for large-scale use in the field. Recently traps and electric nets have become available for field studies on olfaction-mediated behaviour (Costantini et al 1993, Knols & Mboera 1996). One remaining bottleneck is the difficulty of releasing mixtures of candidate odours in concentrations that mimic natural emanations. Studies on pheromone releases of lepidopteran pests have shown that it is of the greatest importance to obtain natural release rates. Combined behavioural and chemical work is needed to develop formulations of kairomones that simulate the natural situation.

The ultimate aim of research on mosquito olfaction is the interruption of mosquito–host interactions in order to reduce the intensity of disease transmission. The identification and study of kairomones will assist in predicting whether these compounds can be used for the manipulation of the mosquito–host interaction, for instance through behavioural disturbance. It is expected that mosquito attractants will be used for the objective monitoring of vector densities, as a safe, reliable and reproducible method for data collection in epidemiological studies. It will be a challenge for the immediate future to develop simple attractive baits for mosquito monitoring and control.

Work on *An. gambiae* has shown that genetically distinct sibling species may also diverge in host preferences. Recent research in Burkina Faso suggests that these differences are olfaction based and encourage further laboratory and field studies to assess their exact nature (C. Costantini, personal communication 1995). Molecular techniques of species identification make a rapid and reliable species identification possible. The observed differences in olfactory behaviour can thus be linked to genetic maps, which will advance the evolutionary studies on important mosquito vector species.

In support of laboratory and field research on host-seeking behaviour, mathematical models of mosquito behaviour have been constructed to help predict response profiles over a wide range of internal and external conditions (Roitberg et al 1994). Such models can be instrumental in decision making about certain steps of research which otherwise can only be based upon the results of empirical studies.

Acknowledgements

I thank B. G. J. Knols and R. Cardé for their useful comments and advice on this manuscript. The assistance of F. Kaminker with editing the text is appreciated.

References

Allen SA 1994 Physics of mosquito vision: an overview. J Am Mosq Control Assoc 10:266–271

Anderson RM, May RM 1992 Infectious diseases of humans. Dynamics and control. Oxford University Press, Oxford

Anton S 1996 Central olfactory pathways in mosquitoes and other insects. In: Olfaction in mosquito–host interactions. Wiley, Chichester (Ciba Found Symp 200) p 184–196

Bentley MD, Day JF 1989 Chemical ecology and behavioral aspects of mosquito oviposition. Ann Rev Entomol 34:401–421

Bowen MF 1991 The sensory physiology of host-seeking behavior of mosquitoes. Ann Rev Entomol 36:139–158

Briegel H 1990 Fecundity, metabolism, and body size in *Anopheles* (Diptera: Culicidae), vectors of malaria. J Med Entomol 27:839–850

Bruce-Chwatt LJ 1985 Essential malariology, 2nd edn. Wiley, New York

Bursell E, Gough AJE, Beevor PS, Cork A, Hall DR, Vale GA 1988 Identification of components of cattle urine attractive to tsetse flies, *Glossina* spp. (Diptera: Glossinidae). Bull Entomol Res 78:281–291

Cardé RT 1996 Odour plumes and odour-mediated flight in insects. In: Olfaction in mosquito–host interactions. Wiley, Chichester (Ciba Found Symp 200) p 54–70

Coluzzi M 1992 Malaria vector analysis and control. Parasitol Today 8:113–118

Coluzzi M, Petrarca V, Di Deco MA 1985 Chromosomal inversion intergradation and incipient speciation in *An. gambiae*. Boll Zool 52:45–63

Costantini C, Gibson G, Brady J, Merzagora L, Coluzzi M 1993 A new odour-baited trap to collect host-seeking mosquitoes. Parassitologia 35:5–9

Costantini C, Gibson G, Sagnon N'F, Della Torre A, Brady J, Coluzzi M 1996 The response to carbon dioxide of the malaria vector *Anopheles gambiae* and other sympatric mosquito species in Burkina Faso. Med Vet Entomol, in press

de Jong R, Knols BGJ 1995 Selection of biting sites on man by two malaria mosquito species. Experientia 51:80–84

Foster WF 1995 Mosquito sugar feeding and reproductive energetics. Ann Rev Entomol 40:443–474

Geier M, Sass H, Boeckh J 1996 A search for components in human body odour that attract females of *Aedes aegypti*. In: Olfaction in mosquito–host interactions. Wiley, Chichester (Ciba Found Symp 200) p 132–148

Gibson G 1995 A behavioral test of the light sensitivity of a nocturnal mosquito, *Anopheles gambiae*, to dim white, red and infrared light. Physiol Entomol 20:224–228

Gillies MT 1980 The role of carbon dioxide in host-finding by mosquitoes (Diptera: Culicidae): a review. Bull Entomol Res 70:525–532

Gillies MT, Coetzee M 1987 A supplement to the Anophelinae of Africa south of the Sahara. Pub S Afr Inst Med Res 55:64–143

Gillies MT, Wilkes TJ 1969 A comparison of the range of attraction of animal baits and of carbon dioxide for some West African mosquitoes. Bull Entomol Res 59:441–456

Grant AJ, O'Connell RJ 1996 Electrophysiological responses from receptor neurons in mosquito maxillary palp sensilla. In: Olfaction in mosquito–host interactions. Wiley, Chichester (Ciba Found Symp 200) p 233–253

Kline DL 1994 Olfactory attractants for mosquito surveillance and control: 1-octen-3-ol. J Am Mosq Control Assoc 10:280–287

Klowden MJ 1990 The endogenous regulation of mosquito reproductive behavior. Experientia 46:660–670

Klowden MJ, Davis EE, Bowen MF 1987 Role of the fat body in the regulation of host-seeking behaviour in the mosquito, *Aedes aegypti*. J Insect Physiol 33:643–646

Knols BGJ, Mboera LEG 1996 Electric nets for studying odour-mediated host-seeking behaviour of mosquitoes (Diptera: Culicidae). Bull Entomol Res, in press

Knols BGJ, Van Loon JJA, Cork A et al 1996 Behavioral and electrophysiological responses of female malaria mosquito *Anopheles gambiae s.s.* (Diptera: Culicidae) to Limburger cheese volatiles. Bull Entomol Res, in press

Laarman JJ 1958 The host-seeking behaviour of Anopheline mosquitoes. Trop Geogr Med 10:293–305

Macdonald G 1957 The epidemiology and control of malaria. Oxford University Press, London

Mafra-Neto A, Cardé RT 1994 Fine-scale structure of pheromone plumes modulates upwind orientation of flying moths. Nature 369:142–144

Mboera LEG, Knols BGJ, Della Torre A, Takken W 1996 The response of *Anopheles gambiae s.l.* and *An. funestus* (Diptera: Culicidae) to tents baited with human odour or carbon dioxide in Tanzania. Bull Entomol Res, in press

McIver SB 1982 Sensilla of mosquitoes (Diptera: Culicidae). J Med Entomol 19:489–535

Muir LE, Kay BH, Thorne MJ 1992 *Aedes aegypti* (Diptera: Culicidae) vision: response to stimuli from the optical environment. J Med Entomol 29:445–450

Pappenberger B, Geier M, Boeckh J 1996 Responses of antennal olfactory receptors in the yellow fever mosquito *Aedes aegypti* to human body odours. In: Olfaction in mosquito–host interactions. Wiley, Chichester (Ciba Found Symp 200) p 254–266

Pickett JA, Woodcock CM 1996 The role of mosquito olfaction in oviposition site location and in the avoidance of unsuitable hosts. In: Olfaction in mosquito–host interactions. Wiley, Chichester (Ciba Found Symp 200) p 109–123

Roitberg DB, Smith JJB, Friend WG 1994 Host response profiles: a new theory to help us understand why and how attractants attract. J Am Mosq Control Assoc 10:333–338

Rubio-Palis Y 1992 Influence of moonlight on light trap catches of the malaria vector *Anopheles nunetzovari* in Venezuela. J Am Mosq Control Assoc 8:178–180

Service MW 1988 Biosystematics of haematophagous insects. Clarendon, Oxford

Takken W 1991 The role of olfaction in host-seeking of mosquitoes: a review. Insect Sci Appl 12:287–295

Takken W, Kline DL 1989 Carbon dioxide and 1-octen-3-ol as mosquito attractants. J Am Mosq Control Assoc 5:311–316

Vet LEM, Dicke M 1992 Ecology of infochemical use by natural enemies in a tritrophic context. Annu Rev Entomol 37:141–172

White GB 1974 *Anopheles gambiae* complex and disease transmission in Africa. Trans R Soc Trop Med Hyg 68:278–299

Ziegelber G 1996 The multiple role of the pheromone-binding protein in olfactory transduction. In: Olfaction in mosquito–host interactions. Wiley, Chichester (Ciba Found Symp 200) p 267–280

DISCUSSION

Curtis: It's important to emphasize that if mosquitoes carrying abnormal genes are going to be released, their effects on fitness have got to be conditional because the mosquito has to survive in the field. We have got to be more subtle, for example by diverting the mosquito from being anthropophilic to being zoophilic, so that it still has a good chance of surviving in the field.

Takken: That's a good suggestion and it might be possible. However, we shouldn't forget that zoophilic mosquitoes will generally feed off humans if they're hungry and nothing else is available. As an exception, this does not occur with *Anopheles quadrimaculatus*, which remains strictly zoophilic. At present we don't know whether it is possible to transfer zoophilic genes into species that are highly anthropophilic, but it sounds an attractive idea and any step that would reduce the biting intensity would decrease the risk of getting malaria.

Guerin: Although I am relatively convinced that kairomones affect the behaviour of mosquitoes, we should be careful about the conclusions we draw. For example, conclusions drawn from analysing blood meals may not be indicative of the true host preferences. We don't really know how strictly zoophilic or anthropophilic they are—they may land on animals first and then decide to approach humans. Has anyone shown clearly that animal odours are selected for in preference to human odours? If not, I would be very wary of sweeping conclusions.

Costantini: We recently did an experiment to investigate the host preference of *Anopheles gambiae* in West Africa. We used a calf and a human as baits, and drew their odours into two odour-baited entry traps (OBETs) (Costantini et al 1993) that were put close together in a choice arrangement, standardizing for the CO_2 output by topping it up at the same level in whichever of the two odour streams had the lower concentration (Fig. 1). In the case of no odour-mediated host preference, one would expect a 50% distribution of mosquitoes between the two traps. This was certainly not the case for *An. gambiae sensu lato* and *Anopheles funestus*, who consistently 'chose' the human-baited trap, whereas other species, such as *Culex* spp. and other anophelines like *Anopheles squamosus*, *Anopheles ziemanni* and *Anopheles rufipes* 'preferred' the calf-baited one. In this area we have both *An. gambiae sensu stricto* and *Anopheles arabiensis*, so it will be interesting to know if there is some quantitative difference in the choice shown by these two species, consistent to what we know about their host selection.

Lehane: But this is an artificial situation in which there is a choice. In the real world mosquitoes are not always presented with such a choice—humans often being the most abundant food source by far.

Gibson: Even in the real world host preference is expressed in a stochastic manner. In *An. gambiae s.s.* a few mosquitoes will take non-human blood meals, presumably depending on the strength of their response to the host cues available. The experimental protocol we have developed with the OBET, however, provides us with an objective measure of differences between species in the likelihood of responding to one type of host or another, which is the closest we have come to an assay for host preference.

Guerin: Another variable that we have not measured is the possible influence of pathogens on the host-seeking behaviour of the vector. Further, is there evidence for pathogen–host interactions that may affect vector behaviour?

Knols: Day & Edman (1983) found that malaria rendered mice more susceptible to mosquito attack at the time when gametocytes (the infectious stages for mosquitoes) circulated in the rodent's bloodstream.

Lehane: Fever has never been shown to make hosts more attractive to mosquitoes. The only evidence I know of in the literature where there's been an unexplained attraction of a blood-feeding insect to an infected animal are chickens infected with arbovirus (Mahon & Gibbs 1982). However, this work was never followed up.

Knols: But if the host is infected with malarial parasite, it is possible that the olfactory signals presented to the mosquito are different because of the influence of the parasite's impact on the physiology of the host.

Gibson: It is difficult to establish whether or not there is a differential attractiveness between sick and healthy people because there is so much biting heterogeneity for many other reasons. Some people are bitten more than others

FIG. 1.   A choice test experiment to study the odour-mediated host preference of West African mosquitoes. Two odour-baited entry traps (upper picture, in the foreground) are put close together, and odours from two hosts, each concealed in one of the tents in the background (in this case a man and a calf, lower picture) are drawn into the traps by fans via the air ducts. CO_2 is topped up from a gas cylinder to ensure the two odour plumes coming out of the traps have the same CO_2 concentration.

because of the design of their houses or their proximity to breeding sites. With tools such as the OBET baited with people in various stages of ill health who may be willing to volunteer, we can determine if there is an olfactory effect and if so, separate that from other close-range effects, such as body temperature and perspiration.

Knols: That experiment would be extremely complex due to the large variations in general health status of individuals and the influence of factors such as diet on the composition of the body odour profile.

Klowden: Lambornella clarki is a protozoan that infects the ovaries of *Aedes sierrensis* and causes a decrease in host-seeking behaviour, so that affected adults don't seek a host (Egerter & Anderson 1989). This parasite replicates in the ovaries, so that the ovaries are full of parasites instead of yolk and the females look like they're gravid. Instead of host seeking, these females engage in oviposition behaviour and oviposit parasites instead of eggs. Therefore, the parasite has manipulated the behaviour of the mosquito host for its own benefit, and behavioural manipulation must be possible because it has been done by a parasite.

Guerin: Given that pathogens may have been a driving force in the development of anthropophily in a certain species of mosquito, this suggests that host preference is something that can be manipulated.

Pickett: But we have to take into account the vast size of the human population. That's why we ought to think about a manageable choice situation which can be engineered with olfactory cues. A 100% behavioural effect cannot be obtained using cues from chemical ecology. Behavioural effects caused by semiochemicals normally only influence a proportion of the population.

Curtis: I disagree. There are many species, such as *Anopheles quadriannulatus*, that never bite humans, and in fact they're less-known to the medical entomologists just for that reason.

Lehane: There is only a small chance of actually fixing that trait within a population and then getting that trait to spread in an area where there are few animals. As I understand it, in Africa where *An. gambiae* is present there are a relatively small number of other hosts available.

Curtis: There are enough to feed *An. quadriannulatus* and *An. arabiensis*.

Gibson: Our behaviour experiments trap many zoophilic mosquitoes, which means there are enough non-human hosts to support these populations.

Lehane: Am I right in thinking that there isn't much overlap between the distributions of *An. arabiensis* and *An. gambiae s.s.* in many parts of Africa?

Curtis: There is an example of overlap in Swaziland where, in the days of house spraying, it appears that *An. arabiensis* was wiped out and only *An. quadriannulatus* remained (Mosbaum 1957a,b). Though not then aware of the existence of the species complex, they decided that the remaining mosquitoes were harmless so they stopped spraying and malaria did not return, or at least not until *An. arabiensis* came back.

Knols: I would like to comment on the issue of driving genes for zoophily into *An. gambiae s.s.* populations. This species is highly anthropophilic and probably responds to human-specific odours whilst host seeking. If we manage to identify these odours than we can release them from ear-tags on cattle, hence making these mammals potential hosts for anthropophilic *An. gambiae.* In this way the host's odour profile would be altered rather than the host specificity of the mosquito.

Brady: There are hundreds of other malaria vectors that are not as anthropophilic as *An. gambiae s.s.*, so it may be possible to convert the African situation to the Indian subcontinent situation simply by lowering the level of anthropophily. Mosquitoes in India manage perfectly well without being so highly anthropophilic.

Lehane: But why is there such a degree of anthropophily in Africa?

Gibson: This is one of the fundamental questions being asked by those unravelling the story of the evolution of the *gambiae* complex. It would seem that *A. gambiae s.s.* became associated with human settlements initially because of the ideal breeding sites created by human cultivation practices. If they were originally zoophilic, like *An. quadriannulatus*, then they had to make at least two major behavioural changes to become the ideal vectors they are today: first they had to develop a preference for human blood; and second they had to 'discover' the benefits of feeding indoors.

Lehane: The problem is determining if there are enough alternative hosts to fill that niche. If the adult mosquitoes have more reproductive success when feeding on human blood rather than animal blood, then that niche is going to be rapidly filled with human-feeding mosquitoes.

Cardé: There should be no reason why mosquitoes that don't get a human meal cannot be drawn to an alternative host. The question facing the female mosquitoes is: if they are fed on an alternative host will they be competitive in the sense of how many eggs can they lay?

Galun: Mosquitoes are probably more likely to lay more eggs when feeding on pig or chicken blood than on human blood.

Gibson: This could become the basis for an unprecedented integrated pest management strategy for vector control. Large areas of Africa do not have cattle because of the disease transmitted by tsetse flies. The programme would begin with eradication of tsetse flies in an area, thereby allowing cattle to move in. The cattle would then help support populations of zoophilic *An. gambiae s.s.*

Klowden: We could take this host diversion issue to the extreme. There is evidence that autogenous egg development inhibits host-seeking behaviour, and I don't know of any autogenous species that are serious vectors. The genetics of autogeny are not well understood, but it could be possible to develop an autogenous strain of mosquitoes, which would eliminate biting during the first gonotrophic cycle. Therefore, the mortality factors after the first gonotrophic cycle would be so great that it may not be necessary to eliminate much more of the population after that.

Bowen: Decreasing vector–host contact is critical to the interruption of disease transmission. However, our knowledge of the host attractants that are involved in these processes is at a primitive level at the moment. For an initial foray into the molecular genetics of olfactory-mediated behaviour, perhaps we should look at other mosquito behaviours that are mediated by single chemicals. One of these is oviposition site location and another is plant feeding. The receptors for at least some of these volatiles and the behaviours are known. It may be simpler to screen mutagenized mosquito populations for those behaviours instead: the assays are fairly simple and quick.

Another possibility would be to screen for repellent insensitivity, which might give us a handle on repellent mechanisms at the biochemical level.

Guerin: We may have to establish short-term and long-term priorities. In relation to the short-term approach we might take our cue from the tsetse fly eradication programmes, where they are being effectively trapped. Once you attract the mosquito you can also take it out. Considering the comment that Gay Gibson made about co-evolution, it might be worthwhile to remove the trapped mosquitoes and not leave them to feed on something else. This trapping-out could have a strong effect locally on a population of the vector.

Brady: But the prevalence of holoendemic malaria is highly stable. There has to be a massive effect on the population mortality to drive the transmission rate off its plateau.

Davis: We don't have a good enough bait to do that. It must be competitive with the natural attractants to be successful in a trap-out strategy.

Pickett: What is the effect of age on attraction? Are small children more attractive?

Curtis: There have been some studies of mothers and babies, who differ in their blood groups, sleeping under the same net. Bryan & Smalley (1978) looked at the blood groups of the mosquitoes and found that there was about a 7 : 1 ratio of mother's blood to child's blood, which is roughly proportional to their skin surface area.

Pickett: Would there be any advantage of thinking about differential protection, so that older people are used as baits?

Curtis: There's some recent evidence from the Gambia that parents are protecting their children by diverting the mosquitoes to bite them, presumably because they go to bed later (Quinones et al 1996). All the children in some villages were vaccinated against rabies, so that the rabies antibodies could be detected in the mosquito blood meal. Most people in the Gambia use nets, and when they were impregnated with pyrethroids, the proportion of rabies antibody contained in the blood meals decreased sharply. When the treated nets were removed, the level of antibody increased. This suggests that it is possible to divert the mosquitoes from adults to children. Adult Africans who live in rural areas are highly immune to malaria, so such diversion is highly advantageous.

Pickett: This type of approach is more within our reach than a large-scale, genetically based modification of mosquitoes.

Cardé: Willem Takken mentioned that we need to understand the behavioural bioassay system in a way that will allow us to identify attractants and develop better repellents. My view, however, is that is we need to have a better comprehension of the insect's actual orientation process. This will, in turn, lead to more intelligent bioassays.

Pickett: But this approach has been relatively unsuccessful for plant protection.

Cardé: I agree, but in that case often the critical decision-making process occurs after the insect has arrived at the plant. Plants will mostly be found eventually if not quickly, and so interfering with host finding may be less effective than interfering with oviposition.

Pickett: I don't disagree with what you're saying from a scientific point of view. However, at the moment we're talking about using knowledge gained over the next four or five years for the development of successful mosquito control strategies.

Hildebrand: In other words, do we know enough? And if we don't, what specific areas of olfaction should we study to be able to move in that direction?

Bowen: There has not been an analogous behavioural analysis in *Drosophila*, and this has not inhibited the molecular genetic analysis of behaviour.

Cardé: But we are not trying to control *Drosophila* by modifying their odour-mediated behaviours.

Boeckh: The careful analysis of orientation behaviour and understanding the behavioural genetics are both important. We know little about what attracts mosquitoes and what alarms them, and the differences between these two behaviours have not been worked out. Each behaviour is governed primarily by a certain series of cues, but we don't know where in this series CO_2, for example, acts, or how important it is relative to the other cues. We need to understand these things before we can attempt to modulate sensory cues.

Brady: Although the upwind approach of a mosquito, which is driven by olfactory cues, is important, if a mosquito chooses not to land and then bite there will be no malaria transmitted. Identifying these landing and biting cues may be epidemiologically more important than knowing the complete series of behavioural events that orients the distant mosquito towards its host.

Bowen: In my opinion we should move forward on more than one front.

Brady: We are making good progress with understanding the olfactory-mediated approach flights, but we know relatively little about the final approach and landing. Epidemiologically speaking, that is what really matters.

Steinbrecht: In that context, I would like to stress the importance of studying olfactory input by electrophysiology. It surprises me that most of this work has been done with the smallest sensilla, the A_3-type sensilla (Bowen 1995, for example), but there are hundreds of longer olfactory hairs and only little information about their function (for references see Steinbrecht 1996, this volume).

Galun: How informative are bioassays going to be in terms of the long-term goal of mosquito control? We may catch many flies within a small area, but this is still only a small fraction of the whole population. Therefore, we also need to know more about the population genetics of mosquitoes.

Lehane: In my opinion traps could never be used to successfully suppress the population of mosquitoes. However, they could be used to interrupt disease transmission. The insect must bite in order to pick up the disease in the first place, it must then take another four or five blood meals over a period of about 11 days before it bites another individual to pass the disease on. Consider a trap in a house that is as efficient as a human. If there are five people in the house, the trap might eliminate 60% of the infected population over that time period.

Davis: We've been focusing on what research can do in a positive sense, i.e. what we can do, but it is also important to consider what we cannot do. We know nothing about the search strategy of a mosquito in a circular, closed arena, i.e. a hut, where it does not have linear flight patterns. We've been talking about odour plumes in terms of a linear approach, but this may not be that useful in this situation. The mosquito may use a different set of cues for these more turbulent wind flow characteristics. Without knowing these cues, we cannot be sure that the traps will be effective. We need to find out what won't work and report those negative results so that we can learn what not to try as well as what works.

Pickett: This is an important issue that needs attention. But we do know at the moment that some people are attacked more than others, which is why the development of a differential attractiveness strategy would be extremely useful.

Dobrokhotov: I do not believe in the success of using mosquito traps because mosquitoes have a different breeding strategy than tsetse flies. Field interventions with traps cannot significantly reduce the number of mosquitoes. At the same time, studying the genetic differences between *An. gambiae* and other zoophilic *Anopheles* species will probably help us to understand the differences in their behaviour. This understanding can then be used for the control of malaria transmission.

Takken: As I understand from John Carlson's presentation, it will not be too long before we can identify genes that code for olfactory behaviour (Hekmat-Scafe & Carlson 1996, this volume). But I would like to take you up on your first point. The tsetse fly is an example of how it is possible to supply simple traps to large parts of Africa, provided they are affordable. People will use anything that may decrease the likelihood of them getting malaria, and interventions aimed at reducing the number of mosquito bites is certainly one of the accepted methods. We shouldn't be too negative about this kind of approach because models of population regulation show clearly how effective these interventions can be.

Gibson: Whether or not it is possible to reduce mosquito populations with traps is open for debate. There is no question, however, that efficient sampling tools are always required; for example, to monitor populations, to study comparative behaviour and to develop alternative methods for malaria control. Social pressure against human-biting catches will grow. In social and practical terms, the use of CDC (Center for Disease Control, Atlanta, USA) light traps alongside human-baited bed nets has already been shown to be an acceptable alternative. All that remains is to standardize this method one step further by replacing the human bait with an artificial one and to make the trap more efficient at collecting attracted mosquitoes.

References

Bowen MF 1995 Sensilla basiconica (grooved pegs) on the antennae of female mosquitoes—electrophysiology and morphology. Entomol Exp Appl 77:233–238

Bryan JH, Smalley ME 1978 The use of ABO blood groups as markers for mosquito biting studies Trans R Soc Trop Med Hyg 72:357–360

Day JF, Edman JD 1983 Malaria renders mice susceptible to mosquito feeding when gametocytes are most infective. J Parasitol 69:163–170

Egerter DE, Anderson JR 1989 Blood-feeding drive inhibition of *Aedes sierrensis* (Diptera: Culicidae) induced by the parasite *Lambornella clarki* (Ciliophora: Tetrahymenidae). J Med Entomol 26:46–54

Hekmat-Scafe DS, Carlson JR 1996 Genetic and molecular studies of olfaction in *Drosophila*. In: Olfaction in mosquito–host interactions. Wiley, Chichester (Ciba Found Symp 200) p 285–301

Mahon R, Gibbs A 1982 Arbovirus-infected hens attract more mosquitoes. In: MacKenzie JS (ed) Viral diseases in South East Asia and the Western Pacific. Academic Press, New York

Mosbaum O 1957a Malaria control in Swaziland. J Trop Med Hyg 60:190–192

Mosbaum O 1957b Past and present position of malaria in Swaziland. J Trop Med Hyg 60:119–127

Quinones M, Drakeley C, Lines J, Hill N, Muller O, Greenwood BG 1996 Permethrin-impregnated bednets divert mosquitoes from children to other hosts. Abstr Brit Soc Parasitol Malaria meeting, Sept 1995. Ann Trop Med Parasitol, in press

Steinbrecht RA 1996 Structure and function of insect olfactory sensilla. In: Olfaction in mosquito–host interactions. Wiley, Chichester (Ciba Found Symp 200) p 158–177

Index of contributors

Non-participating co-authors are indicated by asterisks. Entries in bold type indicate papers; other entries refer to discussion contributions.

Indexes compiled by Liza Weinkove.

Subject index